The Evolution of Naval Mine Warfare

The Evolution of Naval Mine Warfare

JACOB A. MAZUREK

McFarland & Company, Inc., Publishers
Jefferson, North Carolina

The views expressed in this publication are those of the author and do not necessarily reflect the official policy or position of the Department of Defense or the U.S. government. The public release clearance of this publication by the Department of Defense does not imply Department of Defense endorsement or factual accuracy of the material. Use of released U.S. Navy imagery does not constitute product or organizational endorsement of any kind by the U.S. Navy.

LIBRARY OF CONGRESS CATALOGING-IN-PUBLICATION DATA

Names: Mazurek, Jacob A., 1977– author
Title: The evolution of naval mine warfare / Jacob A. Mazurek.
Description: Jefferson, North Carolina : McFarland & Company, Inc., Publishers, 2026 | Includes bibliographical references and index.
Identifiers: LCCN 2026010426 | ISBN 9781476698625 paperback ♾
ISBN 9781476657783 ebook
Subjects: LCSH: Submarine mines—History | BISAC: HISTORY / Military / Naval | TECHNOLOGY & ENGINEERING / Marine & Naval
Classification: LCC V856 .M39 2026
LC record available at https://lccn.loc.gov/2026010426

ISBN (print) 978-1-4766-9862-5

ISBN (ebook) 978-1-4766-5778-3

Front cover images: Mine laying in the North Sea, 1918 NH 2743 (Naval History and Heritage Command); *background* © Shutterstock.

Printed in the United States of America

McFarland & Company, Inc., Publishers
Box 611, Jefferson, North Carolina 28640
www.mcfarlandpub.com

This book is dedicated to the enlisted mine warriors of the world's navies. You do the tedious, dangerous jobs that may never feature in Hollywood movies. Stationed in far-flung corners of your countries' orbits, often well away from the rest of the fleet, you are often referred to within your own navies as "unicorns." You save lives by using mines to prevent battles between fleets. You save lives by removing the mines meant to ensnare your own ships in times of conflict. You save lives by removing the mines that continue to menace merchant shipping in times of peace. You protect the world, one watch at a time, one wrench turn at a time, while others get the glory. I see you, and all that you do for your country.

—Jacob A. Mazurek

Acknowledgments

Thank you to all of those who helped with this book. To my wife, Milena Mazurek, thank you for reviewing every submission prior to sending it to the DoD, the publisher, and even an embassy. Even better, your knowledge and skills with digital imagery far exceed my own. Your redo of my graphics was the better choice for this book. To Don Moody, thank you for the speed of your proofreading that drove my own revision productivity. Furthermore, your deep mineman experience helped shore up points where my own knowledge and research fell short. To Nicole Bishop, thank you for using your specialized knowledge to identify topics that are better left undiscussed. To Jason McKenzie, thank you for your efforts to improve the book's readability. To Anguel Miltchin, thank you for providing your opinion from the perspective of a non–Navy, international reader. To the workers at the Naval Library at the Washington Navy Yard, thank you for helping me to find tons of references and for providing me with reach-back support when I needed it. To the baristas at the coffee shops where I did all of my writing on two continents, thank you for keeping me caffeinated and for allowing me to work. Finally, to the commanding officers of my first two ships, Robert Girrier and Peter Stitt, thank you for accidentally encouraging me to grow a love for naval history.

Table of Contents

Preface

The Story of the Book

The author first gained an interest in writing in elementary school when he represented various schools at regional Young Authors Conferences. He became interested in naval history in 1998 on his first Navy ship in Sasebo, Japan. The specific moment was when he was bored and looking for something to read but could find only the naval history textbook *Sea Power.* Afterward, he developed a taste for the subject and absorbed every naval history book that he could find at used bookstores in Corpus Christi, Texas, and Newquay, UK. In 2003, he set himself the task of reading all of the books on the Master Chief Petty Officer of the Navy's Required Reading List. By the end of his last sea tour in 2009, he had achieved that goal. Together, these books gave him a broad, deep understanding of the course of naval history and the world events that surround that history.

The niche subject of naval mine warfare history first entered his life in 2009 when he was stationed at the San Diego–based Naval Mine and Anti-Submarine Warfare Command (NMAWC). There he was introduced to two old mine warfare history books. The first was *Damn the Torpedoes* by Tamara Moser Melia, which covered U.S. naval mine countermeasures history. The other, *Weapons That Wait: Mine Warfare in the U.S. Navy* by Gregory Kemenyi Hartmann, covered the history of mines. Together, these two books led him to consider the history and heritage of his own community of naval mine warfare. He realized that the subject was never discussed aboard ship, and that most mine warfare sailors had no sense of their own heritage.

Aboard ships, enlisted sailors performed their tasks without ever considering their appropriateness to the tactical situation or whether alternative approaches were available. They did for inspectors whatever checklists and procedures told them to do, without considering the realities of doing those tasks in times of war. They had all heard in bootcamp about the USS *Princeton* and the USS *Tripoli* striking mines in Desert Storm. But they had never heard the story of how those ships got into the position to strike mines. As far as was discernible, minefields were only ever in coastal waters in predefined sections with known borders. Alternatively, sailors might think that mines are randomly strewn around the ocean by shady actors. Or that mines might be encountered in times of peace in pristine, extensive minefields. Sailors knew that mines were laid in the past, but they seldom heard why or how recently mines had been used.

Those two books, especially *Damn the Torpedoes*, were enlightening. While attending mine warfare conferences, the author found himself able to introduce historical examples into discussions. These injections of history helped to add nuance or to offer alternative options. Furthermore, by being a social-yet-midlevel person stationed at NMAWC, he frequently found himself able to learn from members of allied navies. With the primary role of trainer and assessor, he was able to use these many lessons to help teach shipboard sailors many of the lessons that he wished he had known when in their positions.

Then, one day, the Mine Warfare Training Center was looking for volunteers to help rewrite the mineman rating manual. He volunteered at once to rewrite the history portion of that manual. The previous version had been written a couple of decades previously when minemen were not on minesweepers. This entry-level training manual discussed only the history of mines, not those who fought against mines. Over the next several months, he gathered a few more books to help him research this newest iteration of the training manual.

His next occasion for writing about naval mine warfare history came during his last Navy tour in 2016. That year, he started writing essays for the U.S. Naval Institute (USNI). At that point, he had just transferred from a mine countermeasures squadron in Japan. There, he planned naval exercises and delved deep into the science behind the technology. He learned the art of planning, of being among those who must decide how best to use available forces to remove mine threats. Being in that position, he had found himself pushing for greater realism as he read ever more about the realities of mine warfare at sea. What he realized was that there were many techniques used historically that would still have value in certain situations. He began to see that history could give a more realistic image of what could be expected from members of other warfare areas in times of war, when mines are only one concern. At the same time, he recognized that mine warfare is often seen as a niche subject outside of mine warfare circles. His concern was that the only real knowledge that most sailors might ever get about mines would be if they found themselves in minefields.

With these concerns in mind, he researched and wrote a 3,500-word essay titled "Ten Mining Campaigns That Shaped Naval Mine Warfare." This essay appeared on the U.S. Naval Institute's blog. It was referenced in other authors' essays and was linked on the Naval Sea Systems Command's research page. There it was the only reference about mines and the only single-author essay. He went on to submit other essays to USNI for publication. While some were accepted, the only other one to be published was a humor essay titled "Tall Guy Problems in the U.S. Navy." Yet, from the time that he wrote that first essay, he planned to write an entire book on the subject. He began buying and reading every mine warfare history book that he could find. After he retired from the Navy, he attended an MBA program. At the end of his 2019 summer internship in Washington, D.C., he spent four days at the Navy Library at the Washington Navy Yard. During those four days, he took photos of every page that he could find in various sources that mentioned naval mine warfare—more than 600 photos, all told.

He spent a good deal of time in 2019 and 2020 organizing and reading the pages from these photos, even as he juggled his MBA coursework. In 2020, he began writing this book, starting with chapter 1. Writing was slow at first, so that he did not complete the Crimean War portion until a few months before the Russo–Ukrainian war that began in 2022. A year later, he had progressed only up to World War I while also working a full-time job. Then, in June 2023, he and his family quit their jobs, sold almost everything, and moved to Bulgaria with nothing but luggage and a cat. In that luggage were a few information-dense mine warfare history books. He also took an iPad with which he had taken photos of a couple of hundred pages from the books that he had left in storage.

In Bulgaria, he wrote in earnest. His research leaned heavily on the many photographed pages, his few books, the "American Corner" at the Sofia Main Library, and innumerable scholarly internet sources. Writing at coffee shops from Greece to Italy (but mostly in Sofia), he worked his way through World War II, the Korean War, and the Vietnam War. The desert wars were harder to research because fewer books cover the subject. Hardest of all was the last chapter, the one titled "Future History." That history is a greenfield creation requiring 100 percent internet research. While sitting in a country affected by the Russo–Ukrainian war, he scrolled the internet looking for any articles that he could find in English mentioning mines in the ongoing war. His final bit of research involved trying to create as complete a picture as possible of the state of the entire world's mine countermeasures forces. Lacking even an old copy of *Janes Fighting Ships*, he looked through hundreds of webpages to populate an exhaustive spreadsheet. From that process, he was able to distill his final, summarized result.

The last part of the process was due diligence to ensure that he was not accidentally putting out classified or embarrassing information. As someone who had lived and breathed the topic for a quarter of a century, this was a real worry for him. His first step was to provide the final chapter to a local U.S. naval attaché's office to verify that he hadn't accidentally insulted a friendly nation. Next, he submitted the whole document to the Department of Defense to undergo mandatory prepublication review. Half a year later, he received their official release. Finally, a few friends and colleagues volunteered to review his now-completed book for problems with content and readability. Together, they helped identify subjects best omitted from the final version.

The Book's Approach

This book attempts to tell the story of mine warfare (MIW) in a different way. It weaves the warfare area's greater events (and some smaller goings-on) into the fabric of the time periods in which they take place. Nothing happens in a vacuum. Mines were never dropped randomly but were part of wider tactical and operational plans. In at least three cases (two successful), mines were laid as national strategies intended to end wars. Furthermore, mine countermeasures (MCM) were seldom conducted randomly. When minesweepers swept, they did so when and where it was thought that mines would be a problem. This book attempts to explain *why* mines were laid in certain places and how plans and history evolved as a consequence.

Another aspect of this book's approach is analysis, i.e., consideration of why certain incidents and techniques were important. In the early chapters, where few source materials are available in English, a great deal of exposition takes place to help the reader understand MIW concepts. In the chapters on the great wars of the 20th century, where reference materials abound, the goal was to explain the overall reality of mine warfare. That understanding alone can help to inform decisions on how best to equip one's fleet for the slow, tedious, and dangerous work of MCM. As the book turns into the 21st century in the last two chapters, it demonstrates how mines continued to shape history even into the Russo–Ukrainian war that raged as the author wrote this sentence.[1] That last part is especially challenging and subject to future change and additions.

At the time that this book was sent to the publisher, the war was still being actively fought by nations whose people speak languages other than the author's own. The full, propaganda-free tale has yet to be made public. Finally, this book attempts to take a snapshot of the world's MCM forces and reveals MIW's historical lessons that have been repeated throughout its history.

This book is differentiated by the author's profile. Most other MIW history books have been written from one of three perspectives:

1. Engineers who have worked in MIW research, development, and procurement for decades. They concentrate on the how and the who. A part of the shore establishment, they are deeply familiar with system capabilities and concepts of operation and with the various senior officers assigned to the warfare area. These tend to provide single-country, in-depth discussions of equipment and of how peacetime funding and procurement decisions have affected wartime results. Because these matters are so well covered elsewhere, this book will not attempt to replicate them.
2. Officers who have served two or three MIW tours. They concentrate on who, what, when, and where. Among these, the emphasis has been more niche, e.g., a single war or a single ship class. By keeping within a limited scope, they go into a great amount of detail about individual ships, organizations, and commanding officers. Coming from command positions, they often use the style of a narrative report. They describe those entities' actions and decisions in detail. Yet the level of detail often serves to obscure their place in the wider course of history or the purpose and effects of the events. They are at a tactical level, rather than operational or strategic. This often makes their books great for research but not broad enough to give the novice an overview of events over time and across navies. Because such details are exhaustively covered elsewhere, this book will not attempt to replicate them.
3. Historians, either professional or hobbyists, write general MIW history. They also concentrate on who, what, when, and where, though their scope is broader. Usually these are the most entertaining, bringing in elements of the writings of engineers and officers alike. However, what they lack is a deep enough understanding of the subject to analyze and evaluate it. When a writer knows a subject only academically, everything and nothing is important. Just the facts, with all being equal. This is great for informing the general public but can lack the sort of emphasis that can help mine warfare professionals to gain long-term perspective.

This book is different because the author has a much deeper, more personal relationship with naval mine warfare. As such, instead of exhaustive detail, it concentrates on why events happened, what effects they had, and why these events matter today. It attempts to explain MIW as part of a continuum of larger events, rather than as a series of isolated events on tiny ships. Outside the mine warfare profession, people may not care about which units under which officers performed what techniques at a particular time and place. What they may care more about is how a single line of 26 mines sank three battleships, damaged a fourth, and led to a disastrous decision that resulted in over half a million casualties. Leaders may care about the different contexts in which mines have made appearances, and the successes and failures that their contemporaries had in dealing with them. The book also digs into some of the concepts that explain why certain techniques were used in the past and evaluates different approaches used over time. This book is intended to educate and entertain the public, inspire mine warriors with a sense of heritage, and brief decision-makers on the realities and lessons of mine warfare from a long-term perspective. For these latter topics, the lessons learned section of the last chapter alone would suffice for brevity.

This book was written with four audiences in mind:

1. *History Enthusiasts.* Like any longtime mine warfare sailor, the author became accustomed to being the only naval mine warrior that friends, family, and strangers ever meet. As an advocate and storyteller, the author invariably runs into people who are fascinated to hear about the part these niche weapons played in history's great events. For these people, it is important to concentrate on narrative storytelling rather than walls of rigorous details.

2. *Non-MIW Military Professionals.* Even within the wider military or even the wider Navy, few outside the MIW community pay attention to mine warfare. Fewer still understand it. Yet, in a naval war, mines have long been both a threat and a mission for many outside traditional MIW platforms. Naval mines have been laid by many different nontraditional naval platforms. In some navies, they are laid by every class of ship. They are dropped by bombers, patrol craft, helicopters, and fighter jets. Navies, air forces, and even armies have helped lay naval mines. Ships of every description have been sunk by mines, often impacting critical logistics. Mines have caused some amphibious landings to be canceled and have at times led to undesirable land campaigns. Sailors and officers alike can gain from an understanding of the warfare area to help them when cruising through potentially mined waters. When fighting enemies without navies, mines have often been the *only* waterborne threats, the only $5,000 weapon capable of causing $100 million in damage. Many classes of ships (including battleships) have swept for mines. Entire amphibious readiness groups have been devoted to the task of mine clearance. In wartime, when navies expand rapidly, the first commanding officer or chief petty officer roles for many promising leaders are aboard minesweepers. Enlisted sailors of a great many rates can find themselves assigned to MCM platforms. Just as one might read a history of the USS *Indianapolis*'s sinking to better understand the dangers of abandoning ships, this book can help one understand the impact of mines.

3. *Mine Warfare Professionals.* This class of warriors range from boot seamen and ensigns up to grizzled master chiefs, captains, and sometimes admirals. It includes members of the mine warfare shore establishment who work every day, often with limited resources, to man, train, and equip the mine and MCM forces. When struggling to perfect tactics, techniques, and procedures (TTPs) written by unknown strangers, it is difficult to see these in perspective. Why are certain TTPs used and not others? What else has been tried? What other promising alternatives were ignored? For those who have only ever trained in peacetime, it is difficult to grasp the realities of wartime. For those suddenly tasked with solving a problem not addressed by training, have similar situations happened in the past? What factors and problems might be anticipated? For those developing systems, what has been tried in the past and/or by other countries that might help accelerate the latest project? Is it a greenfield innovation, or was there already an operational program that simply lost funding? For those whose duties give them the opportunity to advise their seniors,

what perspectives and anecdotes can they use to help illustrate a point of why certain courses of action are better? Finally, this book is meant to provide a sense of pride and heritage. Often, while mired in the details of maintenance, manning, training, and operations, it is easy to lose sight of one's place in a long continuum of warriors going back to 1777. What stories can senior people tell younger people to help them learn to love what they do? What stories can help them to develop the sort of professional curiosity that leads to advances? What will inspire them to become the next generation of storytellers, the heroes and the writers of the next chapter in naval mine warfare history?

4. *Military and Government Decision-Makers.* The final audience that this book aspires to reach is the high-ranking officials whose discussions and decisions impact MIW for decades. This includes the three- and four-star admirals and their senior civilians who try to project the future direction of naval mine warfare within their respective areas of responsibility. It includes the fleet commanders whose strategies can be made or broken by MIW. It includes the legislative bodies that can make or break naval programs like MIW. It includes the top naval commanders of various countries whose visions for the next-generation fleet can rejuvenate or destroy MIW. Within the United States, chiefs of naval operations with no MIW backgrounds have done both. Lastly, this includes those staffers who draft the documents that broadcast fleet demand. It includes program officers, procurement managers, and those who write the requests for proposals. All of these have the ability to influence the final form of the equipment that mine warfare professionals use to save lives.

Note: There is an Abbreviations List on page 207 to use as a reference for abbreviations used in this book.

About the Author

As an enlisted sailor, the author started working in MIW when he was 19 years old. At the time, he was stationed on one of two U.S. minesweepers in Japan, learning from a crew who had hunted mines in Desert Storm. He sat at sonar consoles for countless hours, scrutinizing screens for anomalies on the ocean floor. Trained as a technician, he worked late nights, often in the dark, troubleshooting and fixing equipment. Stationed overseas, he was motivated by an anxiety that his country would be embarrassed in bilateral naval exercises. If his equipment failed to perform, or if he failed to find mines, it was *he* who had failed to uphold his country's honor. During minesweeping, he worked for hours on a minesweeper's fantail in temperatures ranging from 20 degrees to 130 degrees Fahrenheit (-7 degrees to 54 degrees Celsius). Like any enlisted minesweeper sailor, his cross-training led him to duties that included navigation, radar, planning, communications, paint chipping, crane operations, maintenance management, public affairs, photography, firefighting, force protection, and even time operating a ship's galley (kitchen). Ashore, he assembled naval mines and managed a mine warehouse. Unusual for an enlisted sailor, he served on three different staff commands. In the first, he observed MIW exercises on 16 different ships around the globe and observed five different MCM squadron staffs in action. Back at the office, he read or reviewed most of the U.S.'s and NATO's MIW doctrine, spearheaded the creation of an admiral-level training assessment program, and wrote the history portion of the mineman training manual. In the second staff command, he did the tactical planning for MCM exercises in the dynamic U.S. Seventh Fleet, where he participated in bilateral and multilateral events from New Zealand to northern Japan. In both of these tours, he learned not just from U.S. MIW warriors but also from those of several other countries. His final tour included managing the logistics for the U.S. Navy's globally distributed inventory of MIW weapons and components. In that capacity, he worked with every U.S. MIW operating unit and most of its shore establishment, including research and development (R&D), program offices, and ammunition depots. Upon retirement, following 23 years of service, he earned his MBA at the University of Notre Dame majoring in strategy and in supply chain management. He did his MBA summer internship at Booz Allen Hamilton in Washington, D.C., directly supporting NAVSEA's PMS-495 and PEO USC program office.

It is from that broad and deep experience that the author approached this book.

Introduction: What Is Naval Mine Warfare?

In a time of regional conflict, a warship streams through international waters from one tasking area to the next. The crew is proud, and they are the best at what they do. Troops ashore rest easily knowing that this ship can rain fire on attackers with ferocity and accuracy. No ship afloat can beat them. No smaller raider has ever gotten past them to threaten the ships that they protect. As they cruise gracefully along, this experienced crew scans the horizon and their empty sensors for enemies. They are proud of their part in their nation's defense, and they hope that it is proud of them in return.

The ship shakes violently. From stem to stern, from bridge to bilge, shock waves tear through the ship, whipping it up and down and side to side. Crew members are thrown off their feet. Some, thrown into overhanging angle irons, crash into the deck with broken bones. Those who were topside splash into the water, temporarily too dazed to tread to stay afloat. Water pours into an engineering space and a berthing compartment, causing panic as sailors climb ladders to escape drowning. Electrical systems are tripped offline; pipes and internal communications are severed, and engines are thrown off their mounts. Somewhere, a sailor whose role seldom takes them outside awakens from temporary unconsciousness to find light streaming in through a huge crack in their office's bulkhead and overhead. The proud ship, the best fighter afloat only moments before, is now dying, brought low by a weapon left by an enemy the night before.

Naval mine warfare (MIW) is a subject that seldomly appears in popular culture in an authentic form. Modern navies are known for the exploits of their carrier-based aircraft, their submarines, and the missiles that they can send hundreds of miles to strike far inland. Billions are spent on defending the world's fleets against the same sorts of threats, i.e., aircraft, missiles, and submarines. Those are the weapons of peer competition, of powerful navies able to stand toe to toe exchanging blows. They are the weapons intended to win huge naval battles akin to Trafalgar, Leyte Gulf, or Jutland. Yet, those are not the battles that modern navies fight. For three-quarters of a century, the only ship-to-ship battles have been skirmishes at best. Instead, navies have been fighting asymmetric battles using asymmetric weapons. While powerful navies make frequent use of missiles, and increasingly defend against them, very few guns and almost no torpedoes have been fired in anger during that time. Against the powerful of the world, weaker nations have scored several high-profile coups using other means. Those means have been overwhelmingly cheap, simple mines whose basic designs would have been familiar to sailors of the early 20th century.

Naval mines are a niche subject to the average person and even to the average sailor. When most people hear the term "mine" in a military sense, they will think first of a land mine. In some sense there is a similarity between the land mine and the naval mine. Both represent a hidden danger for passersby, danger that is not shared by the person planting the weapon. If that average person tries to envision a *naval* mine, they typically think of the type of floating spike ball seen in a TV show like *Gilligan's Island* or a film like *Finding Nemo*. While these do represent a type of naval mine, this is an extremely limited view of an extensive class of weapons. This chapter provides an overview of naval mines and naval mine warfare. By first understanding the subject broadly, it will be easier to understand the context of the details revealed in later chapters.

What Is a Mine?

Mines are difficult to define because they fit on a smooth continuum of naval weapons of varying ability to target their victims. At one end are simple moored contact mines that sit until someone hits them. At the other end are the 16-inch shells and harpoon missiles that ships can hurl at each other. In between are quite a few different weapons. Limpet "mines" are placed against ships' hulls, and advanced influence mines can be exploded only by specific types of ships. Remote-controlled mines also allow operators to target specific victims, an ability that makes them slip in and out of different definitions for "mine." Depth charges were tossed into the water near suspected submarines, yet their ability to explode was not dependent on proximity to their intended targets. This model of placing explosives in the water to target specific victims makes them like limpet mines. Torpedoes are actuated in a manner that is like contact or influence mines, i.e., they explode only when a target is close. Often the only real difference has been the motor strapped to the back. Are they mines? And if they are, then how about drone boats? They are essentially remote-controlled torpedoes that can return to home base.

There are a few different definitions for the naval mine. Here is how Dictionary.com defines it: "A device containing a charge of explosive in a watertight casing, floating on or moored beneath the surface of the water for the purpose of blowing up an enemy ship that strikes it or passes close by it."[1] That definition leaves out limpet mines, depth charges, and bottom mines, which do not float. Hartmann's *Weapons That Wait*, a classic on any mine warrior's bookshelf, describes mines as a type of weapon that lets enemies come to the weapon, rather than the weapon going to the enemy. However, that definition leaves out the targeted weapons like limpet mines and depth charges.[2] While this definition is good, the terse phrase "weapons that wait" is itself even better.

Captain J.S. Cowie described naval mines in another way in *Mines, Minelayers, and Minelaying*.[3] He explained that weapons fall into two categories: missiles and mines. Missiles are weapons that are aimed at and propelled at a specific target and that provide their destructive force at the end of their movement. The purpose of combatant ships and aircraft is to aim and fire missiles, which encompass bullets, propelled torpedoes, and rocket-propelled missiles. However, it could also include modern weapons like lasers and sonic devices as well as ancient weapons like swords, spears, bows and arrows, and even ships' rams. Missiles include anything that causes damage at the end of its movement. Cowie's second category is mines, which are a threat after they have stopped moving and the minelayer is well clear. Hartmann further explains that mines are weapons to which the enemy must come, an expansion that would remove some weapons from the mine category.[4] This is a strong definition, as it covers limpets and more traditionally recognized mines but leaves out depth charges and torpedoes. However, it also excludes drifting mines to include the very first weapons commonly recognized as mines from the American Revolutionary War.

The Department of Defense (DoD) definition of naval mines is "in naval mine warfare, an explosive device laid in water with the intention of damaging or sinking ships or of deterring shipping from entering an area."[5] This definition also has its challenges, as it includes torpedoes. In 2020, a proposal was set forward to expand the definition to include in its elements "autonomous, semi-autonomous, sensor-controlled, or directly controlled seaborne device capable of defeating the adversary, providing means to overcome adversary influences, or collecting intelligence, surveillance, and reconnaissance."[6] Such an expanded definition would take into account the possibilities offered by today's drone technology, which are only just beginning to be explored.[7] As shall be seen in the final chapter, it includes the drone boats used in Ukraine to great effect. By that definition, Ukraine's drones may be a form of naval mine warfare. But then, so would batteries of underwater hydrophones.

This book supports a simple definition of "non-propelled, explosive, waterborne weapons that wait." It fits all weapons that traditionally populate minefields. It also fits both limpet mines and depth charges. Torpedoes and drone boats would not fit the definition, as they are propelled toward specific targets. However, it could include autonomous hybrid systems that work as the underwater version of surface-to-air missile batteries. Those weapons autonomously fire propelled weapons at an enemy but do so without being aimed and fired by a human operator. Now, with that definition in place, the next step is to explain how mines are categorized.

Mines are customarily categorized in several different ways. The first category is the actuation method, i.e., what method is used to ensure that a mine blows up when its target is close enough to sustain damage (and not before). A second category is the mine's position in the water, i.e., on the bottom, on the surface, or somewhere in between. Another category looks at what type of craft lays the mine in the water. Finally, mines can be categorized based on how and where they are used. Additionally, certain other types of devices are used in mine warfare to either train mine countermeasures (MCM) forces or, conversely, to counter MCM forces.

Actuation Method

1. *Contact:* Explodes when a ship makes physical contact with some part of the mine. Some common methods are described below.
    a. *Chemical Horn:* Chemical horns are the most common method of actuation. The typical design includes a glass tube filled with a chemical, inside a soft-metal horn attached to the outside of the mine case. When the horn is struck by a passing ship, the glass vial breaks, releasing its chemical into another chemical. The resultant reaction creates enough of an electrical charge to detonate the mine.
    b. *Galvanic Action:* Some portion of the mine, such as an antenna, is made from a different metal than a typical ship's or submarine's hull. When the

passing vessel makes contact with that metal, the galvanic action of the dissimilar metal creates an electrical charge that detonates the mine.

c. *Tilt Rod:* Similar to a horn, a tilt rod acts as a large electrical switch. If a passing vessel moves the switch too much in any direction, a circuit is closed, allowing an electrical charge to detonate the mine.

d. *Other Methods:* Many other approaches were used in the early years to cause mines to detonate. One of the earliest firing devices was nothing more than the cocked flintlock mechanism from a musket, suspended over a gunpowder charge. When the mine hit its target, the mine would be jarred enough to fire the flintlock, thereby sending sparks into the gunpowder.

2. *Influence:* The mine is actuated when a built-in sensor detects the effect that a passing ship has on its environment. Typically, sensor circuits use some method to distinguish between the influence of naturally occurring events and that of actual ships. Later mine circuits could further distinguish categories of ships and other relevant factors, giving the miner a more bespoke targeting capability. Below is a non-exhaustive list of influence types used.

   a. *Magnetic:* Detects changes in the local magnetic environment that fit the profile of a passing ship.

   Note: Frequently, the uninitiated guess that a magnetic mine works like a big magnet, so that it is attracted to the metal hull of a ship and detonates upon making contact. This is not the case.

   b. *Acoustic:* Detects changes in the local acoustic environment that fit the profile of a passing ship.

   c. *Pressure:* Rapid water pressure changes are detected as the amount of water above the mine shifts because of a passing ship.

   d. *Seismic:* Detects vibrations on the bottom of the ocean that are caused by the noise made from a passing ship. This seismic action is like the vibrations one feels when a passing car pumps out a lot of bass from subwoofer speakers.

   e. *Combination:* Two or more sensor types are combined to reduce the chances of mines being spoofed by non-ship sources.

   For a more detailed explanation of how these different types of sensors work, consult Gregory Hartmann's classic book about mines, *Weapons That Wait.*[8]

3. *Remote Controlled (RECO):* Mine is fired by a shore-based operator. The advantage of this type is that it takes advantage of human intelligence to make the firing decision. This makes it ideal for providing protection for working ports, as the mines do not fire unless the operator chooses to make them fire. On the other hand, the operator needs a way of knowing which mine to fire. Several targeting systems were used over the years, including camera obscura, mirrors, two-person lines of sight, etc.

4. *Timed:* Mine is essentially an underwater time bomb. A delay element (mechanical or electronic timer) is set for a predetermined length of time that allows the layer to escape before the mine detonates. This type is far less common and is not normally included in mine classifications. However, limpet mines (a stick-on explosive device with a timer) and certain other weapons fall into this category. Depth charges may potentially fit in this category, with a water-pressure-sensing device acting as the delay element.

Position in the Water

1. *Bottom (Ground):* Negatively buoyant mine case that is affixed to the seafloor/river bottom by its own weight. Except in very shallow water, such mines are normally influence mines. However, there have also been bottom RECO mines in the past, particularly in the 19th century and in the Vietnam War's riverine operations.
2. *Moored:* Positively buoyant mine case that is affixed to the seafloor/river bottom via a cable, chain, or similar device attached to a weighted anchor. The stereotypical floating spike balls are an example of a moored contact mine. Historically there have also been moored RECO mines and moored influence mines.
3. *Drifting:* Positively buoyant mine case that is not anchored but is allowed to drift with the tide and current. By international law, this type is required to disarm itself after an hour because of the danger that it represents to shipping far from the place where it was first laid. This type was usually contact mines. Most drifting mines float on the surface, although some specialized versions that floated below the surface were also invented over the years.
   a. *Creeping Mines:* Positively buoyant mines usually weighted with chain so that they drifted below the surface.
   b. *Oscillating Mines:* Mines fitted with a bellows system that allowed them to hover beneath the surface within a specified depth range. The historical example was the Leon mine of the early 20th century.
   c. *Subsurface Drifting Mines:* Mine is negatively buoyant but is kept from sinking by an attached float drifting on the surface. MCM forces recovering the floats risk coming in contact with the attached mines.

Laying Platform

Mines can be laid from aircraft, surface craft, and subsurface platforms. Each method offers advantages and disadvantages while also having implications on weapon design. The choice of platform depends on the relative importance of factors like speed, volume, and accuracy of mine deployment, as well as protection of the minelayer and the need for secrecy.

1. *Subsurface Laid:* Historically, submarines have been the primary subsurface laying platform. However, certain other subsurface platforms are also used in minelaying. Limpet mines are normally placed against ship hulls by divers, while unmanned underwater vehicles (UUV) have shown the potential for small-scale minelaying applications. Due to the restrictions imposed by submarine design, sub-laid mines are purpose-built to fit the sub's launching systems. Of note, the first mine-laying platform was also the first submarine, which in turn was armed with the first limpet mine. That story will be told in the next chapter.
2. *Surface Laid:* Ships have long been the primary minelaying platforms in most navies. They can be loaded with large quantities of mines and can very efficiently lay vast minefields with reasonably good precision. Often, they are mounted on wheeled cradles (usually mooring anchors) sitting on railroad-like tracks. Mine designs in this category can be more flexible than those launched by subs and aircraft. Where a

sub-laid mine might have to fit into a torpedo tube and an air-laid mine might have to be aerodynamic, surface-laid mines have no such restrictions. Surface minelaying is not solely the domain of minelaying ships. Just about any vessel of any size can be used to deploy mines, if properly rigged.

3. *Air Laid:* Beginning in World War II, nations started to drop mines from aircraft. This gave tremendous advantages to the minelayers, who could lay minefields quickly at long distances from bases of supply. Such mines are almost always influence weapons, though they have come in both bottom and moored configurations. Mines designed to be laid by aircraft will invariably be installed with devices for slowing their descent to avoid damage upon hitting the water. Some exceptions may be the helicopter-laid mines mentioned in chapter 11.

Specialized Mine Category

If the average person was asked how mines are used, they may guess that they are dropped randomly in the water to blow up passing ships. In truth, mines are dropped at specific places for specific reasons, with passing ships blowing up being only a byproduct. Furthermore, certain devices in mine warfare have more specialized purposes.

1. *Anti-Invasion Mines:* No matter how much a fleet may maneuver, if its purpose is to occupy the land of another nation, then at some point the invading fleet must send troops ashore. To do so in volume entails bringing ships near shore through predictable shipping lanes to access hard-won ports. If this proves impracticable, then the likely alternative is a beach assault or a passage upriver. Mines and other obstacles that are made to prevent amphibious landings will specifically target boats rather than ships. Given the smaller size of such craft, anti-invasion mines can be expected to be smaller in size, more densely packed together, and more precise in their targeting. Historically these have been mostly contact mines, though RECO mines were often used in the shallow waters of rivers and harbors.

2. *Anti-Submarine (ASW) Mines:* Anti-surface minefields need only concentrate on the first several meters below the surface through which ships' hulls travel. In contrast to the relatively simple two-dimensional battlespace of surface ships, submarines operate in three dimensions. Thus, ASW mines and minefields are designed to cover the entire submarine battlespace.

3. *Anti-Mine Countermeasures (Anti-MCM) Devices:* Anti-MCM mines attempt to attack MCM forces or to render their efforts ineffective. Some of the earliest methods from the Civil War were little more than mines hidden under mines, so that forces removing one mine would detonate the other mine under themselves. Anti-MCM devices have evolved in their complexity, from cogs that allow sweep wires to pass through them to today's use of sophisticated circuitry and anti-tamper devices.

4. *Exercise (Drill) Mines:* Because of their persistent threat after they stop moving, mines are among the very few weapons for which "service" (i.e., fully explosive) versions are seldomly used in training. To do so is to put one's own MCM forces at risk, as well as commercial vessels innocently passing through the area. Instead, inert versions are normally used to train both minelaying forces and MCM forces (with some exceptions). Often these are little more than the case of normal mines loaded with inert filler instead of explosives. At other times, they may look

nothing like a real mine beyond having similar characteristics when viewed on sonar.

Every so often, the media reports on a mine washed up on a beach somewhere. Beachgoers get scared and call explosive ordnance disposal (EOD) units to blow up the device. However, odds are that unless a naval war has been fought in an up-current area in the past few years, the device is a safe exercise mine that was lost. As a rule of thumb, if the device has an orange-and-white paint scheme, then it is likely either an exercise mine or a sea buoy.

Minefields

In the 2003 movie *Finding Nemo*, the two main characters encounter a submarine minefield in a random spot in the ocean. The mines are new-looking moored contact mines that are supposedly left over from World War II. To a professional mine warrior and historian, this premise makes little sense. Like all underwater objects, sea mines get covered in sea growth after a very short period. For example, the previous paragraph just mentioned exercise mines washing ashore. Despite being in the water less than a decade, the sea growth is of such thickness as to make the mines barely recognizable. Any weapon left over from World War II would have 75-plus years of rust and would be so weighted down with barnacles and seaweed that they would barely be able to float. Furthermore, great pains are taken after every war to remove minefields, especially moored minefields. While there are still well-charted old minefields in the ocean, it is unlikely that any would have been allowed to remain on the path between the Great Barrier Reef and Sydney, Australia, where the movie takes place.

How, then, are mines used, if not placed randomly? The first thing that one needs to know is that a single mine by itself is practically useless. A mine is but a single point in a vast ocean. To be effective, mines are placed together in groups to form a barrier through which passage is difficult without encountering at least one mine. Because there is a cost involved in producing and laying mines, as well as potential danger to one's own shipping, professional navies do not lay mines willy-nilly in random places. Minefields are placed to accomplish a goal and are classified according to the miner's goal.

1. *Protective minefields* are placed in one's own water to prevent enemies from reaching areas that friendly forces want to protect. Thus, they have been historically used in harbor approaches, off potential landing beaches, and occasionally offshore from coastal shipping routes.
2. *Offensive minefields* are used to deny an enemy's shipping and naval forces the use of their own waters. When mines are placed in the enemy's waters, that enemy is limited, delayed, or prevented from safely moving naval and merchant ships into or out of port. For naval forces, this can mean that those in port cannot leave port for either offensive or defensive purposes. Those ships not in port are forced to go elsewhere for repair, refueling, and resupply. With extra miles required to travel, this limits their operational range, decreases time on station, and can lead to vulnerable bottlenecks when available port facilities lack the capacity to handle the increased traffic. Mining a port prevents cargo vessels from loading and unloading cargo. This has the dual consequences of causing supply chain problems for the enemy while also starving it of revenue from the sale of exports.

3. *Defensive minefields* are placed in international waters to channelize[9] shipping to certain areas. Much like a border fence or a city wall, this provides a static barrier. The barrier acts as a force multiplier, enabling available defensive forces to more efficiently focus their resources on protecting areas through which passage is allowed.

4. One final type used for terrorism purposes in the Middle East is the *nuisance minefield*. These are laid covertly across known shipping channels to ambush passing ships. Because these are unannounced and in unexpected locations, they can create outsized panic by ship captains that no longer trust the waters through which they sail. Examples of these will be described in chapter 11. However, nuisance fields are normally small in scale and are laid by limited local forces.

Mine Countermeasures (MCM)

For almost as long as mines have been used, those targeted by them have sought to remove the threat through some form of MCM. Such forces use methods that take advantage of the vulnerability of mines to get at them and destroy them. The tactics, techniques, and procedures (TTP) used depend on the type of mine being countered and the commander's objectives. It is outside the scope of this book to delve into specific TTPs in current use, but some of the general categories are described below.

1. *Mine Sweeping:* Mine sweeping is the pulling of equipment that is designed to neutralize mines over a wide swath. There are two main categories of mine sweeps: mechanical sweeps and influence sweeps.
   a. *Mechanical Sweeping:* Perhaps the most well-known version of sweeping, the goal of mechanical sweeping is to cut the cables of moored mines. Once the cables are cut, the mines float to the surface, where they are neutralized. A lesser-known version of mechanical sweeping was the paravane, a device rigged at the front of surface combatants like battleships to push contact mines to the side and cut their cables for follow-on neutralization. This reduced the risk to ships passing through potential moored contact minefields. A variation of the mechanical sweep is one that cuts the actuation cables for RECO mines.
   b. *Influence Sweeping:* Bottom (ground) influence mines do not have convenient mooring cables that can be cut. What they do have are sensors made to detect the physical forces generated by ships. The goal of influence sweeping is to reproduce those forces far enough away from the minesweeper to satisfy the logic of these sensors to cause detonation without blowing up the minesweeper itself. Two common types of influence sweeps are magnetic and acoustic. The former re-creates the effects that ships have on the local magnetic environment, while the latter makes sounds typical of ships.
2. *Mine Hunting:* The purpose of mine hunting is to detect underwater and floating objects that display the characteristics of mines. Sonar (SONAR = *SO*und *NA*vigation and *R*anging) is the most common tool used for mine hunting, although other methods have also been used. Modern mine-countermeasures ships ("minesweepers") have normally used sonar that is

either hull-mounted, lowered by cable below the ship, or towed behind the ship. Other MCM forces, like helicopters, EOD divers, and unmanned underwater vehicles (UUV), all have sonars of varying capability.

Historically, some other mine-hunting methods used have included visually scanning the environment for mines (mine spotting) and using giant metal detectors to look for metal mine cases. In the American Civil War, "mine hunting" was little more than two soldiers on either side of a river dragging a rope through the water between them to find submerged objects. Another technique was to drag a grapnel along the shoreline to hook the unseen cables going to RECO mines. Usually, these manual mine-hunting methods were supported by infantry and shipborne weapons.

3. *Mine Neutralization:* Once underwater objects have been found and classified as having the typical features of mines, MCM forces seek to confirm their identity. Identification is typically done using visual means like sonar, but laser and touch are also used in some cases. If the objects prove to be mines, the next step is to neutralize the weapons. The traditional definition of neutralization as it pertains to mine warfare is to render the mines incapable of firing. Historically, neutralization was done using gunfire for floating mines, by placing explosive charges, or by disarming them and pulling them out of the channel. EOD divers neutralize by placing explosive charges or by disarming them. Shipborne mine neutralization vehicles can identify and neutralize mines detected on the ship's sonar. Helicopters and unmanned MCM vehicles also increasingly employ innovative new methods for mine neutralization.

Mine neutralization can have a few meanings. One meaning is the version described above in which the job is done by simply damaging the mine enough to make it incapable of firing. Another version, countermining, seeks to cause a "sympathetic detonation" by blowing up the main charge. In World War I, before devices were developed to prevent the technique, countermining could be accomplished by tossing explosives in the water. Often this would cause a whole string of mines to blow up in one massive chain reaction.

Summary

Now that the basic terminology of mines and MCM has been described, we shall go on to the history section. Further chapters will expand upon the terminology of mines and mine countermeasures to use as a foundation of how mine warfare has developed through a series of key campaigns beginning in the American Revolutionary War.

Note: There is an Abbreviations List on page 207 to use as a reference for abbreviations used in this book.

CHAPTER 1

The Famous Pioneers of Naval Mine Warfare

Like any good history, the best place to start talking about naval mine warfare (MIW) history is the beginning. That beginning best starts with talking about MIW's forefathers. It may be surprising that the cast of early MIW pioneers is filled with inventors who are well-recognized for their other engineering endeavors. Bushnell. Fulton. Colt. Nobel. Even Siemens. These are the more famous names that helped raise mines from mere concept to fully fledged weapons systems.

***David Bushnell**—Inventor of the First Submarine and the First Mines*

It was September 1776, at the dawn of the age of the fighting sails. Only two months into its infancy, the fledgling United States of America was locked in its war for independence from the British Empire. At various points along the coast, Royal Navy ships operated with impunity. Having little to fear from American rebels, they could threaten local populaces into submission with the power projection capabilities of their cannons and crews. Such was the situation in New York Harbor, where a fleet of ships sat at anchor preparing to land British troops to attack the American forces defending Manhattan. Admiral Howe's flagship, HMS *Eagle*, sat quietly at anchor in seeming safety off Governors Island.[1]

In the wee hours of morning, a tiny craft eased its way alongside and underneath the *Eagle*.[2] Built like a barrel and resembling two sea turtle shells attached belly to belly, it had been propelled through the water by a single man operating a hand crank. Impossibly small, only 7.5 feet tall and 6 feet wide, it was hardly a thing of nightmares.[3] Yet this tiny vessel packed a powerful punch. It was the world's first submersible vessel used in warfare, one that later historians would call the *Turtle*.[4]

Bushnell's ingenuity was not limited to just the submarine itself. To be useful, a warcraft must be armed, and so he designed a weapon that could punch far above its weight class. His problem was how to use a craft of only a few cubic meters to destroy a ship that was a thousand times its size. Even the smaller ships-of-the-line, such as the 64-gun HMS *Eagle*,[5] were designed to keep fighting even as they absorbed broadside upon broadside of cannon fire from a similar-sized attacker. Fighting ships could expect to leave a battle riddled with holes from "wind to water," a colorful term meaning from the uppermost weather deck and sails down to the waterline. Most such holes represented little threat to the ship's survivability. However, holes at or just above the waterline allowed water to flood into the ship if not checked by damage-control efforts. Left unchecked long enough, the ship could founder and sink.

Therein lay the fatal vulnerability of ships. Any device that could crack a ship's hull below the waterline would be capable of sinking a ship.

The classic, targeted method of creating holes in ship hulls was by ramming. A ram was a metal-capped beam protruding below the waterline from the front of a galley. Rowed into battle, the galley would collide headlong into the side of an enemy ship to create a large hole in its hull. Invariably, this action sank the ship.[6] Even as late as 1945, the submarine USS *Barb*, all ammunition expended, sank a boat by ramming.[7]

Another method was to block a river or harbor entrance with a massive boom consisting of giant logs connected with thick chains. Naturally buoyant, these booms could stop ships dead in their tracks. Unexpectedly running aground on these logs while running at speed, an enemy ship would find itself at the mercy of local forces as water rushed through the seams of its hull, split by the impact. If an approaching ship proved to be friendly, the harbor security forces could simply move the boom to allow unhindered passage.

For Bushnell and his tiny *Turtle*, neither ramming nor cannon fire were options for facing off against a colossal British man-o-war. Instead, he set about creating a brand-new type of weapon built upon a limited amount of earlier research on the use of underwater explosives. He reasoned that he could sink a ship by setting off a large explosion against its hull below the waterline. By using his submarine to surreptitiously place the explosive, the ship's crew would be unaware of the danger until it was too late. His submarine would have time to escape, and the explosive device would remain unmolested until it burst apart in fiery death for its target. That was the plan, anyhow.[8]

There had been earlier work in underwater explosives, some of which Bushnell may have been aware. The Chinese designed a time-delayed underwater explosive device called the "marine dragon king" that used a burning incense stick as a timer. Deployed in the water upstream of an enemy ship, it would explode once the incense stick had burned down to the main charge.[9] Elsewhere in history, there is evidence that the idea of underwater mines had been presented before. For example, the master of ordnance for Charles I received a directive to produce "divers water-mines, water-petards (i.e., limpet mines), and boates to goe underwater" (*sic*).[10] The first device had a similar use to a mine but lacked a targeting feature. The timing method meant that it would probably explode harmlessly rather than sink a ship. As such, it was more similar to a firework than to a mine, exploding when the fuse ran out regardless of what was nearby. As for the second set of devices, the author did not find any references indicating that they were actually produced or tested. Because of issues like these, Bushnell's device appears to be the first documented limpet-type mine used in war, the first to be designed, built, and deployed with the intent of detonating against an enemy ship's hull after being placed by the operator.

Bushnell constructed his limpet mine from between 150 and 200 pounds of medium-ground gunpowder placed in a sealed wooden container. Attached to a ship's hull by means of a hand screw, it would not activate until released from the submarine. Once released, an internal clock mechanism provided a time delay, after which a flintlock mechanism was released to shower sparks on the main charge and detonate the mine.[11]

That was the plan, anyway. Unfortunately, the *Turtle* never successfully pulled off this scheme. It made a total of three attempts at attacking enemy shipping. On the night of September 6, 1776, Sergeant Ezra Lee managed to get the craft underneath the HMS *Eagle* as it sat at anchor off Staten Island, New York. However, he was unable to attach the mine's connecting hand screw to the ship's hull. Several theories have been devised to explain this failure, the most prominent of which is that the screw could not penetrate the copper cladding that the British navy had started using to protect the wooden hull from sea growth. Whatever the case, as dawn approached, Sergeant Lee decided it was best to retreat to avoid detection. As he made his getaway, British forces spotted him and commenced firing. Jettisoning the mine to reduce weight, he made off at best speed. The mine exploded harmlessly and the *Turtle* made it safely back to American forces. Two more attempts were made in the Hudson River, after which Bushnell abandoned the project.[12]

The First Drifting Contact Mines

Following the partial success of the *Turtle*, Bushnell next decided to pursue a less expensive method of using underwater explosions to attack British ships. For his second project, he created simple drifting mines fashioned from sealed, powder-filled barrels connected in pairs by rope. As the barrels drifted downriver, the concept was that rope would snag onto some part of a ship, bringing the barrels alongside the hull. Once the rope was pulled taut, it would release the trigger of a flintlock mechanism and detonate the mine.[13] He called these devices "torpedoes" after a genus of marine animal, the electric ray, known for packing a powerful punch. This name would continue to be attached to mines until engineers began attaching propulsion devices in the late 19th century, creating the need to distinguish between the two classes of devices that today are referred to as mines and torpedoes.

On the night of August 13, 1777, Bushnell's team deployed several of these mines on the Connecticut River against the HMS *Cerebus*. Spying a rope trailing from one of these devices, the crew of the *Cerebus*'s tender hauled in the rope in hopes of discovering something of value. Instead, they pulled the mine's rope taut when it was close aboard. The resultant detonation sank the tender and killed three of its four crew members.[14]

Bushnell's next attack came in December 1777 on the Delaware River against the British fleet anchored off Philadelphia. As a modification, he added weights to his devices and suspended them beneath buoys to ensure explosion below the waterline. Relying on inaccurate information, he released his devices too late for them to arrive at the fleet before sunrise. Discovering one of the devices, the crew of a small boat attempted to recover it, only to have the device explode and kill several crew members. Suddenly aware of the approaching danger, British sailors began shooting at everything afloat in an action memorialized by the poet Francis Hopkinson in his humorous poem, "The Battle of the Kegs."[15]

This was Bushnell's last attempt at underwater warfare during the war. He continued the fight throughout the war in more traditional military engineering roles. In 1790, he took his submarine to France in hopes of gaining fame and fortune. Such efforts came to naught, though he did happen to meet another engineer named Robert Fulton. Since Fulton went on to make his own submarine-like craft and his own mines, it is entirely plausible that his ideas were rooted in his Paris meeting with Bushnell.[16]

***Robert Fulton**—Inventor of the First Viable Steamship and of Moored Mines*

In 1790, while on a trip to Paris, David Bushnell had occasion to meet a young engineer by the name of Robert Fulton. At the time, the young man was supporting himself through the painting of commissioned landscapes and portraits. Fulton had previously lived in Philadelphia and would have been well aware of Bushnell's mines and his submarine. While it is unknown what was discussed, doubtlessly both technologies were included. Over the next few years, Fulton worked on projects to support canal transport, as well as steamship development.[17]

In 1797, following the French Revolution during the reign of The Directory, Fulton began a campaign to sell the French government a submarine. Although it took a few years, his efforts resulted in the construction of the first practical submarine, the *Nautilus*.[18] Powered by the muscles of two crew members, it could submerge for almost 4.5 hours. Armed with a contact naval mine dragged behind the boat, it was envisioned as a vessel for raiding ports. Through various trials, the *Nautilus* proved both safe for its crew and capable of blowing up ships. Unfortunately for Fulton, he was ultimately unable to win sponsorship for his project from a skeptical French government.[19] It did not help his case that they questioned the morality of this surreptitious form of combat.[20]

Note: The use of a submarine to sneak up on an enemy and attack him unseen was not too dissimilar to the guerrilla tactics used by the Americans during the Revolutionary War. American riflemen would sneak up on the marching British troops and fire upon their leadership from behind trees. At the time, European battlefield tactics relied upon the musket, whose inaccuracy necessitated massed fire from rows of soldiers facing each other on open fields. That style of warfare in turn may have been a vestige of the medieval tactic of the shield wall used for mutual defense. At the time, naval combat was fought in a similar manner by opposing lines of ships, all firing away while using their combined bulks to deny their enemies the ability to attack them from a place of safety. Ships large enough to be in such lines were called "ships-of-the-line."

Not finding success in France, Fulton next tried his luck in Great Britain. In 1805, he arranged a demonstration of his first mine. His mine used a 180-pound explosive (similar in size to that used on the *Turtle*) rigged with a timer, suspended from a buoy, and deployed in a manner similar to Bushnell's mine. In that demonstration, he successfully attacked the Danish brig *Dorthea*, causing the vessel to leap out of the water and split in half.[21] While he generated much interest at the time, the Battle of Trafalgar less than a week later effectively destroyed the French naval threat.[22] Attitudes toward his weapon shifted quickly. As Earl St. Vincent observed, the most powerful navy in Europe had no need or desire for the weapon, which could lose them control of the sea if it proved successful.[23]

Having found no luck in Europe, Fulton returned to the United States to continue pursuing his projects. One of those projects, the one for which he is most well-known, was the development of the first commercially successful steamboat. While other steamboats had been made before, various engineering problems made them inappropriate for the demands of routine transportation. His steamboat, the *North River Steamboat* (or the *Clermont*), made the 300-mile round trip between New York City and Albany regularly, making several stops along the way. By solving the problem of traveling upriver, Fulton's steamboat revolutionized the country's commercial riverine transportation system and opened much of the interior of the country. It was the predecessor to the great, ornate Mississippi riverboats much reflected in popular-culture depictions of the 19th century.[24] In 1814, he attempted to convince the U.S. Navy of the potential value of steam warships (as he had tried to convince Napoleon in 1803) but was eventually rebuffed.[25]

At the time that Fulton returned to the United States, the country was still in the early stages of defining its defense needs. One school of thought was that the country needed to have a strong navy that could project power and prestige overseas, as well as fight off invading navies. The other school of thought, embodied in President Thomas Jefferson, was that a powerful navy was unnecessary. This second school's reasoning came from two main points. First, they felt that a large navy would create a sort of aristocracy that would embroil the country in foreign wars because they had the means to do so. The other main point was they felt that a large navy was unnecessary to the defense of the realm. Defensively, the United States enjoyed (and continues to enjoy) the geographic advantage of being located an ocean away from would-be attackers. Any invaders would have to maintain a logistics train several thousand miles long. Upon arrival, they would have to maintain a blockade along a vast coastline riddled with inlets. Any blockade by a navy then in existence would be incomplete. A concentrated enemy fleet could not cover the whole coastline. A dispersed fleet would be vulnerable.

For the Jefferson administration, it made logical sense to create a fleet of 250 single-gun gunboats that could swarm invading ships and overwhelm their defenses.[26] Such a fleet could not start wars abroad but would theoretically be a sufficient force to fight off would-be invaders.[27] Furthermore, such assets would be cheaper than a frigate, easily dispersed, and harder to hit, and the loss of any one of them would not cause significant damage to the overall defense.[28] While action during the War of 1812 proved the ineffectiveness of this approach, it seemed like a good idea at the time. Naturally, this would not be the last time that a peacetime government would embark on a procurement program that proved insufficient in the face of actual war.

It was in the context of this defense-only mindset that Fulton felt he could find a ready champion for his inventions. In 1810, he received funds from Congress to continue his work on naval mines. With these funds, he created two inventions. The first could be described as the world's first moored contact mine. Affixed to the bottom by a weight and an anchor, it was actuated by a ship striking a forked rod sticking out of the top. Anticipating the need to recover these mines at the end of hostilities, he designed a clockwork device that would neutralize the mine and release it from its mooring after a fixed length of time.[29] A century later, moored contact mines would continue to have these basic self-neutralizing capabilities.

Fulton's second mine warfare invention was a "harpoon torpedo." It consisted of a clockwork-actuated explosive device attached by a length of rope to a whaling harpoon. It was carried by a swarm of small boats, from which a crewman would cast the harpoon into the hull of an enemy vessel. As they departed the scene, the mine would be drawn off the back of the boats. Pulled by waterflow of the moving ship, the device would come to rest beside the ship. After a set amount of time (sufficient for the boat crews to escape), the mine would explode and hopefully sink the ship. While it is unknown to the author whether the designers of the 20th century's Harpoon anti-ship missile had this device in mind, it would certainly be appropriate as its namesake.[30]

PLATE IV, *Fig. 1.*

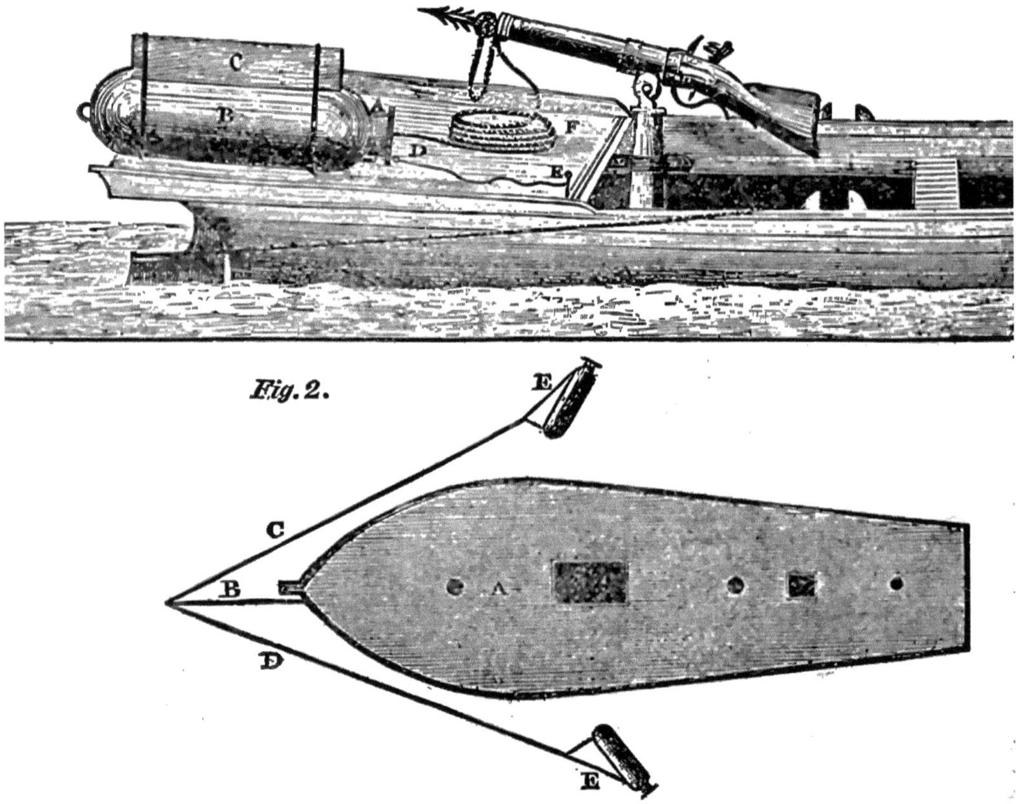

Robert Fulton's harpoon mine from *Harper's Weekly* (1810). After Robert Fulton (artist unknown), torpedo war and submarine explosions, plate IV (Library of Congress).

Later that year, Fulton demonstrated his devices before an evaluation committee. Unfortunately for him, the captain of the target ship for the harpoon torpedo had been directed to design countermeasures. Made from spars holding nets at a distance from the ships, these countermeasures proved sufficient. While Fulton considered it a victory that such extraordinary efforts were required to counter his device,[31] the committee recommended against its adoption. However, they did endorse the value of his moored mine.[32]

While Fulton was unable to get a government contract, his efforts and designs were noted by others who attempted to use the weapons to attack ships during the War of 1812. In July 1813, Elijah Mix attempted to assault the British third-rate ship *Plantagenet* using Fulton's mines. Moored in Norfolk in Lynnhaven Roads, the 74-gun ship represented a threat to the city. After seven aborted attempts, and one premature explosion, Mix ended his efforts to attack the ship. Farther north, forces unsuccessfully tried to use mines against British ships operating off Long Island.[33] The effect of such attacks was to influence where and how the Royal Navy operated near U.S. ports.[34] One additional attempt occurred in the Great Lakes on Lake Ontario. Midshipman McGowan, who is likely to have been involved in Elijah Mix's efforts, left port from Sackets Harbor,

New York, with plans for an offensive attack. Carrying a mine intended to attack the ship-of-the-line *St. Lawrence*, his crew entered Kingston, Ontario. Finding her not in port, they returned to Sackets Harbor. Had *St. Lawrence* been in port and they succeeded in attacking her, this would have been the first offensive mining in a foreign harbor. It also would have been the first and only mining in the Great Lakes.[35]

Samuel Colt—*Inventor of the Six-Shooter and of Remote-Control (RECO) Mines*

Of the inventors discussed in this chapter, Samuel Colt is perhaps the most famous. Colt is known for his revolvers, and his work helped bring multishot weapons to the masses. However, what is far less known is that long before conceiving of his iconic guns, Colt worked with Samuel Morse to develop the insulated wire cables that he would later use to make an electrically fired RECO mine system.

Colt did not invent the concept of blowing up a ship using an electrically fired underwater mine. That was the work of Robert Hare, who used such a device to clear the hazard of a sunken ship. What Colt did invent was the concept of the controlled, networked, underwater mine system. He envisioned a system of an entire field of electrically fired mines, all operated by an observer in a tower. With such a system, friendly ships could pass unimpeded, while enemy ships risked destruction. By using a human observer to provide the firing logic, it had the defensive value of a minefield without the risk to friendly forces.[36]

As a 15-year-old, Colt conducted his first experiments in underwater detonation. Insulating his own wires, he attempted to destroy a raft. He succeeded in giving the raft a good scare and in splashing observers ashore. Having satisfied this itch, he went on to other projects. Nonetheless, as he gained more ideas and information, he added details to the concept that would become what he called his "submarine battery."[37]

In 1836 he had invented his famous six-shot Colt .45 revolver. While multishot revolvers had been available for years, they were hugely expensive, heavy, and difficult to use since the revolving magazine had to be turned and aligned manually. Colt's gun was different. Using a ratchet system to align each bullet with the barrel, the revolver was simple, reliable, and relatively inexpensive. At a time when most firearms had to be reloaded after each shot, the Colt revolver provided a sixfold defensive capability.[38] The government chose not to purchase it for the military. Undeterred, Colt instead sold it directly to the civilian market. Soon his weapon became the hallmark of the frontier.[39]

Early in the 1840s, at a time of tension between the United States and Canada over a border dispute, Colt felt it was a good time to pitch his "submarine battery" concept to the U.S. government. Lobbying politicians rather than approaching the military directly, he secured a $6,000 advance to get started. Raising further funding by selling stock in his newly formed "Submarine Battery Company," he established himself in New York. There he worked with Samuel Morse (of telegraph fame), who loaned him wires and introduced him to Hare's methodology.[40] He was now ready to begin building prototypes of his system.

Beginning in New York Harbor off today's Battery Park, Colt conducted his first demonstration by destroying a vessel with an underwater explosion. Though this enthralled the press and doubtlessly attracted investors, the event was not enough to win a government contract, so he added a follow-on demonstration in Washington. Just off today's Fort Lesley J. McNair, where the Anacostia River enters the Potomac River, he detonated a submerged explosive from shore by sending an electrical signal through an underwater cable. His blast spectacularly destroyed the brig *Volta*, which was anchored for that purpose. Though this engineering feat was hailed a success by the watching crowd (including President Tyler), it also had its detractors who questioned the morality of the proposed system.[41*]

Finally, in spring 1844, Colt was ready to demonstrate his full mining system. Keeping its details a mystery, such as his fire control method for localizing targets, he conducted his demonstration before a gathered crowd of press and politicians. The test was simple enough: An unmanned vessel named *Styx* was released upstream from the minefield. As it drifted over the mines, an observer ashore would actuate the appropriate mine, thus destroying the ship. That isn't quite what happened. With *Styx* passing through the field, one mine fired and then another, with no apparent damage. When a third mine was fired, *Styx* received considerable damage to the bow but continued to float. Subsequent attempts to sink the vessel also failed. Unconvinced of the system's efficacy, reviewing military officials expressed grave concerns about replacing the current coastal defense system with Colt's mines. In the end, the project died in congressional committee.[42]

Hubris is often the enemy of success, and Colt possessed this flaw. That hubris came in the form of believing that he could succeed outside the normal acquisition route. Instead of engaging all the relevant stakeholders to alleviate their concerns and to elicit their support and advice, he went straight to the politicians. Perhaps one of the reasons for this approach was his early failure to get an Army contract for his revolutionary revolver. Whatever the reason, he did not cooperate with the military when developing his weapon system, setting up demonstrations, or trying to sell the idea. Notoriously secretive, he refused to give details about how his system worked to anyone beyond a small number of politicians and reporters.[43]

On the Need to Include the End User When Developing Concepts

The ability of trusted, experienced military members to end or accelerate a weapons development project should never be underestimated. Engineers may devise the most intricate of systems, yet engineering talent is not operational experience. An engineer will know their own system and their own concept, but they may not know the full context in which a system will be used and maintained. Politicians allocate the funds to buy a system, but they too lack such knowledge. Instead, they seek advice from their military advisers, who in turn seek advice from experts from the relevant parts of the military ecosystem.

The cost of a piece of equipment itself is only a small piece of the puzzle. Physical, administrative, and doctrinal infrastructure must be built. If the gear must be deployed from a vessel, then vessels must be configured to the task and procedures devised to ensure safety and efficiency. The gear must be maintained, repaired, deployed, operated, and eventually disposed of by military operators. Tactics, techniques, and procedures used in those activities must comply with existing military standards. The operators must be trained and supported and must be placed within the larger military hierarchy. Furthermore, the capabilities and limitations must fit into the existing tactical and strategic plans, which are often made by those with only a limited knowledge of available non-standard equipment. No one person can plan for everything related to introducing complex systems into the military. Any reasonable objection by any of the stakeholders involved can create resistance that slows the adoption of a system.[44] If the resistance is enough or the problems unresolved, a system can fade into obscurity even after its initial purchase. A system that cannot be used or that is unwanted will never become a cash cow for the inventor. Consequently, Colt's refusal to work within the acquisition process very likely doomed his bid to sell his submarine battery.

While Colt did not obtain a contract for his mine battery system, his concept would soon be adopted in other countries. One such system was deployed during the Crimean War (discussed in the next chapter) in 1854. In Germany, a soldier/scientist named Werner von Siemens used Colt's ideas to develop his own RECO mine system, which was deployed in the defense of Kiel in 1864.[45] That scientist would go on to found the company that today is the massive international conglomerate Siemens AG.[46] Although the U.S. government did not fully appreciate such a system's value at the time, a few decades later it would become a standard part of American coastal defenses.

Note: One of the early complaints about mine warfare was that it was an immoral form of warfare. Reliant on subterfuge, it was seen as the warfare equivalent of a sucker punch or shooting somebody in the back. Unlike a frontal assault on a fort, where both sides assume risk and can defend themselves, the mine is an asymmetric threat. A combatant who encounters mines unexpectedly can neither fight back nor defend themselves. Systems like what Colt proposed were seen in much the same way as roadside bombs are viewed today. Such a system could allow a single operator to destroy a ship or a fleet along with hundreds or thousands of crew members with little risk to the operator. This same moral argument can be applied to today's ballistic missiles or armed drones.

A counterargument is that mines can help to remove the advantage caused by an asymmetry of forces. The United States was militarily weak in its early years. A superior British force captured the nation's capital and torched the White House just a month after Colt was born. Outside of forts and militia, the United States had virtually no defenses against seaborne invasion. Possessing a far larger fleet than the United States, the British could use American waters however they pleased. Like a bully, they could strike almost anywhere with little risk to themselves. In a land with such a huge shoreline, American forces could not be everywhere at once. The strategy for an invader with a strong army and a stronger fleet was obvious: Use their fleet to land the army where the invading force would enjoy overwhelming superiority. This made America's countless rivers, bays, and harbors virtual highways for invaders. The value of a RECO mine system was that it did not require a concentration of naval and land forces to deny an invader the use of waters that did not belong to them. Individual harbors and rivers could be protected by individuals, freeing other forces to concentrate where they were most needed. It gave the weak protection against the strong. Furthermore, by being human-controlled, RECO mine systems were threats only to those who intended harm. For the enemy to avoid destruction, they needed only to stay away. It is in part the moral argument of the preceding paragraph that led to international law requiring that minefields be announced. This allows an adversary the choice to not go where this asymmetric threat existed.

Immanuel Nobel—*Founder of an Empire Built on the First Mines Program*

Of the various inventors listed in this chapter, the name Nobel is perhaps the most remarkable. Swedish-born Immanuel Nobel and his family built a commercial empire in Russia that far exceeds the successes of any of the other inventors mentioned in this chapter. Despite their enormous impact, their story is virtually unknown in the West apart from the prize series that bears their name.

Immanuel Nobel began his career as a relatively unsuccessful engineer and architect living on the outskirts of Stockholm, Sweden. He tried many endeavors in that country, all ending in failure. Possessing a flair for creating ingenious military devices, he had trouble convincing the Swedish military to purchase any but the most mundane products of a rubber company he had founded. However, his interactions with the military as a rubber merchant did expose him to many of the military-related technologies of the time. Such exposure may conceivably have included information about the experiments in naval mine warfare happening in other parts of the world. Whatever the case, he began designing his own variants of the weapon. Unfortunately for him (and fortunately for Russia), he found no champion within the Swedish government. When financial insecurity forced him to bankruptcy, he left home to try his luck abroad. He brought his talents first to Finland and then, at the invitation of a friend, Russia.[47]

His timing in moving to St. Petersburg, Russia, in December 1838 could not have been more fortuitous. Within days of his arrival, he attended a reception where he overheard a conversation about problems being experienced in the development of a Russian naval mine warfare program. The speakers were General Karl Andreevich Schilder, chief of the Division of Army Engineers, and Professor Moritz Hermann von Jacobi, another early pioneer in mine warfare. Hearing of their difficulties, Immanuel described his own mine design and its ability to operate without batteries or observers. A series of demonstrations were arranged. When his designs proved successful, he received a grant and connections that enabled him to build his own factory in 1841. The Nobel family dynasty was launched.[48]

Immanuel's mines were merely the tinder for the family fortune. His sons, Alfred, Emil, Ludwig, and Robert, all learned about engineering and business by working in their father's factory. The factory concentrated on supplying the fledgling transportation industry and the military. A wide variety of weapons came off the line as he sought ever-greater innovation. When the Crimean War threatened, the shop received an order for mines in what may have been the first large-scale government purchase of these weapons. Government contracts arrived continuously, spurring ever-greater investment in capital equipment. Unfortunately, what the tsar gave the tsar could take. When a new tsar came to power, the government's strategy shifted to foreign sources of supply. Domestic suppliers like Immanuel were suddenly left with useless contracts and no recourse. Forced to bankruptcy once again, he left the country and returned to his native Sweden. However, he did not retire in obscurity. Remaining active, he continued working on new projects to include conceiving the product known today as plywood. To make this possible, he also invented the rotary lathe necessary for its manufacture. His retirement was not entirely happy. In 1864, an experiment with nitroglycerine tragically resulted in his son Emil's death.[49] This death may have spurred another son, Alfred, to develop his most famous invention.

Alfred Nobel is famous for two things: TNT ("dynamite") and the Nobel Prize. As a young man, he first learned of nitroglycerine from its inventor, Ascanio Sobrero.[50] At the time, the explosive was highly volatile, making it impractical for field use. As the son of a weapons manufacturer, however, Alfred saw the potential value of the substance and would experiment with it in later years. In 1867, three years after his brother's death, he discovered that by mixing the substance with a neutral absorbent like diatomaceous earth, he could render it stable. His invention, dynamite, would make him a fortune as it was rapidly adopted for weapons of war and for civil engineering.[51] It was used in naval mines for well over a century.

In 1888, when his brother Ludwig died, Alfred had the unique experience of reading his own obituaries written by a press that was mistaken as to which Nobel brother had died. What he read appalled him, for the obituaries widely condemned him as a dealer in death. Dismayed by this description of his life, Alfred rewrote his will to give all of his assets to an endeavor dedicated to celebrating the achievements of all mankind. The resultant foundation would award five annual prizes in the areas of physics, chemistry, medicine, literature, and peace (a sixth prize was added in 1968 for economics). Of these, the Nobel Peace Prize is the most prestigious.[52]

Immanuel's other two sons, Robert and Ludwig, developed Russia's oil industry from a niche enterprise into a major industry. Following their father's bankruptcy and departure from Russia, Ludwig stayed on to manage his father's old factory for its new owners. In time, he built his own factory, where he soon gained some prominence, though he limited his exposure to government contracts. To support one such contract, in 1867 he sent his less successful brother Robert on a reconnaissance trip that led him to the Caspian seaport of Bawa. There he found a veritable gold rush of oil prospectors using ancient techniques to harvest oil from a ground saturated with the substance. Using pits to extract the oil, barrels and carts to transport it, and single-tank stills to refine it into kerosene, the wastage was immense, the product quality poor, and the costs astronomical. Using 25,000 rubles given to him by Ludwig for the purchase of walnut wood, Robert bought a plot of oil-producing land and a small refinery. This impulsive decision was the true start to the Russian oil industry.[53]

After informing his brother of his decision, Robert began improving his fledgling business. By using a drilling technique rather than digging a pit, he could access more oil at a greater flowrate. Rather than hiring hand-drawn carts, he built the area's first oil pipeline to move oil to his refinery. An experienced chemist and engineer, he redesigned his refinery to a multistep system that produced quality on par with the best in the world. In time, his brother Ludwig became involved. The more experienced businessman, Ludwig concentrated on reducing costs, improving quality, and improving the lot of the workers. Among his many innovations was the world's first oil tanker for transporting oil in bulk across the Caspian Sea and up the Volga. He eventually built a network of storage locations, refineries, and railroad tank cars that established the Russian fuel oil industry. Realizing the value of a happy worker, he looked to his workforce's welfare by providing quality housing, education, and recreation facilities. While this approach was wildly unpopular among wildcatters who preferred to keep their workers in squalor, his treatment of his workers provided many benefits.[54] His basic philosophy continues to be used by oil companies today to support workers living in remote locations.

The legacy of the Nobel family's empire is truly astonishing and began with an inventor's chance encounter with two men looking for a working naval mine.

Chapter 2

The Crimean War

In truth, in a book about mine warfare, the Crimean War might not normally warrant its own chapter. Naval mines, or "infernal machines," barely make a blip in the historical record of the period. Popular memory of the war, if any memory exists at all, is from references like "charge of the light brigade" or "the thin red line" (both from the land-based Battle of Balaclava). Nevertheless, the Crimean War was extremely important in shaping the future of naval mine warfare. As the first major nation to employ the weapon in war, Russia was the standard bearer for its use. By using mines successfully in a layered defense, it was able to thwart some of the most powerful navies of the day. This proved to the world's navies the value (if not the morality) of this weapon system. Moreover, neutral observers from the United States, visiting the battlespace of this first industrialized war, brought home reports on these weapons. The knowledge thus influenced how the Confederacy prosecuted its naval campaign during the American Civil War. Furthermore, the layered defensive system used by the Russians in the Crimean War went on to inform how America designed its own coastal defenses less than three decades later. Without Russia's investment in and successful demonstration of these new inventions of Immanuel Nobel and others, naval mine warfare may have never had its same level of impact on world history.

The Roots of the Crimean War

Like far too many wars throughout history, the Crimean War had its base in religion. Three empires, three religions, and one Holy Land. At that time, the Ottoman Empire had become a weak and technologically backward country dominated by other empires. Prior to World War II, the area that is now the nation of Israel was then a part of the Ottoman Empire. In the years leading up to the war, pilgrims flooded this area to pay homage to sites like the Holy Sepulcher. Russia, as the largest Orthodox Christian nation, proclaimed itself as the protector of all Orthodox Christians within the Ottoman Empire. France, as a dominant nation in Western Europe, claimed a similar role for Catholicism. Both nations, France and Russia, had treaties with Constantinople allowing them to exercise that protection. Unfortunately, in an age of empires, neither would be satisfied with anything short of the supremacy of their own rights over that of the other. Following a series of incidents, France forced a decision in their favor by sending a state-of-the-art gunship through the Turkish Straits to threaten the Ottoman capital at Constantinople. In response, Russia invaded Wallachia and Moldavia (modern Romania) and threatened to invade and partition the Ottoman's Balkan territories. Against this threat, Britain and France sent their overwhelmingly superior navies to support the Ottomans.

Desperately, the other Great Powers (Austria, Prussia, England, France) tried to negotiate peace between Russia and the Ottomans. While negotiations were still ongoing, Russia took a step to try to increase its leverage over its southern neighbor. Russia's naval position in the Black Sea was weak. It was evenly matched with the Ottoman navy but had no hope against the French and British navies. Wanting to establish Black Sea naval supremacy in the post-negotiations period (after the Western navies left), Russia staged a devastating naval raid on the Ottoman fleet.

On November 23, 1853, eight Russian line-of-battle ships entered the Ottoman fleet concentration area at Sinop harbor on the southern Black Sea coast. In port, they found 11 Ottoman ships resting at anchor under the fortress guns. Firing explosive shells for several hours, they sank the Ottoman fleet and set the town ablaze. This first battle using explosive shells demonstrated to the world the vulnerability of wooden ships to this new technology. It signaled the need to transition from the highly flammable wooden sailing hulls of old to less-flammable options like steam-powered iron-built ships.[1]

International tensions soon boiled over as news of the attack reached the wider world. Pushed by the British press, on March 27 and 28, 1854, the British and French (respectively) declared war on Russia.[2] Their initial plans were designed to punish Russia. First they sent a combined British/French fleet under Admiral Napier to blockade and raid Russian ports and shipping in the Baltic.[3] Their other action was to land troops at Constantinople and Varna, Bulgaria, to prevent a Russian invasion of the Ottoman capital.[4] When the results of these actions proved unsatisfactory for the resources expended, the Allied forces planned a more decisive campaign. That plan? To capture Sevastopol on the Crimean Peninsula.[5]

During the Crimean War, Russia had both contact mines and remotely controlled (RECO) mines. With RECO mines, the miners had the problem of fire control. For both types, there was the problem of ensuring that mines were laid in an operable condition.

Fire control means firing a weapon so that the projectile or the weapon connects with the intended target enough to cause harm. It is a problem that often must consider four dimensions (range, bearing, height above/below the surface, as well as timing). An explosion that is too far away is wasted. An explosion that happens too early or too late is wasted. To be effective, the person firing a weapon must accurately estimate an adversary's *future* position. They must fire the weapon at the correct time so that the target and weapon arrive at the same position at the same time. With projectile weapons, the bearing along which the projectile will travel is easy to figure out, though it is affected by various ballistic forces. On the other hand, it takes time for the projectile to travel, during which the target can change course or speed.

The fire control problem for contact mines is simple enough. Thickly blanket an area with them and an enemy ship cannot help but run into one. Observation mines fall somewhere between contact mines and projectile weapons for fire control difficulty. On the one hand, travel time is not a factor as the weapons are already in position. On the other hand, the challenge is correctly estimating the target's range as well as the weapon's exact resting position in the water after it was laid. With no visual cues, it is easy for an observer to misjudge the weapon's position relative to the target's position. It begins with knowing where the mine is located. Without GPS, the drop position must be calculated through the never-quite-precise technique of triangulation. Once in the water, the mine's position must be known with accuracy and precision by the shore observer, who must also be able to estimate the target's position with accuracy and precision.

To better understand this problem, imagine that you are burying a treasure in the middle of a huge grassy field surrounded by trees. How do you keep track of where you buried it without placing an obvious marker that will alert others to its presence? How do you relay its position to someone else who comes along later to guard it or recover it? Without closely positioned visual cues, it can be quite difficult to reliably track the burial site's position. This is the problem faced by miners who place mines in a harbor or bay surrounded by water. Various techniques have been used over the years to accomplish this, including bearing markers, camera obscura, multiple observers, etc. Some of these techniques will be discussed in more detail in a later chapter. The problem becomes even harder when mines are laid beyond the sight of land where there are absolutely no visual cues.

Some points should also be made about mines being laid incorrectly. Historically, poorly trained workers operating from crafts of opportunity are more likely to make mistakes. Purpose-built minelaying craft tend to be manned by trained professionals and are designed to simplify or automate the safety device removal and arming processes. On the other hand, it is not always a bad thing when some (not all) mines are laid in inoperable conditions, especially in home waters. First, they are safer to remove. Every interaction between a live mine and clearance personnel includes the possibility of a fatal error. Second, so long as their presence is easily detectable, and other mines are active, they retain their demoralizing effect. The weapon is the minefield, not the individual mine. A minefield can achieve certain objectives (like deterrence) so long as there are enough live mines to appear as a credible risk. Third, mine clearance timelines are not greatly reduced when the armed condition of the mines is variable. Finally, if enough mines are inoperable, then there is the potential for enemy mine clearance personnel or ships' captains to become complacent. If they encounter enough duds, they may stop respecting the mines, which in turn can lead to their taking unnecessarily high risks.

Mining in the Danube River

In 1854, Russian forces gathered in the Balkan possessions of the Ottoman Empire to threaten Constantinople. With their lines of communication stretching along the Black Sea coast, they were acutely aware of the critical vulnerability of their Danube River crossing points. The Danube is quite massive as it nears the Black Sea and is thus a significant obstacle.

Rivers have long acted as barriers for land forces. While on land, armies can choose endless paths to travel from point A to point B. However, where rivers are too deep to cross on foot, options for crossing are severely limited. Crossings are generally limited to either fords (shallow spots), bridges, or ferry points. In all cases, such points are eventually discovered by adversaries. At the same time, these points are bottlenecks because they limit traffic flow. An army that is crossing a river will often amass on both sides where their split forces are vulnerable to attack. Through most of history, that vulnerability was to land-borne attack. However, by the Crimean War, steamships were becoming relatively common. Steamships could travel upriver and could somewhat quickly threaten concentrated land forces with guns far larger than any land-based mobile artillery. Furthermore, troops traveling by river or marching along river roads were especially vulnerable to attack by armed steamships that could at times travel faster than the army's marching pace. This was a relatively new phenomenon for which mines appeared to be an ideal countermeasure.[6]

In 1854, compared even with 1855, naval mines were very much a new technology. Engineers working on the Danube, far from St. Petersburg, would have had little more than rumors and technical concepts to work with to craft what were effectively underwater improvised explosive devices (IEDs). Doing their best with limited information, the Russians constructed their improvised mines at Galati, Romania, and then transported them to Tulcea, Romania. Their goal was to stop British and French ships from moving up the Danube.

Why Tulcea? A touch of historic context and a look at Google Earth reveals a possible answer.

The Danube River is one of the great rivers of Europe. Navigable from Kelheim, Germany, to the Black Sea, it provides access to the Balkans and much of Central Europe. For much of its length, it arcs generally east-southeast. At Silistra, Bulgaria, it abruptly turns north for about 120 miles before turning east-southeast once more near Galati. From there, it is about 50 miles before it branches out into the Danube delta near Tulcea into three main channels that wind their way to the coast. The central Sulina branch is the shortest, and least twisted, and thus the most logical route for shipping of the period. It was there, near Tulcea, that the Russian engineers did their work.

Russia wanted control of the Danube, yet it was under threat by allied naval forces. At the time, they had pushed south from their bases at Galati and Izmail to besiege and capture Silistra. From there, they hoped to rush 90 miles south to capture Varna, the principal port on the Bulgarian coast, before the British and French could use it to land troops. That effort would be greatly aided by having the ability to use the Danube to swiftly move troops and supplies unmolested over the 120 miles between Galati and Silistra.

It was within this context that a small troop of 10 enlisted men under Lieutenant Boreskov set about trying to devise a method for blockading the river using mines. At the time, there was no mining doctrine and only improvised weapons available. While there was probably some brainstorming that occurred with the siege engineers at Galati before Lieutenant Boreskov's men set out, they would have essentially had to figure their own solution on the scene.

The ordnance this team had to work with were double-walled and sealed boxes containing 54 pounds of powder each. Having little prior information, the first test thing that they did was to test to see how far apart these mines had to be to prevent them from damaging each other. When they found that 25 yards was sufficient, they next tested to see if the explosive weight was enough to sink a ship. Their test vessel was an old two-mast ship, which they filled with dirt and whose deck they filled with gabions to simulate a much stronger ship. After failing to sink it with 51 salvos from a battery of shore-based cannons, they ran the vessel over a controlled mine containing 30 pounds of powder. From 1,250 yards, they detonated the mine, instantly reducing the ship to small pieces. Satisfied that their mines were sufficient to the cause, they selected a site where the channel narrowed near a spot suitable for a shore battery.

Now that they knew that (1) the mines could destroy ships and (2) with 25 yards separation they wouldn't destroy each other, the mining team placed their field.

Their minefield was sited at a point of the Sulina where the navigable channel was only about 170 yards wide. Rigging the mines in a checkerboard pattern, they wired them to blow two at a time due to a lack of materials. Like any RECO miner, they had the fire control problem of how to correctly identify when a ship is within a mine's firing range. They solved the problem in the simplest of ways. Using a lookout tower 233 yards from the river as their observation point, they planted two trees on the shore to look like a continuation of the forest. However, when viewed from the lookout tower, these trees bracketed either side of the minefield so that any ship seen on the river between the trees would be over the minefield. This point also fell under the protection of a battery of cannons, providing layered defenses for this section of the river. As an added layer of defense, at night the miners placed four contact mines in the river to catch any allied ships trying to slip through in the darkness.[7]

The fire control problem that was described earlier is made much simpler on rivers than in open waters. In open water, a target's position can be anywhere within a two-dimensional plane. A river is much closer to being a one-dimensional mining fire control problem since the navigable channel is finite in its width. Rivers were historically avenues for moving troops and materials quickly. However, like the real roads of Afghanistan and Iraq, they were often used as killing zones because they make it easy to predict the path of travelers. This has been seen time and again, from Bushnell's kegs in the Delaware River during the Revolutionary War, all the way to the mines intended for

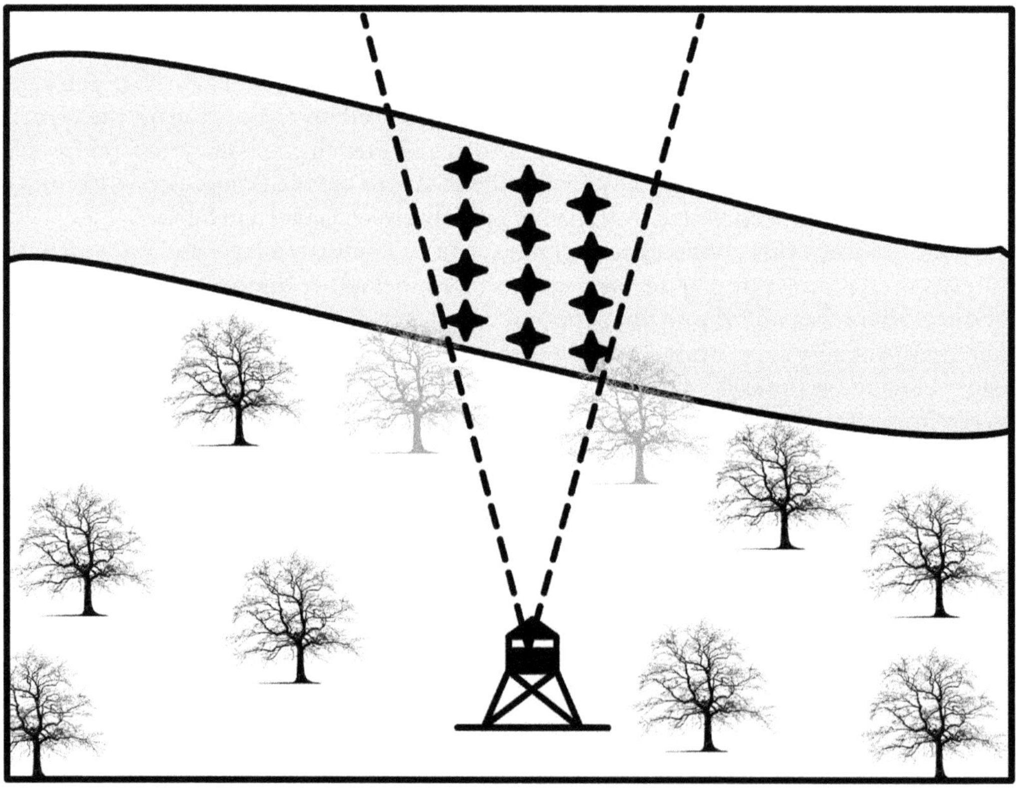

Mine fire control method on the Danube River, 1854 (M. Mazurek, after original drawing by the author).

the river at Umm Qasr during Operation Iraqi Freedom. Without a doubt, rivers will be mined again in future wars, for as long as troops are still transported by boat.

Baltic Campaign

Belying the commonly used name "Crimean War," the war was not fought solely on and near Russia's Crimean Peninsula in the Black Sea. Though the main ire was directed at those directly responsible for the action at Sinop, the allies waged war against Russia in general. Thus, campaigns were fought in most places where Russia met the sea, a domain where Britain enjoyed dominance. Russian and allied naval forces fought in both the Pacific and Baltic as well as the Black Sea. However, it was in the Baltic where Russia's defenses were strongest and where naval mines played their most significant role.

Russia had little choice but to have strong defenses in the Baltic. It was here at the very edge of its vast empire where its power was concentrated. St. Petersburg, the capital city built by Peter the Great, was meant to bring the empire into the circle of great Northern European powers. When the capital was previously in Moscow, foreign visitors to the capital had to travel for weeks or months to progress the 700 linear kilometers from the sea. This distance also made trade exceedingly expensive, thus making it less likely. For an emperor who wanted to be seen as a peer to the sophisticated royalty of France and England, such a capital would not do. He built his new capital in what was previously swampland at the very edge of the Gulf of Finland in recently conquered territory. This gave his court and his traders access to the sea.

When Napier sailed his fleet into the Baltic in 1855, his purpose was to punish Russia. Two of his most important targets were Kronstadt and Sveaborg, which were both series of island-based outer fortresses respectively protecting the maritime approaches to St. Petersburg and Helsinki.[8] Both had layered defenses of forts scattered across several islands, supported by naval mines. Arguably, Helsinki's approaches should have been easier to protect, for the entire harbor entrance is only a few miles wide. Yet its approaches were much deeper, making them harder to mine with available technology beyond cannon and mortar range. St. Petersburg's harbor mouth, at about 12 miles wide, should have been harder to protect. The major difference was that the much shallower water depths in the approaches to that port made them easier to mine with Nobel's weapons.[9] It was because of the difference in water depth that Sveaborg was destroyed, while Kronstadt remained standing throughout the war. However, the allies never passed beyond either fort because of the sheer volume of mines (994 at Sveaborg and 1,865 at Kronstadt).[10]

The fleet that Napier controlled in the Baltic had some 45 or more ships, including 27 ships-of-the-line. To oppose this force, the Russians could muster a similar number of ships to include their own 27 ships-of-the-line.[11] However, quantity is not everything in naval conflicts. Of perhaps greater importance are the tactics, techniques, and traditions that come with experience. For example, in times past, English ship captains exercised their crews at the main guns for an hour or more several times per week using live ammunition. This led to very proficient crews who understood exactly how to safely maintain continuous fire and exactly how conditions changed when firing so long.[12] Other navies that only practiced dry-firing with occasional single live shots could never achieve such adeptness. Efficiency like that is a microcosm of the larger type of proficiency with which a nation long at war approaches future wars. This is where the Russians differed from their British and French adversaries.

Russia has historically been a great land power but had never been a great naval power. Because of the boxed-in nature of its geographic situation, Russia was a literal backwater among maritime nations. Always at a disadvantage in naval conflicts, prior to the Cold War its fleet primarily concentrated on stinging the middle-sized fleets of its closest neighbors. Such battles were usually preemptive in nature, i.e., attacking before their enemies were strong enough to succeed in their own attacks. By the beginning of the Crimean War and even into the 21st century, Russia's entire tradition of naval victories included just half a dozen battles.[13] With such a limited tradition, the Russians would have had little understanding of the true conditions of protracted war at sea. This in turn led to tactics, techniques, and procedures (TTP) that fell under the control of administrators, engineers, and scientists without the insight brought by hard, lived experience. In contrast, their British and French foes had naval traditions of several dozen battles and countless skirmishes across half the globe going back over several centuries. Many of those battles were presumably fought within the professional careers of some of the same sailors who sailed into the Baltic for this latest campaign. Russian and allied fleets may have been similar in size, but they were worlds apart in organizational experience. Thus, Russia rightfully expected defeat if it fought at sea.

With this dose of reality in place, the Russian navy performed a primarily defensive role in the war. In the Baltic, admirals kept their fleets bottled up in port behind

***Merlin* and *Firefly* struck by mines off Sveaborg (Helsinki) during the Crimean War, 1855. John Wilson Carmichael, artist; James Charles Armytage, engraver; and James S. Virtue, publisher (Wikimedia).**

powerful layered defenses, while in the Black Sea the fleet was intentionally sunk to act as a physical barrier. Riga, Helsinki, and Petersburg all enjoyed outer defenses of gun batteries, fortresses, naval mines, and traditional physical obstacles like sunken pilings and anchored booms.[14]

Naval mines proved their greatest worth at Kronstadt, where miners liberally sprinkled the weapons everywhere that it was thought that invading ships might venture. Four of Napier's ships struck mines, being saved from destruction only because the eight to 12 pounds of powder used in the mines was insufficient for the task. Despite this shortcoming, explosions would have alarmed the fleet's crew and given commanding officers pause at the notion of venturing through mined areas.

These moored contact mines, the simplest and most portable among several options, were the easiest to deploy. Crews working from boats could carry several such devices at a time, casting them overboard anywhere where a lead line revealed a proper water depth. Working in this way, a flotilla of small boats blanketed the ancillary areas with these simple weapons, these "infernal machines." Though beyond the protection of fortress guns and easy to counter by allied sailors working from their own small boats, they performed a deterrence function.[15]

Consider the perspectives of Admiral Napier in 1854: Tasked with punishing the Russians, he knew that the best punishment of all would be the pillaging and razing of the Russian capital at St. Petersburg. However, to get there his fleet would need to thread the city's seaward defenses. Those defenses began with a vast quantity of naval mines placed everywhere that a ship might anchor beyond cannon range. Although easily discovered and removed by small boat crews, such work was dangerous and took a very long time. So numerous were these mines that four ships had already received varying levels of damage from them. While the explosive load was low, and quite a few were found to be laid without their safety devices removed, quantity has a quality all its own. He would know that by the time his men had cleared enough of these mines for his ships to operate with reasonable safety, much time would have passed while the perceived danger would cause his fleet's morale to plummet. At that point, the real work would begin.

As ships approached Kronstadt, they would find a mutually supporting line of thick-walled forts bristling with heavy cannons. Unlike at Sveaborg (described later), these fortresses entirely covered their islands, offering no place where single batteries could be neutralized to facilitate the landing of troops. Not that troops could land. Under the fortress guns lay a variety of far deadlier mines, if defecting Russian workers were to be believed. Primarily controlled mines, each would be much more difficult to find and counter than the smaller contact mines laid in the outlying areas.

There were several types of mines laid closer to Kronstadt's fort line. One type was bottom mines of greater explosive weight, laid out in lines across the channel. An operator working ashore could detonate the whole line at once. In such a configuration, fire control was made simple by allowing the operator to sight along a single line of bearing. Anything on that bearing was guaranteed to sustain damage. Another mine type was a triple-cased controlled mine containing 450 pounds of powder and suspended by a buoy. Unlike the small, moored contact mines, these much larger mines were deemed powerful enough to destroy anything attempting to force its way up the narrow channel. Nor was this mine a one-off. By the time Napier's ships arrived in the Baltic, nearly 600 of these mines had been churned out by a local factory, with 11 more being built each day. In addition to being human-operated, they were also given a contact-mine functionality, which allowed them to work even when the operators were distracted. One final type of mine might best be described as a portable controlled mine. Capable of being operated underwater, underground, or even suspended from a tree, it could be quickly deployed anywhere it was needed. With such a weapon, even if Napier's fleet managed to breach a fortress's walls, it would still find it heavily infested with mines capable of devastating his land forces.[16]

To Napier, who had no body of mine countermeasures tactics to rely upon, the heavily mined approaches to Kronstadt would have represented a nightmare. The northern channel was entirely blocked by sunken pilings, i.e., huge logs sticking up from the bottom that would tear great gashes into the hull of any ships attempting to force the passage. That way was not an option. He couldn't assault the forts without first clearing the contact mines, yet the small boats capable of clearing such mines were vulnerable to the fortress guns *and* the human-operated mines. Even if he were to acquire a friendly pilot to sneak his ships past the forts in the darkness through the one channel unobstructed by contact mines (if such a channel existed), there were still physical obstacles. Sprinkled heavily throughout the area were booms made from ships' spars floating in the water between pairs of anchors, a classic naval impediment. A ship attempting to run over a boom might be able to break it but was more likely to find itself stuck. One type of boom heavily used entailed not one but two massive spars bolted together and capable of heavily damaging a ship's hull.[17] Finally, if Napier managed to get at least part of his fleet through this gauntlet of guns, mines, pilings, and booms, he would still find the intact and alerted Russian fleet waiting to finish the job. It may come as no surprise that Admiral Napier, as well as Admiral Dundas the following year, chose to not test the defenses at Kronstadt.[18]

One item of note happened during a major electrical storm in 1855 off Cape Lisy Nos on the northern passage around the Kronstadt fortress when several mines spontaneously exploded.

Upon investigation, it was discovered that the only mines to have exploded were remote-controlled mines. Otherwise known as galvanic mines, these mines required a charge of electricity from a battery located ashore. At the time, such mines were normally left disconnected until enemy ships approached. Once connected, a switch was activated, thus closing the circuit and allowing electricity to flow through the mine's firing circuit. However, during an electrical storm, the atmosphere itself would be charged with electricity. Even without lightning, the long mine cables could act as antennae drawing power from the air itself. Without a clear path to ground, it was thought by the investigating engineers that this induced charge flowed through the mines, causing detonation. To fix the problem, these cables were connected to electrodes placed in the water to encourage these electrical charges to flow in a safer direction.[19]

This was not the last time atmospheric conditions interfered with naval mines. In one infamous episode to be discussed in the chapter on the Vietnam War, several hundred magnetic mines off the coast of North Vietnam spontaneously detonated during a magnetic storm.[20] After all, mines as a technology are heavily reliant on the harnessing of physics. Engineers designing mines or their countermeasures must do so keeping the more obscure effects of physics in mind.

Farther afield at Sveaborg, the allies achieved only moderately more success than they had at Kronstadt. As the fortress protecting the approaches to Helsinki, the capital of Finland, Sveaborg was perceived by both Napier and Dundas as an important target in their campaigns to punish Russia. However, neither admiral could get their fleets past its defenses. Sveaborg (Swedish name), now known as Suomenlinna, Finland, was a series of fortified islands interconnected by bridges. With thick walls and studded with cannons, it was difficult to damage through direct assault by the heavy ships with which Napier sailed. Of particular relevance was a large offshore shoal water area with depths ranging between 10 and 20 feet. This shoal forced the ship captains to choose to either pass directly past the fortress guns at short range or take a more navigable channel that placed the fortress beyond cannon range.[21] For Napier, whose heavy ships were inappropriate for the inshore work required in the Baltic theater of operations, neither option promised much hope of success.

Napier is remembered as being a timid commander, yet his actions reflected both his orders and his available forces. His primary job was to blockade the Gulf of Finland and to take on the Russian fleet if they could somehow be drawn out. However, he was told to not risk his fleet trying to attack heavily fortified Sveaborg and Kronstadt, which were considered to be impregnable. Consequently, after some reconnoitering and intel collection, he limited his actions to lightly held ports and strongholds before impending winter obliged him to return home.[22]

A year later, Dundas brought a larger combined British/French fleet back to the Baltic Sea. Taking on board the lessons of the 1854 expedition, the Admiralty equipped Napier with many mortar barges and gunboats. Mortar barges were shallow-drafted platforms holding huge mortars able to send shells over walls and down into fortress interiors. Gunboats were relatively small boats carrying single, large-caliber guns that could cause outsize damage while operating in the shallows. Carrying their one huge gun longitudinally, they aimed their cannon by pointing the whole boat at the target. This method had the added benefit of making them very small targets when in a firing position.

While there are few detailed accounts about the seaborne defenses around Sveaborg, they can be pieced together from various documents. Within the safer western channel, two ships had been sunk to create an obstacle to free movement. Compared with the previous year, 17 new batteries had sprung up on various islands, promontories, and boats in the harbor. Lines of controlled mines crisscrossed the channels leading into the harbor proper.[23] Like at Kronstadt, booms and iron-bound pilings heavily peppered the area as further obstacles to free and speedy movement.[24] As if the fortress guns were not enough, two warships were moored broadside onto two of the islands.[25] Due to water depth, it is unlikely that controlled mines extended into the northern channel directly in front of the fortress, nor is it likely that such mines would have been used on the shoal for fear of the wires being damaged by passing ships. Nevertheless, it is highly likely that the defenders spread Nobel's contact mines throughout the inner harbor as well as near-shore areas and possibly the shoal. Overall, it was a very hostile environment for the British and French ships-of-the-line.

Unlike the previous year, in 1855 the fleet came prepared for inshore work. Placing 21 mortars vessels among the rocks and islands of the shoal at a range of 3,300 yards, supported by gunboats ranging closer to shore and four large warships behind, the fleet set to work. They pounded the fort with 12- and 13-inch mortars and with heavy guns for some 48 hours, exhausting 100 tons of gunpowder and 5,000 tons of shot and shell. By the time they finished, flames were seen in the fort. Reports on actual damage varied wildly from total devastation to hardly a scratch, depending on who did the telling. On the allied side, not a sailor was killed, though several suffered wounds from their own weapons. Nevertheless, despite their efforts, they could not pass into the inner harbor itself due to the mines and other obstacles. Unable to achieve anything else, the fleet retired.[26]

On July 24, 1855, occurred an event that helped to warn the allied Baltic fleet of the presence of electrically fired mines in the approaches to Sveaborg. Admirals Dundas, Penot, and Seymour were aboard the steamer-frigate *Merlin* in company with escorting vessels as they made a reconnaissance on the port's defenses. As they approached the shore, a whole line of explosions erupted ahead of them. Though doubtlessly the sailors were shaken up, the vessels themselves suffered no damage. The weapons responsible for the fireworks were a line of electrically fired mines whose explosive loads were insufficient to damage hulls beyond two feet away.[27]

One possible explanation for these premature explosions could have been the classic observation mine problem of fire control. The miners may have estimated the ship's position incorrectly. On the other hand, the other explanation is the one that was reported, i.e., a problem with the equipment itself. According to the reports, in the haste of getting the mines in the water, not enough attention and material was used in preparing the weapons for laying. Without proper waterproofing, water leaked into the weapons, shorting the internal switches. The second that their electrical cables were attached to the galvanic batteries as the enemy ships approached, the battery charge went directly to the firing devices. The result was uncontrolled explosions along the entire line of mines, rather than independent, purposeful explosions when actuated by the shore observer(s). Luckily for the trespassing squadron, the closest explosion was far enough away to allow them to escape unscathed.[28]

Roughly 30 miles south across the Gulf of Finland, the Russian defenses at Revel represented a stronger example of a layered defense: well-connected fortifications mounting 400 guns along a very narrow axis of approach, heavily supported by mines. Unlike the port of Sveaborg, whose disconnected defenses could be and were reduced one at a time by attacking forces, Revel's defenses were not nearly so vulnerable to attack and so survived the war.[29]

Thanks in no small part to the naval mines deployed by the Russian Baltic forces, neither Napier nor Dundas was ever able to reach any of the major Russian cities along the Baltic coast.

Black Sea Campaign

Note: This chapter and the analysis below was written in fall 2021, a few months prior to Russia's invasion of Ukraine. Considering that new war, this section gained new relevance.

Yeni-Kale is virtually unknown to the West, yet its location was strategic in protecting the Crimea's supply lines.[30] The Crimean Peninsula juts out from the top of the Black Sea down toward its center. To the west of the peninsula, the sea is wide open and faces other nations, making it important for international trade but less important for the internal movement of goods. The eastern side of the peninsula is a different story. There an arm of the peninsula stretches out toward a similar arm stretching from the Black Sea's eastern shore. These two arms nearly meet at the Kerch Strait and together embrace the Sea of Azov. At roughly only four kilometers wide at its narrowest point and navigable only by shallower draught ships, the strait was easily dominated by fortresses located on either side. Yeni-Kale was the fortress on the Crimean side. With the addition of mines and obstacles, whoever controlled this strait could enjoy an impenetrable layered defense of the type found at Kronstadt located in the north.

At the northern end of the Kerch Strait is the Sea of Azov. At the time of the Crimean War, the Sea of Azov could be considered a Russian lake much like the Mediterranean was once considered a Roman lake. With water travel being the cheapest and fastest way to transport goods, the Sea of Azov was the easiest route to transport troops and supplies headed from the Don River to Sevastopol in the Crimea. The alternative to this water route was an overland route that could add several weeks to the journey.[31] Thus, the Russians would have considered it vital to maintain freedom of navigation in this body of water. Had they succeeded in defending the Kerch Strait, then the war might have had different results.

As it was, Yeni-Kale's defenses proved inadequate to defend against the state-of-the-art force brought by the allied squadron. Surrounded by shallow water, the defenders expected to be able to fend off the shallow boats that could approach them. Instead, they found such boats armed with superior Lancaster cannons capable of firing large-caliber shot against which their older fortress walls could not contend. Even these boats may not have been a concern had the defenders deployed their submarine battery in advance of the acute threat of an approaching fleet and used it upon the fleet's arrival. This did not happen. Either the battery was not fully deployed, it was not made ready for firing, the firing crew panicked and fled,[32] or the shore observation points were captured early.[33] In any case, the weapon system remained unfired.

This failure to properly deploy and use the controlled mine battery at Yeni-Kale had direct consequences on the whole of the Sea of Azov. Prosperous ports, ships, small crafts, and stockpiled supplies abounded all along the coast. The allied squadron laid waste to them all, stopping only when they encountered difficulties at the mouth of the Don River itself. An entire maritime region covering some 15 thousand square miles[34] was denuded of the means to carry on a maritime trade or (to the allies' interest) to relieve the garrison at Sevastopol.[35] Had the Yeni-Kale defenses held during this attack in late May,[36] it is possible that Russia could have sent a strong relief force to take Sevastopol's besiegers from the rear and changed the tide of the war. But history is filled with what-ifs, and alternate history questions such as this can never be answered. Sevastopol fell on September 12, 1855, less than four months later.[37]

The Sea of Azov campaign was a sideline to the main campaign engaged in by the Royal Navy. The West has little memory of the Sea of Azov campaign, where it remains the realm of obscure history books such as this one. Yet it is impossible to believe that Russia did not extract lessons learned from this painful episode in its history.

As mentioned before, Russia has very few points where it can access the world's oceans. It is only its Black Sea ports that are reliably ice-free year-round. When it comes to bulk goods, such as grain, the Black Sea via the Sea of Azov is likely the most economic route for exports. It is there that the great, navigable Don River can discharge the vessels bearing products from deep in the country's interior. Even more importantly, the Don River is today connected to the even greater Volga River via a canal and series of locks.[38] The Volga River provides access to the Caspian Sea in one direction and the nation's capital of Moscow in the other direction. For a country that has been invaded time and time again, the Sea of Azov and its river access represent a critical vulnerability to its international trade and security.

Up until the fall of the Soviet Union, the country had little reason to worry about its Sea of Azov vulnerability. As it was in early 1855, the sea was a Soviet lake with the empire maintaining full control of its defenses and approaches. Half of the Black Sea shoreline fell exclusively under the Soviet sphere of influence. The other half was controlled by Turkey, which, though part of NATO, was not on par with the USSR militarily and was treaty-bound to limit the access of foreign warships. The Montreux Convention of 1936 requires Turkey to restrict access to the Black Sea via the Bosporus for the naval vessels of any nation not bordering the Black Sea. Such vessels are restricted by tonnage and by the amount of time that they are allowed to remain.[39] Together, this meant that Russia could dominate not only the Sea of Azov but the Black Sea as well. The Don-Volga River complex was safe with a 600-mile buffer zone.[40]

All of this changed when Ukraine gained its independence. Though militarily weak at first, Ukraine and the rest of the former Eastern Bloc countries increasingly began turning to NATO for protection against feared aggression from the Russian Federation ("Russia" for brevity). At first, Russia was itself militarily weak and unable to prevent these shifts in its fortunes. Eventually, Russia and its former Eastern Bloc satellites began to craft a new normal. In 2003, Russia and Ukraine signed an agreement giving the two countries co-equal control over the Sea of Azov, with Russia holding the eastern shore and Ukraine holding the western shore. Though the agreement limited uncontrolled access only to Ukrainian and Russian vessels, it left the door open to both Ukraine and Russia being able to control access to the sea via the Kerch Strait.[41] For the first time since the Nazi invasion of World War II, Russia had lost its ability to maintain a buffer zone against attacks on the Don-Volga River complex.

In 2008, NATO agreed with both Georgia[42] and Ukraine[43] to set out paths for those nations' eventual membership. Should such memberships take effect, Russia would have six NATO countries along its European border and no non–NATO neighbors bordering the Black Sea. These countries sought this path to protect themselves from the powerful Russian military with its historically aggressive habit of expansionism. On the other hand, Russia itself sees these moves as a dire security threat. For Russia, a formerly powerful country that had endured multiple invasions of its homeland by European powers, NATO appears to be its most powerful military adversary to date. Having become accustomed to maintaining a buffer zone to give it time to react to invasion, such a situation is very uncomfortable indeed. When that threat was limited to just the tiny Baltic states, Russia could focus on a single roughly 250-mile front. With Ukraine and Georgia added, it would now need to monitor two additional fronts totaling an additional 1,000 or so miles against what it sees as a single adversary.[44] One of those fronts would be a mere 300 miles from the capital.[45] In the event of a war with NATO, it could expect to quickly lose its access to the Black Sea. Should Ukraine allow naval forces into the Sea of Azov, Russia's international trade infrastructure could be substantially diminished.

Then came 2011 and the start of the two-year-long Arab Spring. Dictatorial governments friendly to the Russian government found themselves rocked by major protests, civil wars, and/or regime change. By late 2013, the focus had settled on Syria, a Russian ally that previously could be relied upon to threaten Turkey's southern border in the event of a conflict. By this point, Russia would have felt increasingly isolated as former friends stopped being so friendly.

This is pure conjecture, but it is possible that Russia would have used the spirit of the Arab Spring to its advantage by encouraging (instigating?) Crimean separatists in their efforts to join Russia. This gave the country the equivalent of European Turkey, i.e., an enclave on the other side of a strait that enabled it to control that strait absolutely. By annexing the Crimean Peninsula with the ostensible agreement of its inhabitants, Russia gained several advantages. First, it gained full control of access through the Kerch Strait. This in turn provided a desperately needed buffer zone that increased its ability to protect the Don-Volga River complex and guarantee its access to the sea. To underline this point, it built a bridge across the strait that is too low for large ships to pass.[46] Another advantage it gained was almost tripling its Black Sea frontage to include the major shipping and naval port at Sevastopol. The naval port is 200 miles closer to the bases of the militarily less-capable navies of adversaries in Romania, Bulgaria, and Turkey. This places it in the position to be able to quickly wrest full control of the Black Sea, keeping NATO naval forces beyond the Bosporus. That in turn greatly expands the distance over which ship-launched cruise missiles could be detected and countered before striking strategic targets inside Russia. Finally, by annexing the Crimean Peninsula, Russia gained an unsinkable aircraft carrier. Aircraft based there could dominate the whole of the Black Sea and much of Ukraine. Together, this annexation greatly enhanced Russia's ability to defend its southern border.

If the mines at Yeni-Kale had been properly deployed and used during the Crimean War, then the allied squadron would probably have not been able to force the Kerch Strait. Without being able to pass the strait, the allied naval campaign in the Sea of Azov would not have revealed to Russia its vulnerability in that region. Without the painful historic lesson of that campaign, would Russia have been bold enough to have reannexed the Crimea in 2014? We may never know the answer.

A final anecdote of the war worth mentioning was the mining on the Bug River in Ukraine. As the very first war in which mines played more than a limited, experimental role, the Crimean War revealed a classic challenge for miners. Mines that do not require a human operator are unable to distinguish between friend and foe. How then does one allow for the safe movement of friendly forces while also maintaining the defensive benefits offered by minefields? A report from the military governor of the city of Nikolaev (Ukraine) on October 9, 1855, suggested publishing a warning to local shipping.

The challenge when laying defensive minefields is how to balance operational security versus safety. How that balance is struck depends on the minefield's purpose, the immediacy of its need, and on the local economy's reliance on access to the sea. If a major seaborne assault is imminent, then the minefield might be made strong enough to almost completely block the river. Shipping might be told to avoid the area entirely or may be placed on the tightest of restrictions. For operational security, only certain people would know the specific safe route through the minefield and only fools would attempt passage without their direct assistance. On the other hand, if the minefield is in place as a precaution against a generalized threat of raiders, then the balance might shift toward reduced operational security.

In this instance, the mining most likely took place for the latter purpose. From May through November 1855, a British and French squadron tore up the seaports surrounding the Sea of Azov on the other side of the Crimean Peninsula. Their purpose was to destroy Russia's ability to resupply its Crimean forces with goods shipped via the Don and Volga Rivers from the Russian heartland.

On the other side of the Crimean Peninsula at Nikolaev (now Mykolaiv, Ukraine), the military governor would have been very aware of this raiding campaign. The city lay near the very navigable mouth of the Southern Bug River, which in turn spilled into the Dniprovs'ka Gulf at the mouth of the Dnieper River. Both rivers drive deep into the heart of what is now Ukraine and would thus be potential sources of resupply. Of the two rivers, the Bug would have been far easier to access as the mouth of the Dnieper is a tangled delta of many interwoven, shifting channels.[47] The military governor of Nikolaev would have understood the city to be a logical target for the allied fleet once they finished with the Sea of Azov. To this end, he would have wanted to deter naval scouts from further inspecting the city's defenses. Mines, having proved so successful as deterrents in the Baltic, would have seemed like a logical choice. Since deterrents work only when the enemy knows of their presence, operational security would have been less important. Furthermore, because the threat was only generalized, access to shipping was still a priority. Therefore, the balance was struck: (1) Mine the east side of the river around the Sivers settlement just south of the city where the landing of troops was the most likely, and then (2) tell shipping to stick to the west when heading downriver. The allies would find out and avoid a direct naval raid.

Arguably the most well-known campaign of the Crimean War was Sevastopol. It was there that land forces besieged the port that protected those most directly responsible for the assault on Sinop. While the British navy engaged the Russians everywhere from St. Petersburg to Petropavlovsk, it is army actions that are most closely followed by war correspondents. Thus, it is the campaign at Sevastopol that is most well-known.

Sevastopol's port was situated primarily on the southern shore of a small bay. Its defenses bristled with a vast array of forts and batteries that could subject any attacker to withering fire. Still, a determined enemy willing to weather the firestorm could have entered the bay to directly bombard the city. However, the seven ships of the Russian fleet responsible for Sinop were scuttled across the mouth of the bay and its cannons/crews integrated into the city's defenses.[48] Were the purpose of the Sevastopol campaign to simply neutralize that force, then the Russians satisfied that purpose themselves. However, one does not send armies and fleets to distant shores without engaging in action, and so the city was besieged until it finally fell to the allied forces.

Since this book is about mine warfare, it is worth discussing the use of mines at Sevastopol. Had the city used the same defensive model as used at St. Petersburg, then naval mines would have been deployed. The bay is ideal for the tactic, with all parts being within range of fortress guns. Defense planners would have been aware of the weapons' existence, and had it not been for the sunken fleet, there is no doubt they would have tried to have used the weapons in the city's defenses. However, there seems to be scant evidence of the weapons' deployment at Sevastopol.

In researching this book, the author found a few sources mentioning the use of naval mines at Sevastopol.[49] Nevertheless, there is strong reason to doubt these claims. To begin with, the first of these references was an American one written a quarter century later, while the other two were both written over a century later. All three only briefly mentioned mining at Sevastopol without further citation or details. Contemporary accounts by Russian, British, and American writers make no mention of the presence of naval mines at Sevastopol, though they are quite descriptive of land mines used in the defenses. In the writing style of the period, contemporary sources go into exhaustive detail about the different types, use, and sometimes construction of mines laid at other places. But they do not talk about naval mines at Sevastopol. Other sources go into highly detailed descriptions of the defenses at Sevastopol, complete with intricate maps, but make no mention of naval mines. It would seem illogical that these same writers would fail to discuss naval mines at Sevastopol had they played a role in the defenses.

Is it possible that the failure of contemporary accounts to mention naval mines at Sevastopol had something to do with the prejudices and purposes of their authors? As mentioned before, the Sevastopol campaign was primarily a landward affair, with naval forces having little direct involvement beyond ferrying troops and supplies and a couple of bombardments. A person writing from an army perspective would have little interest in naval matters and thus may dismiss the presence of naval mines as irrelevant. The sources that discuss naval mines tend to have naval origins. So, it is entirely possible that the lack of contemporary accounts of mines at Sevastopol is pure oversight. With that admission in place, the number one argument against their presence at Sevastopol relates to the timing of the siege itself.

Up until the Russian ships were sunk in the harbor on September 22 and 23, 1854, the prince in charge of Sevastopol's defenses had not yet resigned to closing the port.[50] Allied forces did not land in Crimea until September 14, about 30 miles north of Sevastopol in Yevpatoriya, and did not head south until the 19th.[51] When the ships were sunk a few days later, the prince had effectively conceded the inevitability of a siege by both land and sea. At that point, any naval mines used in the city's defenses would have had to have been there already. Since the Russian mine manufacturing program was centered on the Baltic defenses, it is plausible that mines would have been unavailable to be integrated into the Sevastopol's defenses. To do so would have involved transporting the bulky weapons between 1,500 and 2,000 miles—if not more—by cart, kedge, raft, and boat for a total journey that would have had to begin the moment war started to look likely. What mines that were sent would have been diverted instead to places like Yeni-Kale and Nikolaev. The only other viable possibility would be local designs and local manufacturing. This may have been possible if the Russian engineers experimenting with improvised naval mines in the Danube River only a month previously had coincidentally managed to transfer their knowledge to Sevastopol in time for their work to be continued. It seems implausible, however, as no such work was mentioned in a Russian source that otherwise outlined far more obscure uses of mines during that war.[52]

Whether or not mines were present at Sevastopol is irrelevant. The city fell on September 9, 1855, after a primarily land-based siege that lasted 337 days and was incredibly costly in terms of lives and matériel. Action continued in other theaters for a few more months before peace talks began in February. Peace was declared on March 30, 1856.[53]

Conclusion

In the grand scheme of history, the Crimean War created very little permanent geopolitical change. All of the Great Powers involved remained as Great Powers. Russia lost and suffered punishment at the peace table, but nothing that lasted. Its naval power in the Black Sea was destroyed, and it was forbidden to create a new navy in that sea. But after a while, it created one anyway. One side effect was the realization that its position in Alaska, being so far from the Russian heartland, was incredibly exposed in times of war. That realization was part of Russia's reason for selling that territory to the United

States 12 years later. A decade after that, Russia attacked the Ottomans more successfully, freeing the Balkan states and placing them under its own protection.

From the standpoint of the shaping of mine warfare, the Crimean War was immensely important. To begin with, it was the first war in which naval mines were part of the national military strategy. Prior to that, the use of mines in war was at best experiments conducted by inventors and enthusiasts with the help of individual military members. In the Crimean War, however, government sponsorship meant the purchasing and use of thousands of mines, the allocation of considerable resources, and, most importantly, the application of brainpower by hundreds of minds. It was an experiment on a grand scale, the sort that reveals countless lessons learned that could be applied almost immediately. Most importantly, it produced real strategic results, the type that military experts from other countries could use to inform their own military strategies. Three of those experts came from the United States, where naval mine warfare advanced even further less than a decade later in the U.S. Civil War.

CHAPTER 3

U.S. Civil War

On July 7, 1861, the sloop-of-war USS *Pawnee* sat at anchor on the Potomac south of Washington, D.C., near Acquia Creek. Commander S.C. Rowan, her commanding officer, looking out at the river, saw two oil casks drifting past 200 yards away. Less than three months after Virginia joined the Confederacy, his vessel most likely was on a mission to deter blockade runners and to act as a picquet against rebel movement up or across the river. Recognizing that oil casks do not normally pass down rivers in close formation, he decided to investigate further. Soon his crew recovered the strange objects and hauled them ashore.

The casks proved to be more than just loose detritus. For starters, the two objects were adjoined by a 150-foot-long line. Each had a large iron cylinder suspended six feet beneath it by slings and connected by watertight tubes. Packed with powder and with slow-burning fuses passing through the tubes, these iron cylinders represented the first naval mines found during the war. They were similar to mines invented by Robert Fulton five decades prior, with the miners hoping that the casks would spread out and snag on either side of the *Pawnee*, where they would explode when the fuse had burned to the powder. While their principle of operation was crude compared with later devices, they were the harbingers of what was to be the Confederacy's most effective naval weapon.[1]

Introduction

While the Crimean War introduced the naval mine to the world as a useful part of naval defenses, it was in the U.S. Civil War that the weapon truly proved its effectiveness.

Much ink has been spilled in explaining the path that led to the United States Civil War (hereafter referred to as the "Civil War"), its causes, and its major land campaigns. As such, that narrative will not be repeated in this book except where necessary to forward the tale of naval mines and countermeasures during that war. For the off chance that the reader has never heard anything about that war, let it suffice to say that several Southern states separated from the Northern states to retain the institution of slavery. Not desiring such a separation, the Northern states (the "Union") fought to bring them back into the fold. Eventually, the Southern states (the "Confederacy") lost and rejoined the United States. Slavery ended as a legal institution in the United States. with the ratification of the 13th Amendment on December 6, 1865. The practice's after-effects and the resentment caused by it continue to reverberate to this day.

Why is the mine warfare of the American Civil War important to study? The answer is that many of its themes and much of its technology could easily be moved to tomorrow's war and be just as effective. Naval mines are often the technology that a weaker power uses as an alternative to fielding a navy. Unlike the desert conflicts of the past few decades, in areas with river networks, wars are often waged on the rivers as well as on land and at sea. Rivers allow access into the interior by combatant vessels, yet they are shallow and narrow enough that defenders do not need ships of their own to defend them. Simple mines, made with locally sourced materials and deployed from the types of boats any fisherman possesses, can pack enough punch to take down the most heavily armored vessels on the river. This was the case in Vietnam and will again be the case in any future wars fought in areas with navigable river systems. Armored vessels possess capabilities not quite attainable by land or air assets, and as such will continue to have roles in future wars. If a riverine combatant possesses only sea-based mine countermeasures (MCM) systems, they may discover that their boats are sitting ducks when operating on rivers.

Southern Geography and Industry

The Confederacy was made up of the contiguous states of the southern United States. The region was warm most of the year and well-watered by rains, and its economy had developed around cash crops grown in its fertile soil. With navigable rivers driving deep into its interior, it enjoyed easy access to the world's markets for its agricultural products like cotton and tobacco. The North, in contrast, with its colder climate and fewer navigable rivers, had developed an economy built on a balance of industry, food crops, and mercantile trading. Its large population was centered on its industrial cities, in contrast to the South's more rural population. Lacking the easy river access of the South, the North instead had a dense network of railroads connecting its industrial centers. This combination gave it a capacity to produce and transport troops and weapons at a rate that far outstripped that of the South. Ironclad ships and cannons were among the weapons that the South could not hope to produce at the same rate as the North.

When the South seceded from the United States, the Navy and its bases remained the property of the United States. Rather than ships' crews deciding to join the Confederacy and bring their vessels over to the Southern cause, sailors resigned individually to avoid taking up arms against their own states. This left the North with an intact navy, while the south was left with whatever vessels were owned by Southerners, could be constructed with limited industrial capacity, or could be acquired abroad.

In the months immediately following the South's secession from the United States, Union naval forces placed the South under blockade. The blockade's purpose was simple: to starve the Confederacy of sources of supply and funding. Any entity that expends large volumes of materials not produced internally must eventually run out of those materials. The South had ready access to manpower, horses, rifles, and, to a lesser extent, food, but it had very limited heavy industry. This made it perpetually short of the types of resources not normally producible or found on a farm or plantation. Shortages could be shored up by importing equipment and materials from abroad, which is where the blockade came into play. By blockading ports along the length of the Confederacy's coastline, the Union slowed imports to a trickle.

Of particular importance was iron and the products that can be made with iron, like cannons, cannonballs, and ironclad ships. In the entire territory of the Confederacy, there existed only one industrial-sized iron foundry, at Tredegar Iron Works in Richmond, Virginia.[2] This severely restricted what could be produced. Production was further limited by a shortage of iron ore mines and means to transport heavy materials like iron. While the South possessed some railroads, they were not nearly as widespread as they were in the North. With few other sources to supply iron, in some cases the railroad tracks themselves became the iron supply.

The early days of the Confederacy were during a transition period for how fighting ships were constructed. As the war began, naval ships were mostly still constructed from wood as they had from time immemorial. Thanks to engineers like Robert Fulton (discussed in chapter 1), propulsion had progressed to the point where steam power was the norm. Cannon technology had improved to deliver more damage, but the freeboard of ships remained wood. Wooden ships were vulnerable to shore fire, something available to Southern defenders. However, Confederate engineers built a beast that changed that dynamic and thus delivered for the South a real problem that necessitated its future reliance on naval mines for defense.

Rise of the Ironclads

Three days after Virginia seceded from the United States in April 1861, Union forces destroyed the navy shipyard in Portsmouth, Virginia. As part of the destruction, they burned the steam frigate USS *Merrimac* to the waterline when they could not get it out of the channel. The next month, the Confederates raised the sunken hulk and began the process of turning it into the first ironclad in North America. By early March the following year, the new vessel, CSS *Virginia*, was ready for action.[3]

Armored with four-inch-thick iron plate, a low freeboard, and a slanted casemate, and armed with 12 cannons[4] and a ram,[5] the CSS *Virginia* was practically indestructible by anything else afloat. On her first day in action, she took on five Union ships in Hampton Roads and put three of them out of action. However, the North had had ample warning of the CSS *Virginia*'s construction. On her second day in action, she found a surprise waiting for her in the form of the USS *Monitor*, an ironclad with only inches of freeboard and a large turret mounting nine inches of iron plate. The two strange vessels battled it out all day before retiring without a clear winner.[6]

Note: The CSS Virginia *was not the first ironclad, nor even the first ironclad in the Confederacy. However, history remembers it in context to its battle against the USS* Monitor *as well as its successful raid against the wooden warships that preceded that battle.*[7]

Thanks to the vulnerability of wooden ships against ironclad monsters like the *Virginia*, as well as the USS *Monitor*'s ability to take hits, the North started building more and better ironclads. Having far less industrial capacity, the South could not hope to engage in a similar building campaign. This left its port and riverfront cities vulnerable to these floating fortresses that the North was fielding.[8] Unable to produce cannons capable of sinking ironclads, it instead had to consider asymmetric weapons that could punch well above their weight class. Taking lessons from developers like Fulton, Colt, and Nobel, it found its asymmetric weapons in the form of naval mines and submarines.[9]

While the land war between North and South was an even affair of infantry, artillery, and cavalry, the naval war was very much that of the weaker power versus the stronger. It was a guerrilla war, with mines used as waterborne improvised explosive devices (IEDs) against the North's ironclads. With dozens of rivers driving deep into the interior, the North had highways deep into the heart of the Confederacy as far as the fall line. No navy existed to challenge them and there were not enough cannons to stop them. Spars and booms may have delayed them, but such defenses are dangers only to the incautious and are easily defeated once discovered. Naval mines, on the other hand, were dangerous to anybody close enough to do something about them. Cheap and easy to use, they could be placed anywhere in a river where a ship could float and could (and did) sink the strongest ironclads. They turned rivers from highways to horrors, forcing Union commanding officers to slow down and expend much expense and time on mine countermeasure efforts. Yet such measures could only ever be temporary and incomplete. Always there was the danger of missing mines or of mines being placed in their wake, ready to hit them on their return trip. Thus, river traffic slowed to a crawl, giving the Confederates time to set up their landward defenses.

Had the CSS *Virginia* not been built where it was, would the USS *Monitor* have ever been developed? Without the example of their battle, would the North have ever developed the ironclads that they used so successfully along Southern waterways? Without the tactical problem of neutralizing the impenetrable Union ironclads, would the Confederacy's mines and submarine programs have ever been pursued as vigorously? And finally, if not for the extensive, demonstrated success of these programs against armored ships of war, would the world's navies have further developed them to be so successful in all follow-on wars?

Mine Types of the Civil War

While this book generally tries to avoid discussing too much detail about specific pieces of equipment, the mines of the Civil War were unique. With mine development in the Confederacy being practically a greenfield project, inventors were less constrained in letting their imaginations run wild. Consequently, the mines offered development paths seldom seen in other points of history. Here follows a sampling of these mines:

Singer Torpedo (moored contact mine)—One of the more distinctive moored contact mines in history, the tin-constructed Singer was actuated when passing ships knocked their heavy caps off. This in turn pulled a lever that drove a spring-loaded plunger to strike a piston into the chemical primer, blowing up the mine.[10] It was a mass-produced mine, and the examining board recommended it for its combination of simplicity, reliability, and ability to operate independently. It ended up being the second-most-used mine of the war.[11] Does the name "Singer" stand out? That is because its inventor was the brother of the famous sewing machine manufacturer.[12]

Keg Torpedo (moored contact mine)—By far the most common mine of the war, the keg torpedo was similar in construction to Bushnell's keg mines. Both were made from barrels partially filled with powder to which a firing device was connected. In the case of Bushnell's keg mines, the firing device was the cocked flintlock of a musket. In the case of the keg mines of the Civil War, the fuses were either percussion (think of how modern bullets are actuated) or the chemical type designed by General Gabriel Rains. To either end was attached light wooden cones that had multiple operational benefits. Weights were attached to the bottom, which helped keep the fuses on the top where passing ships could strike them.[13]

Barrels made a lot of sense as a mine casing. While hardly ever seen in modern times, barrels were the universal containers of their day. Anything that required watertight packaging, to include gunpowder, was shipped and stored in barrels. Consequently, barrels could be found just about everywhere. In an era where shipping was slow and expensive, barrels were ideal mine cases. Instead of transporting bulky mine cases, Confederate miners could simply ship fuses (possibly in barrels) to be mated up with locally acquired barrels and powder available anywhere. Furthermore, barrels of just about any size could be turned into mines according to need and availability. Being made of wood, they could be set up in several possible configurations by attaching different hardware to the wood. Most importantly, such mines could be deployed from boats or even from shore using the manual means available in the rural areas. This made them cheap and truly anytime/anywhere weapons.

Such flexibility is unavailable in modern mine inventories. The closest to be found is the U.S. Quickstrike series of mines, which mates mine components with general-purpose bombs of varying sizes. Such bombs can be found anywhere that has bombs and can launch strike aircraft. This means that in a crunch, the mines could be dropped by any airplane capable of dropping a bomb, or possibly without even the airplane. However, unlike keg mines, they cannot be assembled with 95 percent locally acquired materials in any zip code and deployed with precision using local transport and local help. That combination of simplicity, cost-effectiveness, flexibility, and streamlining of logistics made the simple and unsung keg mines some of the most useful mines in mine warfare history.

Friction Mines—In an era when people still used fire for lighting and heating and used black powder rifles to put food on the table, the friction mine would have been the easiest to make from household materials. It was a watertight container filled with powder and fired by pulling a chemical fuse quickly through an abrasive, not unlike striking a match. The striking of the match created a flame sufficient to ignite the powder. Submerged beneath the surface, either anchored or drifting, the fuse was attached to a lanyard. At its other end, the lanyard was tied to a harmless-looking float like a small branch or other flotsam. As passing ships came along, they sucked the float into the props, pulling the lanyard and detonating the mine.[14]

Brooke Torpedo and Turtle—The Civil War saw the development of mine countermeasures that expanded beyond the techniques employed by the allies during the Crimean War. One mine countermeasures technique (reminiscent of that earlier war) was two boats pulling a weighted line between them to snag anything sticking up from the bottom. This would have worked well enough against earlier mine types. It was less useful against the Brooke torpedo and turtle combination. These two represented possibly the world's first mine countermeasures countermeasure. Each countered minesweeping in a different way.

Brooke torpedoes were buoyant cases whose bottoms tapered down to a wooden spar attached to an anchor.[15] When sweep ropes encountered them, the rope would simply slide up the case and tilt it over on its spar. Once the force of the line passed, the mine's buoyancy brought it back upright once more.[16]

Another mine countermeasure technique used by some ships was the rake. Attached to the front of ships, it had long prongs arrayed in front of it like a rake. As ships traveled through the water, their rakes scooped up any contact mines in their path at a distance sufficient to avoid being sunk by an explosion.[17]

Turtle mines turned mine rakes into disadvantages. Like their namesakes, their profiles were like turtle shells. Filled with 100 pounds of powder, they sat on the river bottom. Positioned on the side of the Brooke torpedo from which the enemy ships were expected, they were attached to the Brooke torpedo by a 140-foot line. A special type of friction mine, they were detonated by a friction primer attached to the end of that line. If an ironclad's mine rake scooped up a Brooke mine, the force of it pulled the line and detonated the turtle. At 140 feet distance from the bow, this meant the turtle exploded under the ship's engine room.[18] Even if it failed to sink the ship, the force would throw the propulsion equipment into utter disarray, forcing a lengthy and costly repair. In war, it is often to a combatant's advantage to damage but not destroy, to wound but not kill. To kill or destroy takes one combatant out of the fight, while to wound or damage takes the victim plus an escort away from the fight, slows progress, and fills up limited hospital or repair facilities.

David Boats and Spar Torpedoes—Lacking the strength to fight above the water, the Confederates sought to fight Union ships below the surface. Besides the mines that they laid all over their waterways, they also built low-profile or submersible vessels that were precursors to submarines. Called "Davids," these vessels held small crews who manually propelled their tiny craft on what were often suicide missions to attack union ships. Their primary armament was the "spar torpedo," an explosive charge attached to the end of a long pole. The combination spar torpedo/David was an offensive weapon that fell in the same tradition as the explosive device carried by David Bushnell's *Turtle* during the American Revolution (see chapter 1). This spar was held aloft if carried by a surface vessel or affixed to the front of the vessel like a knight's lance if carried by a David. As they reached their target, the crew drove the spar torpedo against the side of the enemy vessel. Ideally the pole was long enough that the explosion happened far enough away that the attacker would not suffer damage from his own weapon.

Electric Torpedoes—Following in the footsteps of Samuel Colt and his submarine battery, as well as the Russians in the Crimean War, the South found a ready use for electrically actuated mines. Similar in firing principles to previous electrical mines, they could at times be as big as boilers containing hundreds of pounds of powder.[19] These would often be laid one at a time, requiring unique fire control techniques to ensure accuracy.

The previous chapter discussed a technique used by the Russians to target ships passing electrical mines arrays, so that any ship on a line of bearing was at risk. In the Confederacy, however, the use of single mines made them point weapons, so that both bearing *and* range were important when estimating a ship's position. Mines are most effective right beneath a ship and become less effective with distance. The trick for the Confederate miners working on rivers that could be a few hundred meters wide or more was ensuring that they did not fire the mine unless a target was close enough to sustain catastrophic damage. To do this, they used cross-bearings. The technique worked thus: Two observers, each equipped with a battery and a telescope, watched along a line of bearing (LOB). To one battery lead was attached a wire going to the mine. Close at hand lay another wire leading to the other observer's battery. As a ship crossed their LOB, they would attach the loose wire to their second battery lead. If the ship left one LOB without ever crossing the other LOB, nothing happened. However, if the ship was on both LOBs at the same time, the common wire was attached to both batteries, completing the firing circuit. At that moment, the mine would detonate with the ship directly overhead.[20]

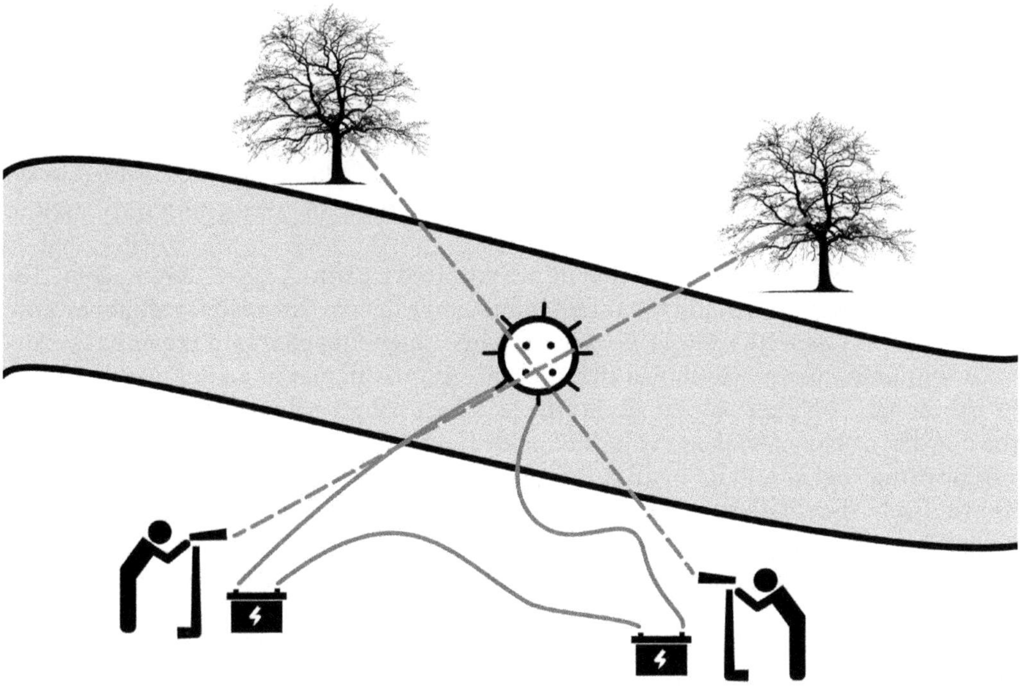

Cross-bearing fire control technique (M. Mazurek, after original drawing by the author).

Electric mines typically sat on the river bottom or harbor bottom, making them impervious to mine rakes and boat-pulled sweeping drags. Instead, it was their wires that made them vulnerable. Find the wire and follow it long enough, and you find both the mine and the miner. Union naval engineers discovered that these wires could be found in a couple of different ways. The first was to drag the bottom with grappling hooks, a trick that was repeated during the Vietnam War. A second technique required troops to patrol both sides of the shoreline ahead of ships. Any wires found could be cut and then followed to the observer station.[21] These techniques served to reduce the risk along rivers when the mines were not guarded to prevent their discovery. For the miner, the wire provided both a benefit and a vulnerability. It provided the miner with the ability to pick their target and to allow safe passage for friendly shipping. At the same time, it required one or two operators to be placed at risk while providing a discoverable means for mine countermeasures.

Frame, Pile, Shell, or Obstruction Torpedoes—As with the defenses at Kronstadt, the underwater defenses around Charleston, Wilmington, and Savannah used physical obstacles as well as mines in their defenses. Piles were logs sunk vertically into the bottom and stuck up high enough in the water column to tear holes in hulls. Frames were like piles. Large, stout timbers, they were affixed to an anchored support so that they stuck up diagonally in a column ready to rake the hulls of passing ships. Obstructions were little more than large objects such as ships sunk in the shipping channel to threaten and slow down shipping. To any of these barriers could be attached artillery shells with mine fuses (a setup known by Rains as a "submarine torpedo mortar battery"[22]). Taken together, a harbor or river so obstructed with a gauntlet of such obstacles and mines was generally considered unassailable by Union forces.[23]

Coal Torpedoes—Coal torpedoes, or "infernals," were not strictly naval mines in the traditional sense. They were perhaps closest to being limpet mines, in that they could be placed on a specific target to cause future damage. The weapons of saboteurs, they were powder-filled iron shells cast into the shape of a lump of coal. Covered in tar and coal dust, they were thought indistinguishable from the coal commonly used to power the steam engines of the day.[24]

The concept of operation for coal mines required stealth. A Confederate agent, disguised as a worker, would slip the infernal into a pile of coal intended for shipboard use (perhaps trains as well?). Ships' engineers, when shoveling coal into the engine's furnace, would inadvertently shovel the infernal into the flames as well. Quickly heated by the tar and coal dust setting alight, the iron case soon became hot enough to set off the powder, causing a violent explosion inside the furnace with a shower of iron shrapnel. Burning coal would fly in all directions to set the engine room alight. Even more destructively, the small explosion broke the integrity of the boiler, setting off a massively violent burst of high-pressure steam and scalding water that could tear the ship apart.[25]

Infernals, though not strictly naval mines, bore a similarity to more traditional mines in one key aspect. It was not the miner who placed the device in the flames and caused damage, but rather it was the ship's crew. A ship sitting pier side, not firing its boilers, was far less at risk of the infernal's devastating effects. Thus, the ship and its crew suffered only when it was actively engaged in supporting the war effort.

Logistics and Development

Mine warfare in the Civil War developed in real time. Unlike later wars, there were no mine stockpiles lying around ready for use. There were a few interested parties who had been tracking the weapons' development and use in Crimea and other places. What they knew was limited to published detailed descriptions of the weapons rather than the more nuanced knowledge of those with firsthand practical experience. Thus, the rather crude mines encountered by the *Pawnee* in 1861. Their inventor, the great oceanographer Matthew Fontaine Maury,[26] designed these weapons as he was seeking a way to make a land-based black powder explosive device float down a river[27]—an interesting concept that may have worked under controlled test conditions but was impractical in real-world conditions and time frames.

Seeing the limitations of such weapons, as well as their danger to friendly shipping, Maury sought to develop an electric mine. Electric mines did not rely on the randomness of current and had their firing logic controlled by a thinking human being. These factors, demonstrated by Colt and by the Russians in the Crimean War, were attractive to this area where rivers were more likely to be used by friend than by foe on most days. The challenge, however, was in the availability of materials and technology.

Electrical engineering was still in its infancy during the Civil War. Telegraphy and other relatively simple technologies existed, but many of the basic principles were yet to be discovered. Batteries were massive and weak, and even the most rudimentary of electrical components were expensive special orders. When Maury thought about designing electrical mines, he was contending with a supply chain problem like what a person of today would have if they were trying to build a jet fighter from materials available at their local hardware store. In the whole of the South, there existed only a few hundred feet of wire, almost entirely uninsulated. The few batteries that then existed in the South were massive contraptions of metal plates and barrels of chemicals capable of generating only rather weak charges. If these materials could be brought together, then theoretically Maury's team would have enough to build perhaps a single electrical mine. Naturally, this stymied the South's development and deployment of the weapon until more materials could be acquired. It was only after better batteries were acquired and telegraph wire was washed ashore during a storm that the South would truly begin its electrical mine program. Even then, when some of the earliest electrical mines were detonated, great care was taken to recover the battery and wires for reuse.[28]

While the lack of electrical supplies delayed electrical mine development, contact mines were similarly delayed. When in 1862 it became obvious that there were insufficient materials to deploy electrical mines in numbers, the strategy switched to improving and deploying contact mines. Here too the Confederates were hampered by a lack of materials, technical knowledge, and production facilities. Thus, most of the remainder of that year was devoted to research and development as hundreds of inventors' schemes were dug through to find viable options using available materials. It was not until December of that year that these efforts bore their first fruit in the first-ever sinking of an enemy combatant by a mine.[29]

The impact of supply chain, production capacity, and basic research on mines and other weapons development and production cannot be understated. Weapons of war have long been some of the highest technology deployed in any nation. Roman roads,[30] medieval castles, ships-of-the-line during the Napoleonic Wars, radar and atomic weapons in World War II, and the stealth drones and hypersonic missiles of today ... all were military weapons at the height of their eras' technology. Consequently, the most dominant technology tends to be extraordinarily expensive. It is in a government's interest to maintain the capability to build these weapons, because war can be unpredictable. If an important weapon is purchased from a partner nation, it can become unavailable if that nation declares war or if that nation is itself attacked. Similarly, if a critical material or component is produced overseas, production can be severely restricted if that source of supply dries up. This was seen during the Covid-19 pandemic when Taiwan and China restricted their production of microchips, thus choking automobile production worldwide. Given these conditions, countries have strong motivation to maintain (as close as possible) a 100 percent domestic capacity for weapons production.

During peacetime, there exists a constant tension between the need to maintain domestic weapons production capacity and the need to fund and produce competing products. Governments cannot maintain unsustainably high tax rates without an existential crisis. Without such a threat, they cannot justify the costs of retaining and replenishing wartime military forces. Instead, they find a lower level of preparedness, one that allows them to hold the line against likely threats until allies and production capacity can be brought online to support a protracted conflict. The result of only low-level weapons production is an industrial base that is not optimized for military production.

For example, a country that wants to be able to quickly build military ships in the event of war would maintain large-scale shipyards in every port. This can work only when there is a high demand for ships of the type and cost made by those shipyards.

Shipyards will pay workers high wages to build high-performance military vessels for top dollar. However, in peacetime the demand for military ships is low while the demand for merchant ships is high. Businesses buying such ships can and will shop for the best deal. In environments where unions and regulations raise the costs of domestic shipbuilding, business has no qualms with choosing overseas shipbuilders that do not bear such burdens. A ship that has 90 percent of the capability but can be bought and maintained for half the cost and built in a third of the time has real business advantages. The predictable outcome is that shipbuilders in high regulatory environments will find it hard to survive and will either reduce capacity or go out of business. With less capacity, they do not achieve the economies of scale or the same level of technological development that can be enjoyed by larger competitors. The end result is an industrial base that is less able to scale up military production without suffering ruinous costs and time delays.

Had the South known that war was on the horizon, it might have built and maintained the production capacity to manufacture whatever mines it wanted. But its economy was focused on the South's strengths, not the what-ifs, and so it started with what it had available. Thus, it turned first to less materially intensive contact mines.

While Maury worked on developing electrical mines in a research and development role, another inventor named Brigadier General Gabriel Rains came into prominence. Rains first began conceiving his ideas for mines during the Seminole Wars when he created rudimentary explosive booby traps. Later, when placed in charge of the Confederate rearguard during the evacuation of Yorktown, Virginia, he had his troops booby-trap the entire town with mines made from artillery shells triggered by friction and impact fuses. While these actions were widely reviled as barbaric and ungentlemanly by the pursuing Union forces, they gained him the attention of President Jefferson Davis. Soon he was placed in charge of the underwater defenses of the James and Appomattox Rivers near the confederate capital at Richmond, Virginia.[31]

Rains took on his new command with vigor. The mines that he designed were natural successors to his land-based artillery shell mines. One early type was the frame torpedo described earlier. Another early type were keg mines that used impact fuses similar to those used on artillery shells but designed to fit in the bungholes of kegs. Rains himself supervised fuse construction, the exact components of which were known only to himself. Soon his keg mines began to appear all along the South's East Coast.[32]

Farther west, Texans E.C. Singer and his partner J.R. Fretwell, along with other inventors, designed a series of practical weapons with which to protect the rivers and harbors feeding into the Gulf of Mexico. Lacking exposure to the efforts of Rains and Maury, they developed them along other lines that took advantage of other known principles for producing sparks, such as friction primers and percussion caps. What resulted were weapons that followed entirely different mechanical principles. These inventors yielded the earliest wins against their adversaries, and later some of the densest arrays of casualties at Mobile Bay. However, the early success of the Union in wresting control of the western waterways outpaced the rate at which these weapons programs could be matured.[33]

This blunting of western mine warfare suggests yet another principle—that war is a race. For the invader, the key is to take territory as rapidly as possible before the defender has time to prepare their defenses. For the defender, defenses must be prepared in advance of hostilities. If not, the dynamic nature of a defensive war places huge pressure on planners, logistics, and field personnel to build defenses while the ground is still held. A spot can be a perfect, impregnable killing ground for the defender, a new Thermopylae, but it remains so only so long as the defender can amass sufficient forces and weaponry to hold it before the invader arrives. Once the invader holds the best defensive locations and controls the best lines of communication, the defender runs out of options, save for risky guerrilla tactics. Likewise, weapons manufacturing and deployment is far easier when doing so in relative safety and virtually impossible when done under the noses of an occupying force. Accordingly, the implied principle is to not pick a fight with an adversary until you are truly ready to protect yourself in areas where it is stronger.

Areas of Operations

Confederate mine warfare saw most of its successes in five main areas, though it also had its isolated victories against single ships in several other locations. Some of the locations where mines were used would spark interest in those who have a connection to those places. For example, in researching this book I learned about a man who used mines to defend the approaches to Port Lavaca, Texas, and the inner waterway leading to Aransas Pass, Texas. As a young sailor, I was stationed on the minehunter USS *Heron* (*MHC-52*) in Ingleside, Texas. I used to take long bicycle rides throughout that area. Had I known at the time that those very waters had once seen hot mine warfare, I might have looked differently at my surroundings. However, for the sake of brevity, I will concentrate on those five areas mentioned above.

Yazoo River, Mississippi—The very first victory for Confederate mining forces came in the Yazoo River in the early stages of the Vicksburg, Mississippi, campaign. A major feature of the Union's strategy included the capturing of the Mississippi River to split the Confederacy. Although most actions were fought in the east, the western Confederate states, which included Arkansas, most of Louisiana, and Texas,[34] provided a steady flow of manpower and equipment. By taking control of the wide Mississippi, the Union could choke off this logistics support and thereby hasten the attrition of Southern strength. Toward that end, Union forces under Admiral David Glasgow Farragut captured New Orleans at the mouth of the Mississippi River and fought their way north up the river. Concurrently, a force of river gunboats under Commander Charles H. Davis fought their way south down the river, winning victory after victory. By early June 1862, the Union Navy controlled every stronghold and crossing point on the Mississippi River apart from a 150-mile stretch between Port Hudson, Louisiana, and Vicksburg, Mississippi. In between lay the mouth of the Red River, providing a vital supply line for men and matériel coming from deep inside the western Confederacy.[35]

Vicksburg lies at the southern tip of the massive Mississippi Delta,[36] a lozenge-shaped alluvial floodplain that lies between the Mississippi and Yazoo Rivers. This area contained flat, rich farmland used for growing cotton in the antebellum period. Despite its richness as farmland, it is dramatically marbled with tributaries to the Yazoo River and is heavily populated with oxbow lakes from where these rivers have flooded and changed course over the millennia. Yet none of these rivers joins the Mississippi. This made the area great for farming but lousy as a route from west to east over the two rivers. Even today, only one minor highway crosses the Mississippi in the entire stretch between the two great interstate crossings at Memphis and at Vicksburg. Just upriver of Vicksburg, the Yazoo River flows into the Mississippi just as the latter kinks into a tight horseshoe bend. Vicksburg sits atop high bluffs at the tightest point of this turn. This advantageous location at the outflow of the Yazoo area just as river traffic must slow to make the tight 150-degree turn naturally led to the establishment of Vicksburg as a trade center. It also made it ideal for defense against the Union Navy. Here on the high bluffs, cannons could plunge fire on passing ships with little concern for corresponding return fire.

Unable to bypass Vicksburg without support by land forces, the Union freshwater fleet maintained a blockade just north of the city at the mouth of the Yazoo River. In preparation for the expected land assault coming down the Yazoo, on December 12, 1862, five Union ships steamed up the river to clear Confederate batteries and reported naval mines. Upon reaching a tight bend in the river, the lead ship, USS *Marmora*, spotted a floating wooden box just as she came under heavy enemy fire. Coming up to assist, USS *Cairo* soon snagged a line connecting two more such boxes. These boxes proved to contain mines made from five-gallon demijohns[37] partially filled with powder and containing friction primers at the end of each connecting line. With two explosions tearing apart her port bow and port quarter, *Cairo* sank in minutes with no loss of life. This was the first time in history that a naval ship was lost to an enemy mine in wartime.[38]

It was not until July the following year that the Union finally succeeded in wresting control of the last two Confederate strongholds on the Mississippi, thus effectively halving the Confederacy. In the weeks following the fall of Vicksburg, Union forces worked to consolidate their gains by pacifying the surrounding areas. One area under contention was Yazoo City, home to a shipyard engaged in the building of ironclads, as well as the workshop where Fretwell and Singer constructed their mines that were popping up in rivers across the South. On July 13, 1863, a Union flotilla under Admiral David Dixon Porter worked its way upriver to Yazoo City. There they engaged Southern troops in heated battle before eventually taking the city. In the process, the armored river gunboat USS *Baron de Kalb* encountered two Singer mines. Suffering explosions at her bow and stern, she sank quickly with no loss of life.[39]

Because of experiences like those of *Cairo* and *de Kalb*, Porter directed his frontline forces to develop protective measures against underwater mines. One such measure was the installation of "torpedo rakes" as described previously. Perhaps the most useful measure was frequent patrolling of vital tributaries by naval and land forces. This prevented the buildup of sufficient men and materials for effective mining efforts and provided the operational uncertainty that would dissuade any attempts at mischief. As a result of his proactive efforts, the Union Navy would suffer only one additional loss to naval mines on the western rivers. That loss, the ironclad USS *Eastport*, detonated a mine on the Red River despite having boats sweeping mines ahead. After she settled to the bottom, Porter chose to destroy her lest he be forced to fight her in the future.[40] This would be the last such loss west of Florida until the bloodletting at Mobile Bay toward the war's conclusion.[41]

Richmond Approaches—Mines had their most strategically important impact on the war on the James River approaches to the Confederate capital at Richmond, Virginia. No other river in the war held such importance for the continuance of the war effort, for to capture Richmond was to knock the heart out of the Confederacy. Besides being the seat of government power, it also provided the industrial capacity that allowed the Southerners to maintain their bulwark against Union movement farther inland. It also provided a source of respite and replenishment for the forces hoping to strike north against the Union's closest sources of supply and industrial capacity. Most importantly, it was a major transportation hub for the marshalling of men and material from throughout the South in the ongoing conflict against the industrial North.

Richmond is situated on the James River at the base of a long, narrow peninsula between the James and York Rivers. Like most of the great inland cities of the American East, it is located at the fall line, that point where the inland piedmont transitions into the wide coastal plains. From this point, ships can navigate clear to the open ocean and thus to the world's markets. This feature made fall line cities important as shipping ports for the agricultural products of the interior. Because rivers tended to narrow at such places, they also were often the closest points to the coast where bridges made continuous overland travel possible. Thus, these were often well-established cities by the mid–19th century. When railroads began snaking their way across the country, the fall line cities were obvious points of connection. As natural transportation hubs, cities like Richmond but also Washington, D.C.; Raleigh, North Carolina; Columbia, South Carolina; Philadelphia, Pennsylvania; and many others were the natural powerhouses of the 19th century. The critical mass of transportation, industry, and people allowed them to develop into the major cities that provide vibrance to the United States to this day.

This strategic position on the James River made Richmond vulnerable to attack from Union Navy forces, as well as blue-clad Union Army forces coming from the north and east. Consequently, considerable effort went into the city's defenses.

For riverine defenses, the Southerners placed gun batteries atop high bluffs at various points to enable plunging fire. They laid electric mines at other points where it was advantageous to do so, which is to say, at points where the channel narrowed or where operators could be easily concealed. Electric types were preferred because the river was used by blockade runners and Confederate ironclads. When one's own side is already short on shipping options, it does not do well to sink that shipping with your own weapons.

The first victim of these Confederate electrical mines along the James River was the USS *Commodore Barney*, a former New York ferry turned gunboat. In early August 1863, a Union squadron headed upriver to test Confederate defenses. En route, they unwittingly came upon the first Confederate mine station. The station's fire control setup was like those described previously, but unfortunately for the mining team, the operator got too excited. Instead of waiting until the ship was directly over the mine, he connected the battery just as the ship was *almost* over the mine. The explosion was massive, throwing a geyser of water 50 feet into the air and lifting the ship 10 feet. Twenty sailors were lost in the explosion, and yet when everything settled once more, the hull remained intact, though some internal repairs were needed. After being towed five miles to Dutch Gap, Virginia, for repairs, the *Commodore Barney* was back in action the next day. Naturally, the miners blamed the ship's survival on the battery having insufficient charge, rather than the too-early mine actuation. But given the nature of how explosive devices work, this seems highly improbable.[42]

Following this Union Navy probe, little else happened on the James River for several months. Then, on May 4, 1864, a joint Army-Navy expedition of several ironclads and gunboats headed upriver to capture the stronghold at Drewry's Bluff[43] at the last major turn downriver of Richmond.[44] The Navy, having experienced 11 major mine strikes by this point,[45] was taking maximum precautions. Small boats were deployed ahead to drag the river, marines patrolled the enemy-held left bank, and ships were ordered to maintain separation to minimize losses from any single mine. On May 6,

Mine explosion at the bow of the U.S. gunboat *Commodore Barney* in the James River. *Harper's Weekly*, dated August 29, 1863 (Navsource Online).

they stopped for refueling a short distance from their target, only to learn from a runaway enslaved person that electric mines had been laid at Deep Bottom only a short distance ahead. Despite this intel, the sweeping parties failed to drag up any mine cables, and so the advance ships were ordered to proceed ahead.[46]

Unbeknownst to the Union Navy, who were looking for wires coming from the enemy-held *left* bank, the Confederate miners had moved their battery setup to the *right* bank. When the USS *Commodore Jones* backed directly over a huge electric mine, the operator completed the circuit by connecting the leads to the battery.[47] Instantly, this mine, containing between 1,000 and 2,000 pounds of gunpowder, threw the ship into the air.[48] There she dissolved in an instant, a raging mass of pieces, smoke, spray, mud, and bodies. Forty men perished, yet somehow an engineer working deep inside the ship survived.[49]

As the debris settled and the smoke cleared, the shore party saw a man running away on the opposite bank. Presuming that only a mine operator would be running through swampy ground so close to the explosion, they shot him. Troops were soon deployed to the opposite bank, where two more men were captured in short order in a trench near a second battery setup. Both were interrogated at length. When they refused to tell the location of additional mines, the exacerbated commander had one of them tied to the jackstaff on the bow of the lead ship. Obstinate at first, he soon sang like a canary as the ship began to approach the next mine upriver. Once that ice was broken, he quickly revealed the other mine locations whose locations he knew. Despite this intelligence leak, a steady barrage of new mines stopped the squadron from passing beyond Bermuda Hundred, well downriver of Richmond.[50]

Note: Today it would be a violation of the Geneva Convention, Articles 17 and 52, for a prisoner of war to be placed in such a situation. Despite this restriction, it is entirely possible that those captured while engaged in the mining of enemy waters would be subjected to such methods by unscrupulous enemies. This would be especially true if they were captured by boat in the water near the mining in question. Such a contingency should be considered by combatant commanders, who in turn should advise sailors and pilots engaged in mining on exactly what they may or may not say if captured and placed in such a situation. In making such decisions, commanders might consider the following in addition to the usual proscription against aiding the enemy: (1) In open-water mining operations, few if anybody involved would know the precise locations of individual mines. With no visual references except long GPS coordinates, and large areas of uncertainty, nobody could be reasonably certain if the patch of water ahead of them is not mined.[51] *For mines in rivers, controlled mine operators and the crew of war vessels would be more likely to recognize known mined areas based on visual references. (2) Per Article 3 of the 1907 Hague Convention, mine danger areas not under surveillance must be announced.*[52] *Thus, the sailor would be following the Hague Convention by reporting mine danger areas.*

Explosion at City Point: Important as the James River was, it was the testing ground for the latest mining technology. It was there that the Confederate Secret Service deployed unusual devices that now may be referred to as improvised explosive devices (IEDs) but were then lumped into the same category as naval mines. One such device caused one of the war's most devastating explosions.

During the siege of Petersburg, Virginia, the Union general established his supply base at City Point (now Hopewell, Virginia).[53] Both cities lay upon the Appomattox River, with Petersburg at the fall line and City Point a few miles away at the point where that river joined the James River. As the last major city before reaching Richmond from the south,[54] Petersburg represented a threat to any Union forces moving to assault the capital from the river. By capturing it, Union forces could push Confederate defense forces back to the capital itself.

On August 11, 1864, a single Confederate captain in civilian clothes carried a simple wooden box to the waterfront. Setting the box on the wharf, he sat upon it and pretended to sleep as he surveyed the scene. Selecting his target and seeing his opportunity, he set off once more with his box. Using a Scottish accent to confuse the German-speaking guard, he managed to get close enough to his target to hand his box to a worker with directions that the commanding officer wanted it placed in the hold. However, this was no ordinary box. Though a common-enough box on the outside, it held 12 pounds of gunpowder and a clockwork triggering device. His target: an ammunition barge in the middle of an offload. His delivery complete, he meandered his way to his observation point, where his partner waited.[55]

Complete devastation followed. Deep in the ammo barge's hold, the clockwork device completed its countdown. Exploding, it set fire to its surroundings. In the resultant flames, a critical mass of powder and ammunition heated to the point of spontaneous detonation. A massive conical blast erupted, raising a mushroom cloud seen for miles and heard as far as Richmond, nearly 20 miles distant. Shock waves smashed the surrounding buildings and vessels, while cooking-off ammunition flew in all directions to start more fires and explosions. General Ulysses S. Grant, who in time would become president of the United States, sat chewing his cigar as ammunition and debris flew through his headquarters. In such chaos, what else *can* one do but stoically wait for fate to reveal itself? Though he survived this disaster unharmed, his depot, his headquarters, and his staff were decimated. Still, despite causing nearly 200 casualties and

$4 million in property damage[56] ($74 million in 2022 dollars),[57] this incident had little long-term effect, as it did not coincide with any other events of import.

The Confederate Secret Service corps found success elsewhere with its imaginative coal torpedo (described previously). Simple in construction, small, lightweight, and innocuous looking, these devices were placed on any coal pile that an agent could reasonably expect to support the Union war effort. Once in position, these devices would eventually be shoveled into the fires heating the boilers that powered the trains and ships of their enemies. While the existence of such devices was unknown to the North, once revealed postwar they were suspected to have caused several mysterious boiler explosions throughout Union lines.[58]

One high-profile explosion later attributed to a coal torpedo was that of the USS *Greyhound*, flagship for General Benjamin F. Butler. Steaming downstream on the James on November 27, 1864, the general and his guests heard and felt a faint explosion. One of his guests, Admiral Porter, recognized the explosion as that of a mine. There soon followed a much larger explosion as the ship's boiler burst apart. With flames and smoke engulfing the vessel amidship, the senior officers joined the crew in abandoning ship. The crew safely evacuated, then watched as the ship burned and sank, taking Butler's horses with it.[59]

As the war dragged on, Union and Confederate forces alike steadily countered the other's increasing threat. To thwart any adventurism by Northern forces, the Southerners continued to build up their underwater defenses to deny freedom of movement upriver of Dutch Gap. Aware of the counterthreat from the Confederate fleet, by October 1864, the Union had begun deploying its own naval mines and underwater barriers. Things came to a head on January 24, 1865, when the Confederate fleet poured downriver intent on attacking their Union counterparts. There followed a disastrous battle between the Confederate fleet, much of which ran aground, and the Union shore batteries. The mission had been to attack the Union fleet with guns and spar torpedoes, yet when the smoke cleared, the Union Navy never saw their enemies. After this, a sort of stalemate settled over the river that would last until the war's end.[60] The only remaining mine loss in the area was the Confederate flag-of-truce steamer, which was sunk by a mine that had broken free of its mooring. Whose mine it was, Northern or Southern, was never determined.[61]

Charleston, South Carolina—Aside from Richmond as the capital, Charleston was perhaps the second-biggest symbol of Confederacy for both the North and the South. It was there on April 12, 1861, that an armed attack on Fort Sumter by South Carolinian troops turned the secession from an unfortunate situation into a hot state of war. Perhaps no other city in the Civil War experienced such enthusiastic pursuit of undersea warfare.

Charleston Harbor is situated along the South Carolina coast, where several major and minor rivers and other waterways join to flow to the ocean between coastal barrier islands. Upon navigating past these islands into the harbor, ships must pass through a turn into the inner harbor, where they find themselves surrounded by land on all sides. At the time of the Civil War, this harbor was well protected by a multitude of forts and artillery positions, so that no enemy vessel could enter without being pounded on all sides.[62]

In April 1863, when the city of Vicksburg was still under siege, a force of nine union ironclads entered Charleston Harbor, where they were collectively struck with cannon fire hundreds of times. In response, they could return only 139 shots total. The marked disparity in rates of fire notwithstanding, they proved that their vessels could take a beating and remain afloat.[63] Despite such resilience, they could not maneuver freely due to the plethora of barrel mines (some fake), frame torpedoes, booms, rope obstructions (meant to foul propellers), and three massive electric mines built from boilers. Furthermore, low-profile David boats mounting spar torpedoes threatened any enemy ship sitting idle at night. Combined with wide shoal water and a low-lying shoreline, the harbor was an unhealthy place for Union ships to dwell.[64]

It was due to the known mine threat that the USS *Weehawken* led the way pushing a minesweeping barge known as a "devil." In the process, she managed to set off one of the smaller percussion barrel mines with only minor damage to herself. The more daunting damage came from the 53 shells that hit her during the 40-minute ordeal, though she still managed to leave under her own power. Luckier still was another vessel, the USS *New Ironsides.* Taking a position farther out, she dropped anchor almost directly atop one of the massive iron boiler mines. This electric mine, 127 feet^3 and packing 3,000 pounds of powder, was big enough to sink a few ships at once had they been in tight formation. Luckily for the *New Ironsides*, nothing happened when the operator attempted to actuate the mine. Later it was discovered that the cable had been broken when it was run over by a supply wagon where it crossed the beach.[65] This was neither the first nor the last time that a mine operations department was screwed over by supply.

The presence of so many mines in Charleston gave the fleet pause, filling them with caution and limiting their freedom of movement. Fearing the effects of these weapons, the fleet switched from an all-out assault strategy to a siege position instead. Naturally, the armchair strategists of the press lambasted the fleet commanders for their caution.

Emboldened by their weapons' success in driving back their adversaries, Confederate mining forces increased their efforts. Soon mines were laid farther into the harbor, finding homes across the mouths of both the Ashley and Cooper Rivers (which pass the city to the west and the east, respectively).[66] These mines threatened Union and Confederate vessels alike. Twice Confederate transports, first the *Marion* and later the *Etiwan*, struck mines at the mouth of the Ashley River. These defenses in place, the minelayers set their weapons even farther afield in the Stono River. Winding its way from Lighthouse Inlet, the Stono River flows to within a couple of miles of Charleston. From there, a Union force could have conceivably attacked Charleston from the west across the Ashley River where its defenses were weaker. By mining the Stono River's mouth, the miners shored up this vulnerability to Charleston's western flank.[67]

Such defenses soon proved their worth on the night of August 16, 1863, when one of these mines drifted near the *Pawnee* (mentioned earlier). With a sudden report heard from the miners who had just finished minelaying, the mine destroyed the *Pawnee*'s launch with only minor damage to the ship. Made aware of the danger, the fleet responded by recovering and disarming what mines they could find, while also taking measures to minimize the danger of mines that they may have missed. As though taking note from the measures taken to defeat Fulton's mines a half century earlier, the *New Ironsides* surrounded itself in a shroud of nets suspended from spars. The fleet also took to questioning captured Confederate watermen, who proved unaware of the details of the city's underwater defenses. Then, as now, mining plans were kept under close guard.[68]

Static mines and obstructions weren't the only underwater weapons threatening the Union fleet off Charleston. The Davids were weapon systems that were almost exclusively unique to Charleston. Spurred on by a $100,000 prize for the sinking of the *New Ironsides* and a $50,000 prize for the sinking of a monitor, a team had built a few of these cigar-shaped, low-profile boats.[69] During development, a few of these had flooded with water or toxic gas, bringing death to all onboard.[70] Undeterred, a team under Lieutenant

Glassell attacked the *New Ironsides* on October 5, 1863, with a spar torpedo extended from the bow of a David. The damage to *New Ironsides* was substantial but not catastrophic, while the David suffered damage of its own. Some of its crew abandoned ship and were soon captured, while the David drifted away to be eventually brought back to port.[71] Later, the hand-powered *Hunley* would attack the USS *Housatonic* in a similar manner, resulting in the sinking of both vessels. History would remember this as the first sinking of an enemy ship by a submarine.[72]

Mines continued to threaten Union forces off Charleston for the remainder of the war. In the final weeks of the siege, the union fleet under Rear Admiral John A. Dahlgren entered the harbor to support Major General William T. Sherman's landward assault on January 15, 1865. Having learned from experience about the underwater threat, Dahlgren made efforts to protect the fleet from mines. Each ship was rigged with assemblies of 50-foot spars and nets to act as torpedo catchers, while he preceded the fleet with two monitors and a fleet of small boats dragging for mines. Consideration was given to deploying divers as well, but this proved untenable.[73] Despite his precautions, the monitor USS *Patapsco* struck a mine and sank between Forts Sumter and Moultrie at 32 45'12"N, 79 51' 58"W, where it remains to this day at a depth of 40 feet.[74] Two months later, after the city's fall, two more Union ships struck mines in Charleston Harbor with only slight damage.[75]

St. Johns River, Florida—Perhaps the least well-known campaign of the Civil War is that of the St. Johns River. Starting in the lakes and swamps in the center of the Florida Panhandle, the St. Johns River is relatively narrow as it wanders its way north. Once it reaches Palatka, the river widens substantially as it continues to flow north to Jacksonville. There it narrows again and veers east to empty into the Atlantic at present-day Naval Station Mayport.[76]

Action along the St. John's River began as early as 1861. Union troops supported by gunboats patrolled the river throughout the war, at various points capturing and abandoning forts and cities along its length. On February 7, the Union occupied Jacksonville, followed by Palatka on March 10, 1864.[77] This gave it effective control of the river, allowing it to reinforce and supply troops attempting to wrest control of the surrounding countryside.

Responding to this renewed threat, the Confederate Army began its own campaign to mine the river to deny its Union adversaries the use of this vital waterway. To aid in this campaign, the Confederate general in charge of Charleston's defenses sent a mining expert south along with enough mines and mine components to be able to construct 300-plus naval mines. Centering their operations some 12 miles south of Jacksonville, they managed to sink four ships between Beauclerc's Bluff and the mouth of Cedar Creek.[78] Due to these efforts, the Union Army transports *Maple Leaf, General Hunter, Harriett A. Weed*, and *Alice Price* all sank to the bottom.[79] These losses led to Union forces adopting countermeasures including mine exploitation (recovering mines to see how they work), posting mine lookouts, continuously dragging the river for mines, and steaming in single-file convoys.[80] Most of these countermeasures would be repeated in wartime clear into the late 20th century.

Mobile Bay, Alabama—In covering the naval portion of the U.S. Civil War, general naval history texts generally mention a relatively limited number of events. They mention the *Monitor* vs. *Merrimac* battle, the activities of the raider CSS *Alabama*, and Admiral Farragut's breaking through into Mobile Bay, Alabama, while shouting "Damn the torpedoes! Full speed ahead!" Farragut's quote was meant to inspire young sailors to be willing to charge into danger regardless of the obstacles. Yet this legend seriously misrepresents how the events unfolded. The reality is that (1) Farragut's orders were not mere bravado but based on an understanding of the battlespace, and (2) the real bloodletting from the mines at Mobile Bay happened a few months later during mine clearance operations in the inner harbor.

For a solid understanding of how the initial breakthrough took place, one needs look no further than the introduction to Dr. Tamara Moser Melia's book *Damn the Torpedoes.*[81] It is a seminal work and is a staple for the library of any mine warfare professional. As an author, I cannot do a better job than she did with the resources she had available as a historian for the Naval Historical Center. As such, I will only attempt a summary of the action.

In early August 1864, a couple of months after the fighting on the St. Johns River in Florida, a joint Army/Navy expedition gathered outside Mobile Bay for an assault on the city. At that point, Mobile Bay was one of only two major ports left under Confederate control, the other being Wilmington, North Carolina.[82] Capturing the port would complete the Union blockade of the South and hamper the flow of war matériel into the region. This in turn would weaken the states in the Deep South, making them more vulnerable to follow-on invasion.

The outer defenses to Mobile Bay were formidable. Although the bay itself is huge, up to 19 miles wide, a long barrier peninsula and a barrier island leave it with only two entrances. Of the two, the navigable one facing the Gulf of Mexico is only about 2.5 miles wide. During the war, heavily armed coastal fortresses dominated the shore on either side of this entrance. In between lay three rows of mechanical contact mines, with the only cleared path lying at the eastern extremity directly under the guns of Fort Morgan. This Charybdis and Scylla situation left hostile naval forces with the choice of death from below or death from above. Once inside the harbor, attackers would find at least one large ironclad waiting for them.

Farragut's first task was reconnaissance of the battlespace. He sent a team under his flag lieutenant to survey the underwater battlespace over a few nights preceding the assault. Diving from darkened boats propelled by muffled oars, they explored the area to the east of the easternmost buoy for reported mines. Finding none despite considerable intelligence from deserters, refugees, and others,[83] they further examined the mines to the west of the buoy. Discovering mines aplenty, they drilled holes to sink some, dragged some out of the way, and examined others to find that many would likely be inoperable due to sea growth. These underwater mine countermeasures survey operations completed, they reported their findings.[84]

With reconnaissance completed, Farragut devised a strategy tailored to the situation. The Union fleet would enter the harbor through the marked channel to avoid the mines. In order to minimize the impact of Fort Morgan's guns, the ships would enter in pairs, with ironclads on the fort's side to help shield the less-armored vessels tied beside them. Going at high speed in paired column,[85] they should have minimal exposure to mines and fortress guns alike. By steaming in column, if the lead ships were to detonate mines, they would have succeeded in clearing the path for the follow-on ships.

The battle began with the fleet arranged as described above, though with the addition of the USS *Tecumseh*, which had requested to be the lead vessel, along with a cutter rigged with an improvised mine catcher. Things went wrong when the Confederate ironclad CSS *Tennessee* came into view. Seeing this vessel as the primary threat, the *Tecumseh*'s captain began edging toward her as he led the column into the bay. Passing to the west of the marker buoy despite orders, the ship struck a mine and vanished in less than a minute, taking most of her crew with her. *Brooklyn*, the wooden ship accompanying her, began to back down while engaging the fort's guns. Seeing the danger of the fleet being bogged down under the fortress guns because of *Brooklyn*'s actions, Farragut called from his perch in the *Hartford*'s rigging, "Damn the torpedoes! Full speed ahead, Drayton!" He knew that he could safely make such a call because of his flag lieutenant's work over the previous few nights. Despite several ships' hulls scraping sunken mine cases on their way into the bay, no other ships suffered mine damage that day.[86]

Once in the bay, the Union Navy engaged the heavily armored *Tennessee*. While attacking the middle of the Union line, her armor allowed her to stay in the fight much longer than the relative numbers would suggest. Ultimately, however, she was reduced to immobility and surrendered. There followed sieges on the two forts guarding the entrance, though these soon capitulated when it became apparent that support would not be forthcoming. With the fall of the forts, the Union had control of the lower bay, completing the blockade of the Gulf of Mexico ports east of the Mississippi. This, however, was only the beginning of Union troubles in Mobile Bay.

Cleanup Operations: Taking control of the lower bay was one thing, but taking Mobile itself was a far greater challenge, one that Farragut did not have sufficient forces to achieve in August 1864. With hurricane season in full swing and other campaigns taking higher priority, Union forces could do little more than maintain a blockade of the upper harbor. Meanwhile, the Confederate forces under General D.H. Maury, nephew of mine scientist M.F. Maury (discussed previously), did not sit idly by waiting for the inevitable attack. Over the next half year or more, they bolstered their defenses, including mining the inner harbor and the Blakely River, whose outflow formed the bay. Few naval actions took place during this lull, as the Navy did little beyond maintaining the blockade and removing mines as they found them. In December, the light gunboat USS *Narcissus* struck a mine while on picket duty, but most action did not happen until March 1865 and later.[87]

With only one month remaining in the war, the Union Army had finally fought its way through the Mobile landward defenses. The Navy forced its way into the inner harbor intent on engaging the batteries protecting the city, as well as the fortress at Spanish Fort on the other side of the Blakely River estuary. It was in this area that the Navy found mines in great numbers.[88] Before the war's end on April 9,[89] four more Union ships struck mines on the Blakely River while attacking Spanish Fort and Fort Blakely. Another four would be lost in the month that followed before peace once more came to the United States. With 10 vessels lost to mines in the vicinity of Mobile Bay, this campaign was by far the most successful Confederate mining operation of the war and the costliest for the Union.[90]

Impact of the War on Mine Warfare Development

The U.S. Civil War introduced the world to naval mines as serious weapons of war. While they were present in the Revolutionary War, the War of 1812, and many others, they were primarily just the occasional one-off nuisance. In those early days, fleet commanders had no need to prepare for mines in the water, but only to respond on the rare occasion that they appeared. Little damage came from mines, and usually they went entirely unnoticed by their intended targets.

When the Crimean War came along, mines had become part of a national defense strategy. Russia showed the world that mines could be used as a part of an integrated harbor defense. They still did little harm beyond damaging a ship here or there, but their presence was sufficient to deter amphibious landings at the ports where they were active. Still, the war was primarily an affair of armies, with naval forces providing more of a supporting role. Russia did not depend on the sea to keep its economy running, and as such little was decided by the absence or presence of the navy beyond the threat to St. Petersburg. Russia was a naval power that chose to not use its navy, so mines were of secondary importance.

Things were different in the Civil War. The Confederacy was an example of what a nation with no navy at all can do against what was then the most powerful navy in the world. Unlike the Crimean War, where little was at stake for the invaders, the Civil War was one that the Union *had* to win for the survival of the nation and the nation's soul. That impetus gave rise to a different strategy, one where any cost would be borne to take the day. Furthermore, the unique geography of the South made possible a far more pronounced naval role than that of the allies in the Crimean War. Being geographically isolated with an enormous coastline, numerous natural harbors, and long navigable rivers, and only one border with a neutral country a thousand miles from the action, the South was in effect an island nation. It was both dependent upon the sea and prone to attack from the sea. With rivers flowing deep into its interiors to major population centers at the fall line, it was uniquely vulnerable to the shallow-draft, heavily armored gunboats of the era. Thus, if this pseudo-nation with only a small, blockaded navy was to survive, it had to equalize the balance of power on the water.

Mines were the great equalizer. At a time when Union ironclads roamed rivers and harbors with relative immunity from cannon fire, the Confederacy proved that they were vulnerable from below. In the previous chapter, the Russian use of incendiary shells to destroy the Ottoman fleet at Sinop showed the world that the days of wooden ships were numbered. At Hampton Roads, the battle between the *Monitor* and the *Merrimac* showed the world that armored ships were the nearly invincible wave of the future. This in turn spurred the Great Powers of Europe and elsewhere into a naval arms race that led to the great battleships that dominated the seas a half century later. But most importantly to the development of mine warfare, the Confederates showed the world that such armored behemoths could still be sunk, if only from below.

During the war, some 47 or so vessels were sunk or damaged by mines or other technology then classified with mines ... far more than by any other cause. This, for the first time, established the legitimacy of mines as weapons that could neutralize the powerful battleships of their rivals. Industrialized nations began in earnest to use the examples from the U.S. Civil War to develop the types of underwater weapons that would be commonplace by the First World War. Moored contact mine designs were refined from the shallow-water devices of the Civil War to become mass-produced, spherical, deep-water devices. Controlled mines were developed into complex coastal defense weapons. Drifting mines were transformed in two directions. In one direction, the weapons were morphed from the improvised weapons sent down-current in previous wars to become highly sophisticated weapons dropped to slow followers. In a different direction, this idea of offensively sending explosives underwater to attack enemy fleets merged with the spar torpedo concept. These ideas evolved into the self-propelled torpedoes of the First World War and later. Furthermore, the idea brought on by mines and the Davids of using stealth to hide from powerful fleets until the moment of attack led directly to the first real interest in submarine development.

Had it not been for the Civil War's example of mines as foils to armored ships, the navies of today might look far different.

Table 1. Vessels Damaged or Destroyed by Mines in the Civil War

		James River, VA	Hampton Roads, VA	Roanoke River, NC	Cape Fear River, NC	Georgetown, SC	Charleston, SC
1862	Feb						
	Dec						
1863	Feb						
	Mar						
	May						*Weehawken*
	Jul						
	Aug	*Commodore Barney*					*Pawnee* and launch
	Sep	*John Farron*					
	Oct						*New Ironsides*
1864	Feb						*Housatonic*
	Mar						
	Apr		*Minnesota*				
	May	*Commodore Jones*					
	Jun						*Etiwan / Marion / Shultz*
	Aug	Ammo Barge / *Lewis*					
	Oct			*Albemarle*			
	Nov	*Greyhound*					
	Dec			*Otsego / Bazely*			
1865	Jan						*Patapso*
	Feb				*Shawmut*'s launch / *Osceola*		
	Mar				*Thorne*	*Harvest Moon*	*Bibb / Massachusetts*
	Apr						
	May						
	Jun						*Jonquil*

North Edisto River, SC	Wright River, GA	Ogeechee River, GA	St. John's River, FL	Mobile Bay, AL	Yazoo River, MS	Port Hudson, LA	Red River, LA
	Susquehanna's launch						
					Cairo		
		Montauk					
						Richmond	
					Baron De Kalb		
Memphis							
			Maple Leaf / Gen'l Hunter				*Eastport*
			H.A. Weed				
			Alice Price				
				Tecumseh			
				Narcissus			
				Althea / Milwaukee / Osage			
				Rodolph / Ida / Sciota / Cincinnati			
				R.B. Hamilton			

Chapter 4

Russo–Japanese War

Note: In a normal book about naval mine warfare history, the obvious thing to do is a lot of talking about mines. With the Russo–Japanese war, however, there is not much material available. Most of the naval action involving mines centered on one place for a limited amount of time. If a writer wishes to simply write a bland recounting of events, the few notable mining events in that war are hardly the basis for a dedicated chapter. However, in a book about impactful *mine warfare campaigns, the Russo–Japanese War deserves a seat at the table for at least three reasons. To begin with, it was the first war in which offensive mining took place. This tactic would appear again in nearly every naval war that followed. Second, a single minefield would ultimately determine the course of the entire war, a war that led to the enormous world events that dominated geopolitics for most of the century that followed. Third, mines laid in this war drifted in neutral shipping lanes for years afterward, leading to the international rules that govern the use of naval mines today, over a century later. Thus, where the paucity of direct mining events prevents a purely narrative approach, the bulk of this chapter will instead concentrate on discussing and analyzing the events leading up to—and following—the actual mining. Hopefully this device will make the chapter interesting to you, the reader.*

Introduction

Sunday, April 13, 1902: An open-ocean minefield changes the course of history, leading to a new world order and to the first international rules on the use of naval mines.

At the beginning of the 20th century, the world was in transition as empires were coming to the ends of their frontiers. Africa had been formally divided by the European powers at the Berlin Conference of 1884–1885.[1] The countries of North America were consolidating their gains, while the European powers no longer found room for gains anywhere in the Americas.[2] Seemingly, the only places left for expansion were the lands controlled by other empires.

China had recently gone through the Boxer Rebellion in which attacks against foreigners led to the landing of troops by eight nations. The loss of the Chinese Imperial Army to these troops led to a fearsome indemnity and the stationing of foreign troops in several Chinese cities. Thoroughly weakened, the Qing Dynasty was left with little leverage against foreign influence. In this environment, each of several nations gained special privileges and controlled spheres of influence within ostensibly sovereign Chinese territory.[3]

During this same period, Japan was coming into her own as an important regional power. As part of the reforms of the Meiji period, the country had built a modern army and a modern navy. Highly trained based on the best Western tactics, systems, and equipment available, she was a peer competitor for any potential regional rivals. In 1894/1895, the nation won the first Sino-Japanese War. In winning, it gained a foothold in the important base at Port Arthur in the Manchu region in the northeast corner of China.[4] Furthermore, China agreed to cease exercising power over the emperors of the Korean Peninsula. With China out of the way and out of Korea, Japan installed a government friendly to its own cause.[5]

Russia had other ideas. Taking advantage of China's weakness, it infringed on Japan's interests in three ways. First, it sponsored a coup that removed Korea's new pro-Japanese government. Second, with the support of Germany and France, it forced Japan to return Port Arthur to China. Third, using its leverage as the guarantor for the huge financial indemnity owed to Japan, Russia compelled China to grant several concessions. Among these concessions were the leasing of Port Arthur to Russia, as well as the right to build railroads connecting Port Arthur and Vladivostok to the Trans-Siberian Railroad. By controlling the areas surrounding these railroads, Russia could exercise de facto control over northeastern China.[6] For a newly reformed country like Japan, these actions were both humiliating and a threat to its own imperial ambitions.

Fleet Comparison

In 1904, the size of Russia's main naval battle line of battleships far exceeded that of Japan. Whereas Japan could field six battleships, Russia would ultimately commit 18 battleships to the Pacific theater (though three were smaller coastal battleships). However, this numerical superiority did not tell the whole picture. To begin with, the Japanese fleet represented the latest in technology and interoperability. Built almost entirely in British shipyards, they were all similar in capability and speed. This force was further supported by a strong shipyard capacity, able to repair battle damage and build the smaller naval vessels needed to support the fleet.[7] Their sailors were almost universally literate and trained in the best tactics, techniques, and procedures (TTPs) then available. It was a fleet ready to fight above its weight class.[8] With a recently built world-class naval base at Sasebo on the Tsushima Strait, this fleet was within easy striking distance of the Korean Peninsula and the entire Yellow Sea.[9]

By contrast, the Russian fleet was a museum of battleship development. Built across a hodgepodge of different countries' shipyards over a range of time, the fleet suffered from a wide disparity in capabilities. These ships were dispersed across the globe, with seven of them initially installed in the Pacific at Port Arthur, while the other 11 started the war as the Baltic fleet. More battleships were stationed in the Black Sea, where recent wars with the Ottomans compelled them to remain in place. For those ships at Port Arthur, no shipyard facilities existed, making it difficult to repair battle damage.[10] They were crewed with peasants led by the officious and often drunken men who then made up Russia's officer class. Poorly trained, illiterate, and harboring a certain amount of resentment over their unequal treatment, these Russian crews were inadequate compared with their Japanese counterparts.[11]

Naval mines technology had progressed somewhat since the handcrafted, small-batch devices of the American Civil War. After a few decades of use in various wars around the world, mines were now metal-built, mass-produced weapons of war created to be laid using purpose-built minelaying systems. Mines had become part of the integrated coastal defense systems of navies around the world.[12] New fortifications were being built with retracting cannons, searchlights, and bunkers for the storage, deployment, and firing of mines.[13] Any ship's commander considering a bombardment knew that he could possibly have to contend with mine barriers intended to keep vessels like his own from approaching within gun range. Mines had become accepted as viable weapons for defense.

The Conflict Begins

In a foreshadowing of Pearl Harbor, the Russo-Japanese War began with a surprise attack. With a strategic situation as precarious as Russia's, with most of its Pacific fleet at a single port and reinforcements half a world away, a successful attack could prevent Russia from interfering with Japanese designs in the region. Without a fleet at Port Arthur, Russia would potentially need to withdraw from the Manchu region entirely.

On February 8, 1904, 10 destroyers from the Japanese main battle fleet under Vice Admiral Togo entered Port Arthur under cover of darkness. Forewarned of the potential for war, Russian Admiral Stark had ordered measures to blunt an attack but had not ensured that his orders were being followed to the letter. Thus, despite his orders, some ships had not yet implemented his precautions. Taking advantage of the Russian fleet's lack of preparedness, the Japanese destroyer squadron launched a flurry of torpedoes that damaged two battleships and a cruiser before retiring back to sea.[14] The next day, Japanese forces attacked the small Russian naval force at Inchon. A week later, the Japanese began their invasion of Korea, unopposed by Russian ships.[15]

As at Pearl Harbor, the admiral in charge of the attacked base was fired. Admiral Stark was relieved of command, to be replaced by a dynamic new commander, Vice Admiral Stepan Makarov.[16] The choice was a good one. An inventive warrior-scholar who had made the world's first successful mobile torpedo attack during the Russo-Turkish War of 1877–1878, he had extensive experience in the Pacific theater of operation.[17] Immediately launching into a series of reforms intended to improve the morale and material readiness of the fleet, he showed an aggression that had all the potential to throw a wrench into the machinery of Japanese plans. Where Stark favored caution in his defenses, Makarov favored aggressive patrolling and using offense as the best form of defense.[18]

Mines at Port Arthur

Mines, at the time, were used to keep enemy ships away from friendly ports. This is the essential element of a *defensive minefield* as discussed in an earlier chapter. It takes advantage of the fact that only the defenders know the safe way through the minefields, while attackers would be ignorant of this key information. If they did not know that mines were present, they would likely stumble into the minefield and suffer damage.

If they did know of the presence of mines, they would be deterred from launching an assault. The effect was (and still is) a sort of one-way valve, keeping the port operational while also reasonably safe from seaward attack.

For Japan, this situation was unacceptable. Its strategy required the neutralization of the Russian fleet, either by destruction or by otherwise preventing it from leaving port. The traditional method of blocking a fleet in port was through blockade, an approach involving closely patrolling the harbor approaches with smaller ships while maintaining a battle fleet close at hand. However, aggressive Russian sorties directed and led by Makarov created a hazard to patrols. At the same time, the limited size of the Japanese fleet meant that the few battleships would be tied up and unavailable for service elsewhere. With numerical inferiority, the Japanese admiral, Togo, could ill-afford to allow even a single ship to return to Sasebo for refitting. Doubtlessly cognizant of Russia's ability to reinforce its fleet within half a year, Admiral Togo needed a new tactic to neutralize the Russian fleet. He needed to do this quickly, to give his fleet more freedom of movement.

His first attempt to solve this problem was to try to sink five merchant ships in the harbor mouth. With a relatively shallow harbor mouth subject to a huge tidal range, Port Arthur was vulnerable to such tactics. Unfortunately for him, his efforts to do so on February 24 ended in failure as shore and ship fire destroyed all five vessels before they reached the chokepoint.[19]

On April 12, Togo's fleet executed a new tactic. Working in the cover of darkness, a Japanese minelayer planted a field of 48 mines off Port Arthur. Today such a tactic seems mundane and obvious, but at the time it was revolutionary. Consider the implications. Throughout history, a fleet commander had complete freedom of movement into and out of his own port, barring the presence of an opposing fleet. Having steamed into and out of that port innumerable times, every hazard was known and charted. He could be certain that his planned course would be a safe one. Togo's tactic, the first use of what is today known as "offensive mining," changed that certainty.[20] Offensive mining places new hazards in waterways previously known to be safe, in much the way that a sunken ship could create a new hazard. However, unlike sunken ships, it is hard for a shore observer to notice mines going underwater. Moreover, sunken ships create a hazard only at points where the water is too shallow to allow a ship to pass over the sunken one. Naval mines, on the other hand, can be placed in deeper water, with moored mines being restricted only by the length of line that can affix it to their anchors and influence mines being restricted by the limits of their sensors. With offensive mining, fleets faced a persistent threat outside their own ports. That threat, and the even scarier thought of its hidden nature, could keep a fleet bottled up in port even when no enemy fleet was present.

On the morning of April 13, 1904, Admiral Makarov knew nothing of offensive mining. That morning, in response to a skirmish of destroyers, he steamed out of port on the battleship *Petropavlovsk*, accompanied by the battleship *Pobieda* and four cruisers. Giving chase through the fog, his force encountered five Japanese battleships. Recognizing that he was outnumbered, he directed his force to return to port. As they approached the supposed safety of home, they passed directly over the Japanese offensive minefield laid the previous night. *Petropavlovsk* struck two mines,[21] kicking off a cascade of sympathetic explosions that sank the ship in two minutes, bringing Admiral Makarov and 677 crew members[22] down with it. Shortly afterward, the *Pobieda* also sustained mine damage, though she was still able to make it into port.[23]

Russian mines on the beach near Port Arthur, 1905. Courtesy of Boris V. Drashpil (Naval History and Heritage Command).

Russian mining of the approaches to both Port Arthur and Vladivostok began shortly after the attack. Supported by the coastal fortifications and the guns of the ships in port, their bolstered defenses deterred further attack by Japanese forces. With this now formulaic combination of weaponry in place, the Russians enjoyed a place of refuge unassailable from the sea. Any attempt to take the port required a landward assault by first taking the Liaodong Peninsula on which the city lay.[24]

Other than improving their defenses, the Russian response to Japan's offensive mining attack was doubtlessly one of panic. Admiral Makarov's replacement, Rear Admiral Wilgelm Vitgeft, opted for a much more cautious approach than his predecessor, refusing to take his fleet to sea. Instead, he moved many of his ship's crews to man the forts ashore and he set his minelayer, the *Amur*, to work laying additional fields. The *Amur* laid at least one minefield farther out to sea than his orders allowed, resulting in the sinking of the Japanese battleships *Hatsuse* and *Yashima* on May 16.[25] These sinkings significantly shifted the balance of forces. Where Russia now had three damaged battleships and three good ones against Japan's six good battleships, the latter was now down to four. Togo's response, on May 30, was to order the ostentatious laying of 1,000 more mines around Port Arthur to frighten the enemy into staying in port.[26]

Fleet Actions

In the interim, Japanese land forces invaded and captured the northern end of the Liaotung Peninsula, thus cutting Port Arthur off from resupply. Isolated and under siege by both sea and land, decisive action was needed to change the Russian situation.

On June 20, Tsar Nicolas ordered the Baltic fleet to deploy to the Pacific to relieve the beleaguered fleet at Port Arthur. A mere three days later, perhaps spurred by telegraphed news of the intended relief force, Rear Admiral Vitgeft steamed out of Port Arthur with his fleet. With his damaged battleships repaired, he enjoyed an advantage of six battleships to the Japanese's four. Unfortunately for Russia, he soon lost his nerve and led his fleet right back into port, with the battleship *Sevastopol* sustaining mine damage along the way. There they would stay until Japanese howitzers, now approaching the outer defenses, began raining shells directly into the harbor.[27]

On August 10, the newly promoted Vice Admiral Vitgeft led his fleet once more to sea with plans to escape to Vladivostok. They would never make it. The four Japanese battleships found them and engaged. In the battle that followed, Vitgeft was killed and nine of the 10 total battleships involved were damaged (with no sinkings). All Russian battleships escaped, with five making it to Port Arthur and one escaping to Kouchou, where it was interned. The ones making it to Port Arthur were doomed.[28]

While the First Pacific Squadron sat in Port Arthur awaiting its fate, the Baltic Squadron, now called the Second Pacific Squadron, made ready for sea. This force was made up of seven battleships: four dreadnought battleships barely off the ways, one ship with only a year in service, and two pre-dreadnought battleships.[29] This motley squadron of obsolete ships, combined with newer ships crewed by inexperienced draftees and reservists, finally made it to sea on October 15. Once at sea, they engaged in a series of mishaps that would make any professional mariner cringe with despair. One error was the Dogger Bank Incident. Knowing that most Japanese warships were built in Britain, they feared attack while still in the North Sea. As they passed through those waters, the battleships mistook British fishing trawlers for a Japanese fleet. In a show of excessive force, the battle fleet engaged the fishing fleet with a withering slew of shells, nearly sparking war.[30]

Back in Port Arthur, the First Pacific Squadron loitered in port. There they remained until December, when Japanese troops were within line of sight of the harbor. Now that the Japanese artillery spotters could direct fire, they soon had all but one Russian battleship sitting on the bottom of the harbor. The very last battleship, the *Sevastopol*, escaped to sea, where it was chased and eventually forced to scuttle to avoid capture.[31] Now unopposed at sea and with plenty of warning of the approaching Russian replacement squadron, Togo took this interlude to refit his fleet at Sasebo.[32]

Meanwhile, the Second Pacific Squadron was working its way around Africa. Upon arriving in Madagascar, they learned of both the loss of Port Arthur and of the brutal suppression of protesters in St. Petersburg known as "Bloody Sunday." With morale already low, a wave of unrest swept through the fleet and led to a mutiny on one ship. On April 8, the fleet passed Singapore, giving the Japanese plenty of warning of their pending arrival. At Cam Ranh Bay in Vietnam, the fleet was ordered to await reinforcements from a newly formed Third Pacific Squadron. Assembled from a medley of older ships previously rejected by the Second Pacific Squadron's admiral, these four pre-dreadnought battleships added little to the fleet's firepower while taking much from its speed of advance.[33]

On May 14, the now combined Second and Third Pacific Squadrons finally left Vietnam for their final run to Vladivostok to join the remains of the naval forces there. Most never made it. On May 27, they met the Japanese fleet. Forewarned of the Russians' impending arrival, Togo had his ambush set. His force of four battleships and 11 cruisers[34] swept in on the Russian squadron as it passed through the western passage of the Tsushima Strait. His action plan was well planned and thoroughly discussed with his commanders. By contrast, the Russian admiral had no tactical plan at all and had done little to train his fleet en route. In the battle that followed, five of 11 Russian battleships were sunk and the squadron was broken up. Of the ships that remained, some scattered to neutral ports, some surrendered, some scuttled, and the last battleship was sunk in a running gunnery duel. Of the 53 total Russian ships that left Cam Ranh Bay, only one cruiser and two destroyers reached Vladivostok. The loss was devastating and decisive. Peace soon followed.[35]

Postwar Political Fallout

The consequences of the war on subsequent world history were immense. It was the first war in modern times in which an Asian country beat a European country. This changed the balance of power in Asia while also destroying European illusions of dominance as destiny.

For Japan, the prestige brought on by the resounding victory helped to quell the unrest then fomenting at home because of the heavy military spending. With Russia out of the way, Japan now enjoyed absolute regional dominance and a foothold in Manchuria. From there, it was positioned for its annexation of Korea in 1910 and its invasion of the rest of China in the 1930s. With China in hand, it had the platform from which to launch its invasion of Indochina and the Pacific Islands. These actions put the Empire of the Rising Sun in direct conflict with the United States. When the United States sanctioned Japan by cutting off her oil supplies, the latter attacked the American fleet at Pearl Harbor and invaded the Philippines.

For Russia, the war and its result were gasoline to the flames of unrest already burning, leading directly to the Russian Revolution of 1905. Prior to the war, Russia had been an autocratic regime struggling with the social challenges of modernizing and industrializing while still retaining the repressive tactics from its days of serfdom. Unrest grew as war spending caused a recession that led to massive inflation without compensatory wage increases. With the war drawing away more and more of the troops that the Russian rulers traditionally used for internal suppression, the omnipresent dissidents became bolder. When a peaceful protest for reform turned into the massacre known as Bloody Sunday, unrest flashed across the country. This unrest skyrocketed after the disastrous loss at Tsushima and came to a head soon after the peace treaty was concluded in Portsmouth, New Hampshire, on August 29, 1905. While a government can always expect some level of domestic unrest when waging war, it had better win to justify the costs in lives and money. The total humiliation suffered by the tsarist government had the opposite effect. Protests exploded throughout the empire. These included several mutinies in the naval fleet led by sailors resentful about the poor leadership that caused the death and imprisonment of so many of their comrades.

Faced with the unrest that continued to rage, on October 30, 1905, Tsar Nicholas II signed the October Manifesto. This document promised a constitutional monarchy and the creation of Russia's first popularly elected legislature. Much of this promise was later repudiated and the repressive tactics returned. The resultant resentment, along with the continued, demonstrated government incompetence, would ultimately lead to the Bolshevik Revolution during the First World War. Such was the start of the Soviet Union. The revolution drew the new Union of Soviet Socialist Republics out of the war, allowing Germany to concentrate resources on the Western Front. Peace in the east also meant that German prisoners of war, nourished on revolutionary ideas of their recent captors, returned to their homeland. These returning prisoners helped inspire domestic unrest in Germany as local socialists pushed for their own revolution. On October 29, 1918, a general mutiny in the German fleet sparked a revolution throughout Germany that forced the emperor to abdicate on November 9, 1919. Two days later, the country signed the armistice that ended World War I. The follow-on German Revolution of 1918 to 1919 ended with the forming of the Weimar Republic. It was within the framework of that republic, and the deprivations suffered by it as a result of the hastily signed armistice, that the Nazi party came into power.

It is impossible to say what would have happened had Admiral Togo not sunk Admiral Makarov with the world's first offensive mining attack. Nonetheless, a case can be made that had it not happened, World War II, USSR, and possibly even World War I may have never come into being. Such is the butterfly effect of history.

The Hague Conventions

During the late 19th century, the nations of Europe were engaged in an arms race. Technology had rapidly progressed since the Napoleonic War, the last major European war involving fleet engagements between naval powers. True, there were naval actions such as the Russo-Turkish War and the Crimean War discussed previously. But such actions were generally small in scale and asymmetric in the relative strengths of the forces involved. On land, the technology for killing soldiers had advanced to now include new and horrifyingly efficient means of death. Recognizing that the morality of warfare had changed since the days when a soldier fighting in line had enjoyed more than an even chance of survival, intellectuals began to advocate for limits. With enough pushing on the public discourse, they eventually gained official attention in diplomatic and military circles.

In 1899, Tsar Nicholas II of Russia (the same emperor discussed in this chapter) sponsored peace talks at the Hague in Brussels. These talks brought in diplomats from many nations with the purpose of establishing rules for the conduct of war and the use of certain weapons. Many rules were agreed to concerning land forces, but little attention was paid to naval forces. With little recent use of the new, revolutionary naval technology, diplomats would have lacked deep insight into the many ways that this technology could be used and misused. Based on history, offensive naval warfare would have consisted of relatively evenly matched battleships fighting in line (ostensibly similar to the ships-of-the-line of old). Defensive naval warfare would have included coastal fortifications and maybe some mines in a combatant's own harbors and rivers. Self-propelled torpedoes were available, and had been used with some success, but may still have been thought of as short-range weapons. These being generally seen as appropriate uses for such weapons, there would have been little drive for increased regulation.

By 1907, when the Geneva Convention reconvened, the situation was far different. While much of the death from the Russo-Japanese War was from the land campaign, the actions at sea were shocking. It was the first meeting of fleets of battleships, with

Russia having lost not just one but two fleets to Japan's total of six battleships. It was the first use of open-ocean mining and the first use of offensive mining. At least three battleships and four cruisers, as well as other ships, had been lost to these mines. Freedom of movement had been lost throughout the region for belligerents and neutrals alike as mines broke loose and drifted throughout the Yellow Sea. Torpedoes were no longer just a short-range weapon representing grave risk to the user. Over 300 of them were fired during the war between Japanese attacks on Port Arthur and the three main fleet actions. Naval war was shown to be no longer a simple fight-or-flight affair between relative equals. Now it was an asymmetric contest where superior explosives delivery technology could kill large numbers of sailors who had little ability to defend themselves.

After such events, combined with increased antagonism among European Great Powers, the 1907 Geneva Convention paid much more attention to naval warfare. One of these areas was submarine mining. The Hague Convention VIII of 1907 set forth rules regulating (though not abolishing) the use of naval mines and torpedoes. Though these rules were far from perfect and were written when only contact mines existed, they have persisted in their applicability to this day. In what follows, this author's analysis of the articles specific to weapons regulation (vice convention administration) appears in italics.

Text of the Hague Convention (VIII) on Submarine Mines, 1907[36]

Convention (VIII) relative to the Laying of Automatic Submarine Contact Mines. The Hague, 18 October 1907.

(Preamble.)

(List of Contracting Parties)

Inspired by the principle of the freedom of sea routes, the common highway of all nations;

Seeing that, although the existing position of affairs makes it impossible to forbid the employment of automatic submarine contact mines, it is nevertheless desirable to restrict and regulate their employment in order to mitigate the severity of war and to ensure, as far as possible, to peaceful navigation the security to which it is entitled, despite the existence of war;

Until such time as it is found possible to formulate rules on the subject which shall ensure to the interests involved all the guarantees desirable;

Have resolved to conclude a Convention for this purpose, and have appointed the following as their Plenipotentiaries:

(Here follow the names of Plenipotentiaries)

Who, after having deposited their full powers, found in good and due form, have agreed upon the following provisions:

Article 1. **It is forbidden**—*(emphasis added by author)*

1. To lay unanchored automatic contact mines, except when they are so constructed as to become harmless one hour at most after the person who laid them ceases to control them;

In short, drifting mines must self-neutralize within an hour. Several interesting drifting mine designs had appeared by this time, with the idea of releasing them into harbors in the manner of fire ships of old, to wreak havoc with the incoming tide. If a nation were complying with the agreement, this tactic would be less available except in very distinct circumstances. That in turn made drifting mines less valuable assets in a nation's mines inventory. Nevertheless, drifting mines were used with success only a decade later, as will be shown in the next chapter.

2. To lay anchored automatic contact mines which do not become harmless as soon as they have broken loose from their moorings;

In short, moored mines must self-neutralize if their mooring wires are severed. The fact that this was included suggests that prior to the convention, this was not a universal feature of moored mines. Likely neither the Russian nor Japanese mines at Port Arthur had this feature, accounting for the reason why mines floating around the Yellow Sea would continue to be a hazard to shipping. Theoretically, if all the world's moored contact mines truly did become harmless upon breaking loose from their moorings, then successful mechanical minesweeping should cause all swept mines to become harmless. However, only a fool would trust this to be the case and move among them without caution. The challenge is to ensure that all cut mines are neutralized before the sweeper loses track of them. Integrated training for mechanical minesweeping should always go beyond the procedural practice of a ship or helicopter just putting out a mechanical minesweeper and driving around with it, followed by another unit taking all the time in the world to neutralize a single mine. Historically, realistic training would involve several exercise mines being cut in a short period of time. This adds the dimension of dealing with a now dynamic threat, which in turn affects the time that can be spent neutralizing individual mines. It also requires more complex inter-unit coordination, contact tracking, and balancing the speed of sweeping against the speed at which available neutralization assets can keep up.

3. To use torpedoes which do not become harmless when they have missed their mark.

In short, torpedoes must self-neutralize at the end of their runs. There is some possible point of contention here if one were to follow the strict wording of this clause. Suppose that a submarine lined up a shot at a ship in a convoy so that if it missed its mark, then it would hit another ship just beyond. This was a common tactic in World War II. Technically, once missing the intended mark (the first ship), then the torpedo should have become harmless before reaching the second ship. Practically speaking, however, this would have been impossible to achieve. In reality, the implied interpretation of this requirement is that it self-neutralizes when it stops moving.

Art. 2. It is forbidden to lay automatic contact mines off the coast and ports of the enemy, with the sole object of intercepting commercial shipping.

Offensive mines are not supposed to be laid solely to target commercial shipping. Note that there is a lot of gray area built in here. If a single armed vessel lies within a port, or if military cargo is transported into a port, the port becomes a valid target for mining since the interception of commercial shipping would no longer be the sole object of the mining. But a port not guarded is vulnerable to commerce raiding and other attacks. Another challenge is that a navy that does not spread itself among many ports at once risks having nothing left in the aftermath of a surprise attack, or in the modern age, a tactical nuke. Thus, in war it is conceivable that there would be few ports capable of handling ships that would not also have a naval presence, thereby making the port susceptible to mining against the off chance of intercepting military shipping. If the goal here was to protect commercial shipping, then it should have been better worded to prevent other types of threats to commercial shipping.

One consideration is related to mercantile competition that existed in 1907 among the several empires. Prior to the codification offered by this article, it may have been conceivable for a nation to lay mines in neutral waters along a competitor's trade lanes in times of peace for the purpose of damaging their trade. This article makes it clear that such actions would be universally considered an act of war. As will be seen in chapter 11, such actions still happen.

Art. 3. When anchored automatic contact mines are employed, every possible precaution must be taken for the security of peaceful shipping.

The belligerents undertake to do their utmost to render these mines harmless within a limited time, and, should they cease to be under surveillance, to notify of the danger zones as soon as military exigencies permit, by a notice addressed to ship owners, which must also be communicated to the Governments through the diplomatic channel.

Belligerents must neutralize their mines and/or tell others of the mines' presence if they do not maintain a constant presence in the vicinity of the minefields. Again, note the level of gray areas in this wording. A country's "utmost" in clearing mines could be very little if they have not developed a robust mine countermeasure capability. A "limited time" based on "military exigency" is vague, potentially leading to a delay of months if the "military exigency" means waiting for the mines to blow up their intended targets. It also allows justification for a delay if a belligerent frequently returns to reseed the minefield. Furthermore, the duty to notify of "danger zones" also lacks specificity as it does not provide a size limit for this zone. Thus, a danger zone could conceivably encompass every inch of water in which mining is possible.

Art. 4. Neutral Powers which lay automatic contact mines off their coasts must observe the same rules and take the same precautions as are imposed on belligerents.

The neutral Power must inform ship owners, by a notice issued in advance, where automatic contact mines have been laid. This notice must be communicated at once to the Governments through the diplomatic channel.

Same rules apply to neutrals.

Art. 5. At the close of the war, the Contracting Powers undertake to do their utmost to remove the mines which they have laid, each Power removing its own mines.

In other words, countries that lay mines must remove their own mines from their own territorial waters and neutral water. Generally, no nation is likely to want to leave mines where they are a threat to their own shipping, if they thought that they were safe from external assault. The problem can come from a defeated nation. If a country is left with no navy, no control of their own territory, and possibly destruction of any records of their mine-lay plan, then they may not have the capability to clear their own mines. They may not even be able to reimburse another nation for doing it for them. Thus, a weaker nation, though having substantial reasons for mining their own waters, does so at the risk of poisoning their own waters against future use.[37]

As regards anchored automatic contact mines laid by one of the belligerents off the coast of the other, their position must be notified to the other party by the Power which laid them, and each Power must proceed with the least possible delay to remove the mines in its own waters.

After peace, belligerents must inform their former enemies of the locations of offensive minefields. Note that there is no requirement to teach their former enemies how to neutralize those mines, as there is no requirement for a nation to already know how to clear their own mines before laying them. Thus, unclearable minefields could be laid, potentially spoiling an enemy's waters for decades after peace returns. As we shall see later, this problem contributed to Japan having to work for over 25 years to clear U.S. mines from its home waters following World War II.

There are several more articles, but they are all to do with the applicability of the agreement and its future administration.

Impact on Shaping Mine Warfare

The Russo-Japanese War was a watershed moment in naval mine warfare. Despite the mining portion being restricted primarily to the waters around Port Arthur for only about a year, its consequences are profound. As the direct cause of the Hague Convention VIII of 1907, it led to international rules that have governed mine warfare to this day. That by itself would be sufficient to warrant the war's place in the annals of naval mine warfare history, but it was also the setting for the first uses of offensive mining and open-ocean mining. These tactics were used extensively in World War I just a decade later and in every major naval war that followed. With such a tactic available, most future mines developed around being able to support it (the exception being shallow-water mines and RECO mines). Finally, and possibly most importantly, the success of that first offensive mining campaign enabled a cascade of events that may have contributed to both world wars and the Soviet Union to come into being. Few mining campaigns have had such an impact on mining and world history. Not bad for an event of which few in the modern era have ever heard.

CHAPTER 5

World War I
Dardanelles Campaign

It is summertime in the second decade of the century. Years of regional tension have resulted in the larger countries of the region building up their militaries with the best weaponry available. A grievous offense perpetrated by an individual causes a burst of indignation and rhetoric that builds into all-out war. Conflict breaks out as each side calls on the help of their larger allies. Large countries trample the rights of smaller countries to gain military advantage.

Amid the turmoil, one of the belligerent countries closes the major strategic strait that it controls. Its adversaries soon discover that it is not mere words keeping the ships from passing through this important waterway. Mines in vast numbers festoon the waters, ready to sink any large vessel foolish enough to force the strait. Fortified shore-based weapons systems wait at every point, ready to destroy anyone who tries to remove the mines.

Over the next three weeks, the coalition forces operate as close to the minefields as possible to seek out and destroy the defenders' mobile shore-based weapons. Meanwhile, an ill-prepared minesweeping force attempts to reduce the minefields while operating under withering shore fire. By the time the campaign is over, 700 sailors are dead, the strait remains closed, and mines have sunk or heavily damaged five major warships. There follows a protracted ground campaign of attrition in mountainous terrain, whose horrors eventually lead the coalition to abandon the effort.

While this account sounds strikingly like a near-future nightmare scenario in the Middle East, it describes a historical situation in Europe just over a century ago.

Political Situation

How did the assassination of one man lead to a global conflict that left 38 million killed and wounded? The answer lies in Europe's political situation leading up to the war.

The 1914 map of Europe looked far different than it does today. The Russian, German, and Austro-Hungarian Empires all shared borders with each other. Together they covered an area now occupied by at least 19 countries.[1] Each of the "Great Powers" (Britain, France, Germany, Russia, Austria-Hungary, and the Ottoman Empire) allied themselves with other Great Powers and offered protection for smaller nations within their spheres of influence. An arms race was on as each country sought to build a military to match or exceed those of its neighbors.[2]

Germany was a young country only a few decades old.[3] The six-centuries-old Austro-Hungarian Empire was a salad bowl of ethnic groups, led by 84-year-old Emperor Franz Joseph. Several small nationalist groups in the Balkans brooded under the domination of the Austro-Hungarian Empire, which in turn struggled to maintain control. Russia, though huge, was plagued by slow economic development and extreme unrest from its industrial workforce. The Ottoman Empire, suffering from a depressed economy, had recently been through a series of political coups and regional wars with Russia and other nations. In Bulgaria, recent territorial losses in the 1912–1913 Balkan War left the newly emancipated country bitter against its neighbors and the Great Powers.[4]

With an intricate network of alliances and a powder keg of international tensions, the stage was set for a cascading war that would embroil the whole of Europe. All that was missing was a spark of pretext.

Europe Descends into War

A teenager provided the spark that launched Europe into World War I. On June 28, 1914, a 19-year-old Bosnian Serb named Gavrilo Princip assassinated the heir to the Austrian throne, Archduke Franz Ferdinand. He and five other young assassins from a group known as Young Bosnia carried out the attack as part of a program to wrest the Southern Slav (*Yugoslav*) regions from Austro-Hungarian control. While the assassination took place in Austrian-held Sarajevo, Bosnia, Austria believed that significant material support came from the Serbian revolutionary group Black Hand, based in neighboring Serbia. After Serbia rebuffed initial calls for an open investigation, Austro-Hungary responded with a heavy-handed ultimatum aimed at shaping public sentiment toward declaring war.[5]

Note: It is worth mentioning that the situation at this point had strong parallels to the 9/11 attacks and the events that led to the United States declaring war on the Taliban in Afghanistan. In both cases, a small cell of a terrorist organization carried out an attack within the territory of their victim. In both cases, the aggrieved nation asked for an investigation of the organization's host nation and was rebuffed. In both cases, the aggrieved nation declared war on the grounds that the host nation was harboring the terrorist organization.

After the Austrian ultimatum, a chain reaction of activated alliances led to World War I:

1. Austria threatened war against Serbia.
2. Russia honored its treaty to protect Serbia.
3. Germany honored its treaty to protect Austria from Russia.
4. France honored its treaty to stand with Russia in the event of war with Germany.
5. Germany declared war on France, attacking across Belgium.
6. Great Britain honored its treaty to protect Belgium's neutrality.[6]

Within eight days, nearly all of the Great Powers, with their global colonial empires, became locked in war. Only the Ottomans remained neutral.[7]

The war officially began on July 28, 1914, when Austria declared war on Serbia. At that point, Turkey did not have any alliances that should have pulled it into the oncoming war on either side. Within the empire, there were both pro–British and pro–Germany factions.[8] In the end, they joined the German side of the war in reaction to early decisions made by Great Britain.

As of summer 1914, Great Britain was nearing completion of two modern battleships for the Ottomans. The Ottomans had already sent two crews to bring the ships home. However, seeing the inevitability of the rapidly developing war and the need for ships, the Royal Navy requisitioned the two ships for their own use on August 2. That was a mistake. The expected battleships were sources of great national pride for the Ottomans, who viewed this action with great bitterness. In response to the insult, they completed a secret alliance to side with Germany against Russia that same day.[9]

Lesson Learned: If you hope to keep a powerful nation friendly or neutral, avoid taking high-handed actions when they are still deciding where they stand.

Goeben *and* Breslau

When Germany declared war on France on August 2, it had only two ships in the Mediterranean: the SMS *Goeben* and the SMS *Breslau*. The larger of the two, the battlecruiser *Goeben*, outclassed anything that the British then had in the Atlantic. It was faster and had superior range and firepower compared with any of its counterparts. That meant that it could keep out of range of its opponents while simultaneously keeping within range of its own weapons. However, a lack of friendly ports made it hard to replenish its coal stores in wartime and remain independent. Additionally, it was plagued by the limitations of engineering plant issues typical of spending long periods in subtropical waters. Holding such an unsustainable position, these ships had to do something quickly before time or the enemy caught up to them.[10]

Knowing this, Britain First Sea Lord Winston Churchill directed his three battlecruisers and four cruisers to find and contain the two German ships. Over the few days preceding the British entry into the war, he sent a flood of guidance to the on-scene commander. The Mediterranean fleet was directed to prevent the German ships from attacking an expected French convoy returning troops from Africa. At other points, it was also directed to prevent the German ships from escaping through the Strait of Gibraltar and to find and destroy the German ships. Finally, it was told to preserve its own ships in case it encountered a superior force. The British admiral was uncertain about which objective took priority, and so he attempted to do all at the same time. He kept his three battlecruisers in the western Mediterranean to prevent a breakout and to protect the convoy. He sent his four cruisers to track down the German ships and engage them. What remained unclear was what "superior force" meant. Were the British ships the superior force with their greater numbers? Or was the *Goeben* the superior force with its ability to pound the British ships from a standoff distance? Eventually, the lack of clarity led to a huge toll of ships and men.[11]

Lesson Learned: When tasking on-scene commanders, prioritize your objectives. If multiple directives are necessary, summarize all priorities still in effect. Be clear on what you mean when giving your commanders leeway to not *engage.*

While the British were unclear on their primary objective, the Germans had no doubt. On August 4, Great Britain declared war on Germany. Recognizing that time was limited, the German commodore, Wilhelm Souchon, ordered his ships to bombard the important French-Algerian ports of Bone and Philippeville that same day.[12]

Lessons Learned: If it appears that your opponent's position is weak and untenable, expect them to try to be as lethal as possible as soon as possible. Expect an attack somewhere important, and soon.

The bombardment was followed by a chase as four heavy cruisers under Rear Admiral Ernest Troubridge tracked the two German ships east across the Mediterranean. On August 5 and 6, the Germans refueled for 36 hours in Messina, Sicily, while the British squadron blocked the northern exit route. At approximately 1700 on the second day, the Germans broke out through the southern exit, brushing aside the single British light cruiser, HMS *Gloucester*, which was monitoring that route. While *Gloucester* was hopelessly outmatched, she continued to shadow the German ships as they made a feint toward the Adriatic, then changed course south to head toward Turkey. The pursuit continued until 0255 on August 7, when Troubridge gave up the chase. He reasoned that he could not hope to catch the German ships, and even if he did, he could not hope to beat them. This prudent decision, though in line with Churchill's orders to not engage a superior force, gave the Germans time to get closer to their objective.[13]

Toward the evening of August 7, *Gloucester* engaged the *Breslau*, which had dropped back to feign deploying mines across the pursuing ship's path when *Goeben* experienced an engineering casualty. After a short skirmish, Souchon called off the action and the two ships sped toward their destination. When *Gloucester* slowed to avoid the "mines" believed to have been dropped by the *Breslau*, the two German ships made good their escape.[14] By the 10th, they arrived at the Dardanelles Strait. Two days later, the ships were "purchased" by the Turkish government, and Souchon was made the commander of the Ottoman fleet.[15]

Lessons Learned: (1) In a pursuit scenario, an onboard drifting-mine deployment capability (complying with the 1907 Hague Convention) might help to slow pursuers, especially for platforms that otherwise lack anti-surface capability. Lacking that, the appearance of mining can also help to slow pursuers. (2) Ship bridge teams should be cognizant of the effect of unexpected mines/mining in their path. Preplanned responses (PPR) should be reviewed to ensure that they cover such situations. Rules of engagement (ROE) should address what a military vessel should do when it witnesses active mining in international and territorial waters at different levels of conflict.

Over the next few months, the Ottomans became increasingly hostile to the Allied interests. The Ottoman Empire closed the strait and negotiated with Germany for possible entry into the war. On October 29, 1914, Souchon used the Turkish fleet to attack four Russian Black Sea ports. The Ottomans blamed Russia for the incident and declared war the same day.[16]

The Dardanelles Campaign

Under pressure to strike a quick blow in support of their ally, a British squadron bombarded the outer forts of the Dardanelles two days before declaring war on the Ottomans. Although the Ottomans sustained minimal damage, the raid induced them to review and update their defenses. When the Allies returned, aerial reconnaissance revealed that the defenses had been strengthened with 324 mines laid in 10 lines, protected by heavy guns and mobile howitzers. An integrated communications network and mobile searchlights further strengthened the defenses.[17] Thanks to the English raid, the Dardanelles became a much harder nut to crack.

Lessons Learned: (1) Small-scale victories can mean large-scale defeats later. Every defeat is a chance for one's adversary to learn about their own weaknesses, so the first attack must be the hardest blow. If a place is strategically important, do not teach the enemy the biggest lessons about themselves by mounting unimportant raids. Make the first victory a decisive one. (2) If there is a danger that mining may have occurred, it is possible that moored mines and mine lines would be visible from the air. This is not to say that aerial spotting is 100 percent effective, as not all mine lines were detected at the Dardanelles.

The Allied assault began on February 19, 1915.[18] The force contained several fishing trawlers manned by civilian crews placed in military service and supported by numerous British and French cruisers, destroyers, and mostly obsolete battleships. Initially, the trawlers swept at night while supported by cruisers and destroyers. By sweeping at night, it was hoped that the minesweepers would be harder targets. In reality, the Turkish searchlights proved more than capable of finding and illuminating their targets.[19] Furthermore, the current was heavy and constant, forcing the trawlers to sweep at perhaps two knots. Such slow and well-illuminated targets made excellent practice for the Turkish gunners. Under heavy bombardment, the civilian crews could not cope and were forced to flee time and again.[20]

On March 10, the Allies tried a different tactic. Up to that point, the strong current worked against the boats dragging heavy gear. Under the new plan, they pushed at best speed past the first mine lines and then deployed their sweeps while being pushed by the rapid current through the mine lines. This might have had some success, except the Ottomans fired a tremendous barrage against the unprotected trawlers. With shells splashing all around them, the minesweeping crews panicked and rushed back down the channel, pushed by the current. Some crews forgot to deploy their sweeps in their desperation to return to safety.[21]

Note: It should be highlighted here that the minesweeping trawlers were able to pass over the first set of mines. Likewise, at least one Allied submarine would later pass underneath. These were both possible because of the state of moored contact mines tactics at the time. Up until that point, most mining had only been to defend against raids by large warships like battleships and cruisers. The number of mines available would have been limited, thus requiring efficiency in their deployment against their primary targets. Setting mines close enough to the surface to threaten smaller targets would result in wasting the weapons. A mine too close to the surface would be easily spotted and easily countered. If the goal was important enough, they could be countered by simply preceding the battle fleet with a series of unimportant, concrete-reinforced ships manned by volunteer skeleton crews. Indeed, such sacrificial ships then existed in the German navy in the form of Sperrbrechers.[22] *Mines too close to the surface would also hamper one's own waterborne communications, such as supply boats and motor launches for crossing the channel. At the same time, mines laid too deep would threaten nobody except submerged submarines. Since submarines had not yet played a major factor in war outside harbor defense in the U.S. Civil War, they were not yet seen as a significant threat. With all these factors in mind, the Dardanelles defenders would have laid their mines shallow enough to threaten battle cruisers but deep enough to avoid wasting them on ships that could otherwise be handled by coastal defense artillery. Later in the war, when submarines became the greatest coastal defense threat, tactics and equipment evolved to consider maritime defense in three dimensions rather than a strictly two-dimensional problem of defending against surface combatants.*

Lessons Learned: (1) When clearing mines in a contested environment, man your ships with trained, disciplined crews. In mine countermeasures (MCM), as in any other warfare area, choose units capable of operating in the expected environment or use them in a way that mitigates the effect of the environment. (2) Frequently assess the environment before arrival (if possible) and continuously reassess the environment to determine if your plans need to be changed. (3) When the environment has the potential to seriously impact your operations, seek creative solutions to work with the environment rather than against it. (4) Reserve forces operating their own equipment can be good for rapidly increasing MCM forces. Fishing boats converted for MCM purposes have been used in the past for home waters. Reserve units outfitted with mine-hunting drones can also be useful for rapidly expanding mine detection capability during overseas contingencies.

Leadership under such complex battle conditions can be difficult. Admiral Carden, the leader of the expedition, suffered a nervous breakdown on March 17 from the strain of nearly two weeks of unsuccessful nighttime sweeping operations. His replacement, acting Vice Admiral John de Robeck, implemented a new plan devised by his predecessor.[23]

The rising sun on the following morning revealed a much stronger force than seen in the strait previously. This time, the minesweepers arrived escorted by a force of 18 battleships. Aboard the minesweepers, volunteer naval personnel helped to galvanize the morale of the crews. As this force squeezed into the strait, the battleships engaged the shore batteries and forts in a torrential downpour of shells while the minesweepers worked on clearing the fields. Under heavy fire, the Turkish gunners concentrated their attacks on the battleships, allowing the minesweepers to operate in relative peace. After several hours of fighting, the shore gunners began running out of ammunition and were forced to temporarily evacuate their positions.[24]

Lessons Learned: (1) Frequently reevaluate your plan for effectiveness. If your plan is not yielding the desired results, revise your plan. (2) Priorities change from kicking little guys around when big guys are punching you in the face.

26 Mines, Four Battleships, One Battle Cruiser

This gargantuan attack may have truly succeeded had it not been for an idea conceived by Turkish Lieutenant Colonel Gheel.[25] During the first week of the campaign, he noted that the larger Allied ships preferred to maneuver in the slower waters of the southeastern side of the channel in Erin Keui Bay (or Gulf of Erenköy). On the night of March 7, he ordered the *Nusret*, a 131-foot minelayer, to deploy 26 mines in a line running parallel to that shore.[26] Earlier sweeping in this area had not revealed any mines, so no further mining was expected. Likewise, sporadic Allied aerial spotting in this strait failed to detect any mines in that area. For these reasons, the Allies did not suspect that a new mine line awaited them.[27]

Turkish minelayer *Nusret*, responsible for sinking or grounding five capital ships, 1912 (Wikimedia).

Fresh from the March 18 battle, the battleships once more turned around in Erin Keui Bay. This time, they ran over the line of 26 mines. As the French battleship *Bouvet* made its turn, it was rocked by an explosion that caused her to capsize and sink in less than two minutes. Soon afterward, two more battleships, *Irresistible* and *Ocean*, struck mines and eventually sank. The battlecruiser *Inflexible* struck another mine and was heavily damaged while the French battleship *Gaulois* had to ground herself following a mine strike.[28] In this way, the tiny minelayer *Nusret* managed to sink more battleships than sunk by the direct fire of all other Turkish and German surface ships throughout the rest of the war.[29] Such a success arguably made it one of the most cost-effective combatants in history.

Lessons Learned: (1) Guard against the possibility of new mines being seeded. (2) OPSEC: Vary your routine. Examine your own forces' activities for habits that make them vulnerable and find ways to reduce those vulnerabilities. (3) Observe your enemy's habits for opportunities to strike them unexpectedly. Take advantage of those opportunities. (4) Offensive MCM: The surest way to keep an area safe from mines is to prevent the enemy from laying them. Where possible, find and strike their minelayers early in the fight to minimize the number of mines that they can later deploy. (5) A well-laid minefield can be extremely effective and can be a game changer.

Consequences of the Dardanelles Campaign

While the loss of so many battleships is enough to give anyone pause, in reality it did not seriously hurt the Allied tactical situation. The lost ships were all obsolete, scheduled for scrapping. There were still more than enough battleships remaining to keep fighting, and more were on their way. Turkish ammunition supplies were severely depleted. Believing that the breakthrough was imminent, the Ottomans had already begun evacuating the more important assets from the capital of Constantinople. All that was needed was one more day's effort and the strait would have been laid open to the ravages of the Allied fleet.[30]

De Robeck did not know how close they were to victory. Unfortunately for the Allies, the sinking of the battleships took the wind out of his sails. Believing it impossible to force the strait, he ordered no more assaults. Four days later, he and the army commander decided that a shore campaign was the only way to silence the guns that were preventing the minesweepers from working.[31] More than a month later, the Allies conducted their first landings at Gallipoli (the European side of the Dardanelles). By that point, the Ottomans had resupplied and had massed their forces to repel the foreign invaders.[32] In the land campaign that followed, 302,000 of the 489,000 Allied troops put ashore (62 percent) were casualties of battle or disease. On the Ottoman side, 250,000 of the 315,500 troops deployed (79 percent) were casualties.[33] The little *Nusret* with her 26 mines had caused such a fear in the Allied commanders that half a million troops fell to prevent further ship losses.

Lessons Learned: (1) If the enemy is down, do not give them the chance to regroup and resupply. Hit hard the next day, and the next, until the enemy has nothing left to throw at you. (2) On the reverse side, if a heavy, drawn-out engagement looks probable, start your resupply train as soon as possible and keep it going until the engagement is finished.

As with all great events in history, the action at the Dardanelles on March 18, 1914, was a major turning point that could have gone in another direction. What would have happened had the *Nusret* mining never occurred, or if it wasn't so effective, or if the Allied commanders had greater resolve? In all these scenarios, the implication would be that the Allied attack would have continued the next day. With the Ottomans having expended most of their ammunition, it is reasonable that they would have been unable to continue defending against minesweeping without resupply. Once the Allies realized this fact, they would have stationed battleships to engage any supply convoys. Mine clearance would have taken perhaps another week, and then the Allies would have been through the Narrows, the Dardanelles, and into the Sea of Marmara.

It is possible that the Ottomans would have hurriedly put together secondary defenses. This might have included bringing whatever parts of its battle fleet that were available into the Sea of Marmara to engage the Allies in pitched battle. This would have been most effectively accomplished as the Allies were still in the bottleneck of the Dardanelles. They could have furthermore brought to bear drifting mines and their unused inventory of unsweepable Leon oscillating mines.[34] However, the Ottoman battle fleet was quite small and would have been no match for the Allied battle fleet then arrayed against them. The Allies would most likely have won the engagement, entered the Sea of Marmara, and steamed to the Ottoman capital of Constantinople.

There is some debate on what might have happened next. Constantinople is (and was) an important city, the key to controlling the only seaborne route between the Black Sea and the Mediterranean. Capturing it may or may not have knocked the Ottomans out of the war. On the one hand, capital cities like Constantinople are often the political, cultural, and economic heart of an empire. By taking it or even threatening it, the Allies may have been able to convince their not-altogether-committed foes to drop out of the war. After all, their cause was limited and to that point only a limited amount of blood had spilled on the Ottoman side.

On the other hand, the empire was still huge, and Constantinople lies at its very edge. The government could have accepted the loss of the European portion and simply shifted to Ankara in the center of the Anatolian peninsula. After all, it had already lost a lot of territory over recent years and continued to survive. Beyond the range of ships' guns, they would be immune from further threat from the ship-rich, manpower-poor Allies. After licking their wounds, the Ottomans could then have continued the fight by any means available. This would continuously create a drain on Russian, French, and British manpower that might otherwise have been spent fighting the Germans. Despite that threat, the Allies would be able to maintain control of the entire strait. This would allow them seaborne communication to resupply Russia and to put warships into the Black Sea, while also giving them the ability to prevent Ottoman troops from entering the Balkans en masse.

One thing that happened for certain was that with the strait closed, Russia had no access to the wider world economy. Its exports dropped by 98 percent and its imports dropped by 95 percent.[35] This would have had a devastating effect on its economy. Jobs related to the import and export industries would have been cut. Jobs related to supporting the workers in those industries would have also been cut. The price of goods normally imported, or made with imported inputs, would have shot up dramatically. This in turn would cause inflation at a time of higher unemployment. For a population already agitated from the Revolution of 1905 and its aftermath, this would certainly have created additional unrest. Of particular importance to the military was that one of the desperately needed imports to Russia was weapons. At the time, Russia had not fully industrialized and was not making enough guns to arm all of its own troops. Without the ability to defend themselves, let alone assume technological dominance, Russian troops suffered appalling losses.[36] For their families back home, such losses would have been further demoralizing. Considering the economic and military impact of the strait's closure on Russia, it is probable that it extended the war. It prevented the Russian government from gaining the support of its population and was at least a contributing factor in bringing about the Bolshevik Revolution of 1917.

Whether the Ottomans stayed at war or not, an Allied victory at Constantinople would probably have kept Bulgaria from going to war on the German side six months later. Bulgaria initially had no reason to enter the war, being at diplomatic odds with its neighbors. However, if the Ottomans lost control of the strait, Bulgaria would have had the opportunity to expand its borders to include the capital of its former colonizer. But since the Ottomans still held the strait, Bulgaria had little to gain from joining the Allies. That fight would have meant a drawn-out land campaign against the much larger Ottoman Empire. Help could not be expected from the Serbians, who were fighting their own desperate campaign against the invading Austrians. The probable result would have been a loss of territory.

The Bulgarian situation would have changed with the Ottomans removed as a threat. Had the Ottomans become neutral, Bulgaria would have had no threat and may have maintained the status quo. Had the Ottomans joined the Allies, Bulgaria may or may not have joined as well. On the other hand, the Ottomans would have found it disastrous to remain aligned with Germany. Without control of the strait, the Ottomans were less able to defend their European territory. Under those conditions, Bulgaria would see joining the Allies as a golden opportunity. In one fell swoop, the Allies would have had a motivated partner to finish pacifying that territory. Bulgaria's price would be that same territory to include Istanbul. That would have made Bulgaria a much greater player in subsequent regional politics. Furthermore, with Bulgaria on the Allied side, Serbia could be reinforced by Bulgarian, Russian, and ANZAC troops. This in turn would put powerful pressure on Austria from the south. With Serbia, Bulgaria, and Russia all on the Allied side, Romania would have found little advantage in joining Germany and Austria. Instead, it would either have stayed out of the war or would have joined in the impending fight to capture Austrian territory. In either case, the strategic situation on the southeastern front, and consequently, the postwar Eastern European map, would have been far different.

Conclusion

The Turkish victory at the Dardanelles during World War I is perhaps one of the most remarkable turnarounds in naval history. Few such events exist in history where a very minor player wrested victory from the jaws of overwhelming defeat. Had the Allies succeeded in breaking through the strait in the face of the mines and guns, they may well have driven the Ottomans out of the war. It was not to be. Thanks to 26 naval mines laid in an unusual spot, five capital ships were lost and the campaign abandoned. This led to around 550,000 casualties in a grueling land campaign. The strong Ottoman defense on land helped to draw Bulgaria and Romania into the war on the Ottoman side, rather than the side of the Allies. All because of 26 well-placed naval mines.

A strategic strait near a continental crossroads, protected by layered defenses of mines and shore firepower, defied the best efforts of Western powers. History, it seems, stands always ready to rhyme with itself.

Chapter 6

Mine Warfare in the North Atlantic, World War I

Introduction

Where the Russo-Japanese War introduced the world to offensive and open-ocean mining, World War I brought both tactics to the level of a science. In all previous wars, belligerents typically laid minefields in the immediate harbor approaches of either their own ports or those of their enemies. World War I was different. Germany and the United Kingdom (later with the aid of the United States) blanketed much of the North Sea with hundreds of thousands of mines. This copious stream of mines gave each country extensive experience in assembling and laying mines, while kick-starting research and development. Such mining in turn forced rapid development of mine countermeasures techniques to restore desperately needed freedom of navigation. By the time the war finished, the techniques for both laying and removing moored contact mines had matured to a level not to be substantially improved upon until the Vietnam War some 50 years later. It also saw the introduction of the world's first influence mines, a weapon that would see significant service in later wars.

The last chapter, on the Dardanelles, discussed in detail the lead-up to the war and the early actions in the Mediterranean. This chapter will focus on the rest of the major mine warfare actions of the conflict, which were centered primarily on the North Atlantic.

Mine Posture at the Beginning

When it came to World War I at sea, mines and submarines played a role in shaping the battlespace that far exceeded their cost. For years prior to the war, the various Great Powers had competed in assembling powerful surface fleets. Governments poured their pride, esteem, and budgets into building goliaths that could go toe-to-toe with one another in heated battle sufficient to put Admiral Horatio Nelson's accomplishments to shame. Possessing massive battleships, they were ready and willing to absorb the awesome power of shells over a foot in diameter slamming into their waterlines. Only some recognized the lessons from the American Civil War and the Russo-Japanese War that even the biggest battleships were no match for a fishing trawler dropping a well-placed naval mine. All the topside armor in the world would soon prove wasted in a war where most attacks came underwater in the form of mines and submarine-launched torpedoes. Those who were ready for this new way of fighting started the war with an advantage that would be overcome only through ingenuity and considerable sacrifice in men and matériel.

Germany—Germany came into the war with the best preparation for undersea warfare. Despite building up a powerful prestige fleet, it would have known that it was precipitously vulnerable to the experienced fleet of the United Kingdom (UK). As a nation, Germany was still young. Long a land-based power, its entire North Sea frontage was a mere 120 or so nautical miles wide as the crow flies. Alternatively, Germany had quite a lot of frontages on the Baltic Sea from which she could operate. However, the three narrow passages between the Baltic and the North Sea were easy for the UK to blockade.

Meanwhile, Germany's two most likely enemies each had different types of naval vulnerabilities. Russia's problem, as discussed previously, was an even worse version of Germany's naval situation. This presented an opportunity for Germany to conduct blockades of her own. The UK had the opposite problem. With nothing but coastline and countless ports, it was impossible to fully protect all ports at once. Heavily dependent on maritime trade, it wasn't necessary to engage its main battle fleet to cause it harm. Germany could not hope to blockade the whole of the island with any sized navy. However, the lessons learned from the Russo-Japanese war indicated that such a blockade could be conducted with the simple expedience of planting mines in an enemy's harbor approaches. Why risk a direct battle against a larger enemy possessing centuries more experience at sea than one's own navy? From Germany's standpoint, the difficulties of containing the fleets of its neighbors could be easily resolved by vigorously developing a robust mining capability.

The enthusiasm of some Germans for mine warfare development began before the country itself existed as a nation. As early as 1846, during the Schleswig-Holstein War, Werner von Siemens designed and laid a field of RECO mines in the Port of Kiel. This was believed to have deterred their Danish adversaries from bombarding the city.[1] The following year, von Siemens founded today's telecommunications giant bearing his name.[2] In 1878, a Prussian designed the first chemical horn (vital for reliable, independent mine operation).[3] Two years later, during the Franco-German War, mines were laid to protect three rivers feeding into Helgoland Bay.[4]

With these past successes in using mines to defend against stronger maritime powers, Germany was well inclined to develop mines into a major warfare area. In the decades prior to World War I, they built war plans based on undersea warfare. War plans indicate capability needs. As such, much effort went into developing submarines with minelaying capabilities, as well as contact mines that could be safely laid by those submarines. The contact mines were fitted with Hertz horns and were attached to their anchors via a doubled length of wire.[5] Constructed with typical German high standards of workmanship, they would have been pricier than most mines on the market. They could also be deployed by a variety of vessels, from battleships to auxiliaries to *Unterseebooten* (U-boats) designed with minelaying capability.[6]

United Kingdom—The United Kingdom, as the world's dominant naval power, had little interest in preparing for mine warfare prior to World War I. In the view of most members of the British Admiralty, a mine was an underhanded weapon that might have

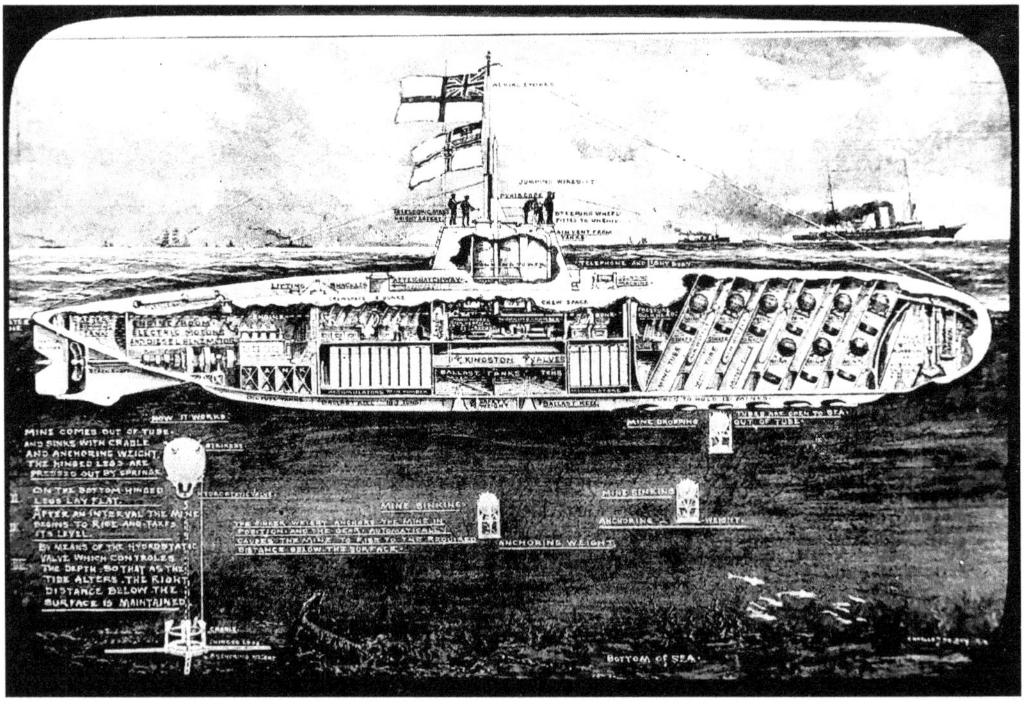

Drawing of German minelaying submarine UC-5, 1916. Mine tubes on the right (Naval History and Heritage Command).

value in harbor defense but whose use would only limit the British fleet's freedom of navigation.[7] While they did make some effort at developing mines and building a stockpile of about 4,000 of these weapons, most of their efforts were in minesweeping technology. The mine stock consisted of unreliable local designs, as well as Swedish-made Leon oscillating mines and some moored mines of German manufacture.[8] Minesweeping was in the form of wire passing between two vessels of opportunity and suspended from heavy "kites." Once the mines were separated from the moorings, they would float to the surface, where they could be sunk with gunfire.[9] The country was hardly prepared for the type of war Germany would wage.

United States—Despite an extensive history of developing and using mines within the United States, little MIW development or stockpiling took place in that country between the Civil War and World War I. As a continent-sized nation separated from potential adversaries by entire oceans, it was seen as having little to risk from foreign navies. With isolationist sentiments prevailing, it also saw little reason to build an offensive mining capability.[10]

In 1885, Secretary of War William Endicott commissioned a board to study the fortifications of the United States and to make recommendations. This resulted in the building of fortresses near several major seaports. At least one of these, Fort Washington, can still be visited just south of Washington, D.C. These forts featured sloped walls, glacis, retracting rifled guns, and controlled mines fired by operators working from fortified observation stations. The mines were connected in branched clusters and were individually selectable. They were deployed only in times of conflict and saw service in the Spanish-American War, World War I, and World War II before they were finally disestablished at the start of the Cold War. By that point, close-in shore bombardment was no longer seen as a plausible threat.[11]

Outside of these localized defenses, little was being done to seriously develop contact mines or minesweeping other than incremental work by individual officers. The entire result of efforts to build a comprehensive U.S. mine warfare program was the assignment of a single officer to the Naval War College in 1912. In 1913, with war looking increasingly imminent, the U.S. Navy adopted the British minesweeping manual and issued it to the fleet. The following year, it took the step of fitting out a grand total of four ocean tugs with minesweeping capabilities.[12]

Had the U.S. Navy joined the war at the beginning, it would have been essentially helpless against the threat of mines. As it was, the war had been going on for nearly three years before it finally entered the fight. Perhaps because eventual entry was seen as inevitable, the U.S. Navy spent those years improving its readiness. Its minesweeping tactics and equipment were adopted almost entirely from the Royal Navy. While few ships were assigned permanent duties as minesweepers, several ships were fitted for collateral duty minesweeping. In November 1916, it conducted its first large-scale MIW exercise off Sandy Hook, New Jersey, where fleet units cleared a field of some 200 moored contact mines. For a country not at war, this was a sure sign that its navy expected to soon be engaged in heavy levels of mining and MCM.[13]

Note: Exercises featuring large-scale contact minefields reflect historical reality. Even in modern times, any future war near the coast, including the present Russo-Ukrainian war, would likely involve large numbers of mines. While some of these would be mixed fields, almost any minefield would likely contain enough moored mines to represent a logistical challenge for those removing them from the sea-lanes.

Russia—Like the German navy, and with a wealth of practical experience, Russia saw mines as a major weapons system. In the years since the Russo-Japanese War, it continued to build a stockpile of mines, with new designs as late as 1912. Understanding the value of offensive mining, it fitted out all cruisers and many destroyers for that task.[14]

Germany vs. the United Kingdom

Some countries in some wars lay mines only periodically as a series of limited events at a limited number of places to achieve limited objectives. Germany was not one of these countries. The *Kriegsmarine* began laying mines from the very beginning, and seemingly never stopped.

German mine warfare in the first couple of years of World War I began with surface vessels. Surface combatants, as well as converted merchant vessels operating out of the Heligoland Bight ports on the German northwestern coast, made mining raids along the UK coast. Surface minelayers had the advantage of being able to drop huge numbers of mines within a few hours and could come in force with armed escorts. Merchantmen converted for minelaying practiced the tradition of the *ruse de guerre* by flying neutral flags or even brandishing the color schemes of British ferry lines. So decorated, they could simply pass through British shipping lanes dropping mines in their wakes.[15]

At the beginning, the Royal Navy (RN) still considered its Grand Fleet of battleships to be the king of the high seas. The public expected its navy to engage the German fleet in a pitched battle similar to Lord Viscount Horatio Nelson's famous 1805 victory at Trafalgar. But such a battle cannot happen if one's enemy never leaves port.[16] Within the first few months of the war's beginning, the RN was already beginning to understand that the heyday of the battleship was in its twilight. Already their supremacy was being challenged by weapons deployed by the much smaller submarines and even smaller vessels ... anything capable of delivering a weapon that explodes underwater. The foot-thick armor of battleships could stop shells a foot in diameter. But it meant little against explosions delivered from beneath by torpedoes and mines. Submarines were already showing success at sinking ships with torpedo attacks with the sinking of three cruisers by one submarine on September 22, 1914. With U-boats running rampant, the RN sought ever-safer protected shelter from their predations. While they knew that all German surface combatants could lay mines, the fear was that the weapons would be used tactically during battles and skirmishes. They were not, however, quite yet thinking of mines as a widespread area-denial weapon. History, after all, had no precedence of such a threat. That would soon change as Germany began a campaign of steady mining.[17]

The mining strategy was set on course the evening before war began between the two nations. On the day that Germany was to respond to the British ultimatum that it leave Belgium's territory, the converted excursion steamer *Königin Luise* slipped to sea with 180 mines. Painted in the color scheme of the neutral Hook of Holland ferry service,[18] it delivered the German response in the form of mines dropped in the Thames approaches at dawn on August 5, 1914. Discovered in this endeavor, the vessel was soon under fire by the British destroyer *Lance* steaming in company with the light cruiser *Amphion*. In short order, *Königin Luise* lay sinking while *Amphion* took prisoners aboard. Soon thereafter, while investigating another vessel painted in the same color scheme, the *Amphion* struck one of *Königin Luise*'s mines. Famously, these first two vessels sunk by either side in the war had effectively sunk each other.[19]

On October 17, the Germans once more attempted to mine the Thames approaches using four old destroyers. Discovered before they could accomplish their mission, all four were sent to the bottom.[20] Not long after, a fishing trawler working in the vicinity of one of them, the sunken destroyer *S119*, fortuitously hauled up that ship's jettisoned classified documents. Included in these documents were the code books for Germany's highest-level interservice and diplomatic communications. This boon allowed the Admiralty intelligence to decode the most important German messages.[21] The

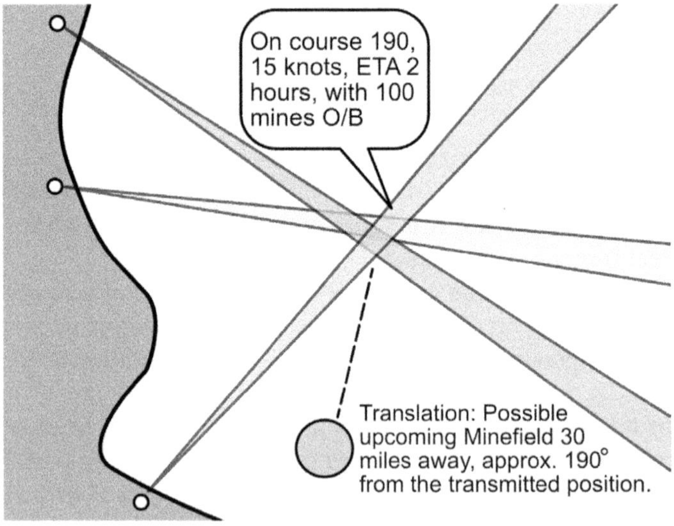

Principle of radio direction finding (M. Mazurek, after original drawing by the author).

other two major code books would be found in other fortunate captures. These, along with radio direction finder stations, helped the RN to anticipate German naval activities.[22]

Within a week of the sinking of the four destroyers, the Germans struck yet again, this time far from the Thames. On October 22, the converted liner *Berlin* laid some 200 mines near Tory Island off the northern coast of Ireland. At the time, the RN Grand Fleet was using the area as its main anchorage, perhaps in hopes of keeping far from the raiding submarines. A few days later, it claimed its first victim, a merchantman, followed on October 28 by a far more valuable prize. Not informed of the minefield, a squadron of eight dreadnought battleships departed the anchorage for gunnery practice. Soon one of these, the *Audacious*, experienced an underwater explosion believed to be a torpedo. It wasn't until nearly eight hours later when reports of other mine strikes in the area started reaching the Admiralty. Twelve hours after the initial strike, the *Audacious* suddenly capsized and exploded, its crew having been already evacuated.[23] Fearing that the Germans would understand that their fleets were now nearer to parity, the sinking of this powerful dreadnought remained a state secret throughout the war.[24]

Surface mine raids continued through 1914 into 1915. German surface raiders laid further minefields along the northeast coast of England. The efforts to counter these mines became the training ground for the UK minesweeping forces. Among other things, they learned lessons such as the following: (1) It is better to sweep moored mines at high tide. (2) When conditions aren't conducive to sweeping, it is best to anchor (in a cleared area). (3) Sweepers will get destroyed, so have a deep bench. (4) An enemy's offensive minefield also works as a defensive minefield. (5) As long as set routes are cleared regularly and followed, it isn't necessary to clear every mine in the sea. (6) First try different types of vessels in the conditions where mines will be found, and then rapidly build a fleet of cheap sweepers based on the best candidates. For the Royal Navy, the shallow-draft, twin-screw *Hunt* class was the winner for offshore sweeping, while the *Dance* class of tunnel tugs (3.5-foot draft) proved best for inshore sweeping.[25]

By the end of 1914, two things became clear to the Royal Navy. First, it needed to do some mining of its own to help blockade the surface raiders. Second, it was sorely lacking in minelayers. Where Germany had built nearly all of its surface combatants with mining capabilities, the traditional British aversion to mining meant that they had not devoted much attention to the capability when designing their own ships. Some old cruisers of the *Apollo* class were

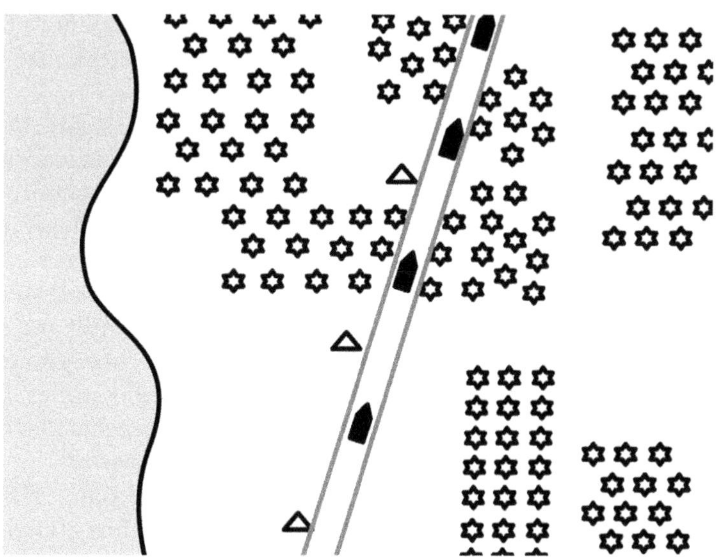

Principle of cleared routes through simple minefields (M. Mazurek, after original drawing by the author).

converted for the purpose and laid the first UK offensive minefield in the Heligoland Bight area. More ships were retrofitted for mining, and the Heligoland Bight minefield grew as British and German navies alike added to the field. Combined with heavy British patrolling, Germany soon found it too hazardous to use surface raiders for mining. Instead, it turned to a different solution.[26]

Submarine Minelayers

In July 1915, the Royal Navy learned of the existence of the U.C. I-class submarine. Only 111 feet long, they seemed made for short runs to mine the British coast. Carrying 12 mines in six tubes, these tiny craft could lay small nuisance minefields. Along with the shift from surface mining to submarine mining came a home-base shift. While the surface raiders had operated out of the ever-more mined and patrolled Heligoland area of Germany, U-boats increasingly operated out of Zeebrugge, Belgium. The U.C.-I's were soon followed by the U.C. II-class submarines in the latter half of 1916, capable of deploying 18 mines anywhere in the European theater of operations. Furthermore, they had three torpedo tubes for sinking ships through more direct means. Together, these submarines were responsible for sinking 1,800 ships during the war.[27]

The advantage of the Zeebrugge base was its proximity to the Thames Estuary and to the Dutch border. Since it was situated across the mouth of the English Channel from the Thames Estuary, boats operating out of here could quickly reach most of the British south or east coasts. Transiting in the dark through lightly defended areas, when submerged they were relatively invulnerable to even the most heavily armed RN warships. Its location about 20 nautical miles from the Netherlands meant they could take advantage of that neutral country's safer waters to avoid RN patrols and minefields. Upon arriving in their assigned areas, they could mine harbors both major and minor, torpedo a few merchants or fishing vessels, and return home … all inside one to three weeks. Through these means, the German navy laid more than 100 minefields in and around the British Isles and the French coast. See the map above for details (darkest areas).

While there was a pause in U-boat operations against shipping in 1916 as a result of an American ultimatum,[28] they returned with a vengeance in October of that year as the first of the UC-II boats were released. They would continue until nearly the end of the war before the U-boats were limited to operations closer to home. In 1918, the UC-III class of long-legged U-boats were introduced; they were capable of operating off the U.S. East Coast. The UK, for its part, continued its own system for countering this mining with small minesweepers, sometimes purpose-built and sometimes fishing trawlers converted to the purpose. Operating from ports throughout the islands, they checked their areas of responsibility daily for the presence of mines. Battle fleets and squadrons sent fleet minesweepers ahead during the rare moments that the fleet moved. Any mines discovered were cleared as time and resources allowed.[29]

Before long, minesweepers became a routine sight off harbor mouths throughout the island. Their activities, and the persistent, white-knuckled efforts of their crews, became a daily routine. All these interactions with mines were hazardous. Over the course of the war, Great Britain and her allies lost 13 minesweepers and 242 gunboats/picket boats to mines. Of the latter, 193 were fishing vessels supporting the fleet.[30]

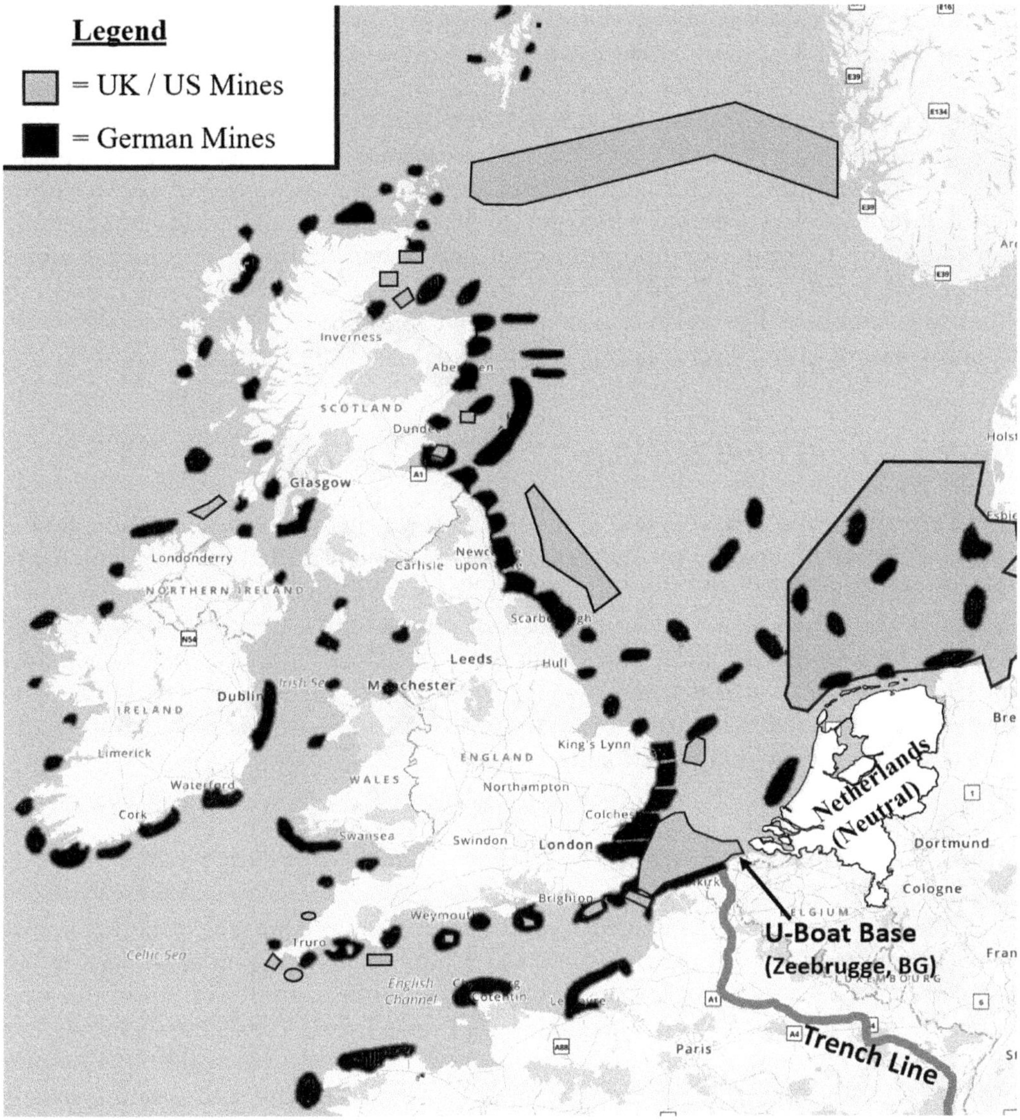

British, German, and American mines laid during World War I (from H. Taprell Dorling, *Swept Channels: Being an Account of the Work of the Minesweepers in the Great War.* London: Hodder & Stoughton, 1935, heavily modified by the author).

Paravanes

One particularly interesting World War I development in mine countermeasures is the paravane. Looking at the concept, it appears to have developed as a natural extension of the Oropesa sweep. Whereas the latter is towed behind a small vessel, the paravane was towed from the *front* of a large ship. It consisted of a weighty device hanging from a chain from the bow of a capital ship or a merchant. Serrated wire cables streamed from each side of the device, leading to otters with controllable depths. As the ship passed through a moored minefield, any mines near the ship's bow would be pushed outboard by the serrated cables. As the ship continued steaming forward, the mechanical action of these serrated cables would cut the mine cables at a distance from the ship. The mine cases would then float to the surface, where they could hopefully be destroyed by deck guns. These devices would later be used in World War II and even make an appearance during Desert Storm when the United States recommissioned the USS *Iowa*.

Unfortunately, paravanes are useless against bottom influence mines. If used today, they would doubtlessly continue to be effective against purely moored contact mine fields. But when could one use a paravane? Besides the fact that minefields are generally rare in our mostly peaceful world, the prevalence and proliferation of influence mines make it risky to assume their absence from any minefield. However, if a ship finds itself stuck in a minefield, having an escape route that you know is clear of at least contact mines is safer than no clearance at all.

United States Enters the War

Throughout the first few years of the war, U.S. President Woodrow Wilson did his best to maintain his country's neutrality. While there was a cordial trading relationship with the British Empire, the two nations were not quite as close as they are today. Concurrently, there was also a friendly trading relationship with Germany, as well as a large German American population. Having no staunch military alliances with either side, the United States had little to gain from joining the war and little interest in doing so. Its merchants continued as best they could to maintain trade with whatever willing partners were to be had, including the belligerents.

A Conflict Between Systems of Ethics

What eventually led the United States to joining the war was the refusal of one of the belligerents to respect their neutrality as a nonbelligerent (similar to what led to the War of 1812). In these early years of open-ocean undersea warfare, there was serious contention about what constituted legitimate means of naval warfare. As the world's longtime dominant naval power, the United Kingdom subscribed to a traditional system of chivalry in warfare. Under that system, opponents were to declare themselves to their enemy, fight them toe-to-toe, and let the strongest force take the day. The victory was nearly always the one with the skill and technology to strike the hardest blows and the discipline to take their opponent's hardest blows without fleeing. One could usually predict the outcome by counting the number of guns present while accounting for the relative skill of the fighters. It was warfare made for a technologically advanced, disciplined society with a long tradition of victory at sea. Naturally, anything that short-circuited their power as the stronger force was thought uncivilized. As the inheritors of English culture, this same ethic was also present in the United States, though it was not as strongly held.

Germany followed a different approach, one that perhaps hearkens back to the guerrilla methods used by their ancestors fighting Rome in the forests of Germania. Under that ethic, a weaker fighter can use guile to neutralize a stronger enemy's advantages in size and strength. Such guile often takes the form of the surprise attack, of sneaking up on one's opponent and attacking them with devastating force when they least expect it. It rewards the one with the most cunning, not the most strength, and punishes those not continuously ready to defend themselves. Applied to war at sea, where giant battleships roamed, this meant that clandestine attacks by submarines and mines were both valid means. Both could attack those goliaths at their weakest points, while also being protected from counterattack.[31]

One goal of the German navy was to drive the UK out of the war by hurting it domestically. Against an industrial island nation, the obvious approach was to prevent resupply (an approach later used by the United States against Japan, as will be discussed in chapter 8). Because of the number of British ports and the numerical advantage of its navy, a full blockade was not possible. Instead, they leveraged naval mines to perform the blockading function while limiting risk to their own forces. They also used the German submarine, the U-boat, to directly attack shipping bound for the United Kingdom.

U-boats were well armed but vulnerable. They could clandestinely approach the largest of surface ships and fire mobile torpedoes to deliver devastating blows. They were also armed with deck guns to sink merchant vessels and fishing boats. However, their protection was their ability to remain unseen. Submerged, they had a chance to defeat a battleship. Surfaced, they had little hope of survival. This placed them directly at odds with the Royal Navy's ideas of chivalry. Surfacing to provide notice gave time for guns to be brought to bear and for reinforcements to arrive. Smaller vessels were safe enough to be dealt with by surfacing nearby and threatening. Crews could then be allowed to board life rafts before the U-boat destroyed the vessel with gunfire. If the target was a war vessel, or a huge ship capable of outrunning the U-boat, victory was assured only by firing torpedoes at the moving target without notice. While U-boat captains took some pains to identify ships to establish their legitimacy as targets, geography alone was deemed sufficient. They announced a war zone area and declared that any vessel found in that area would be considered in material support of the UK war effort and thus a legitimate target. The same zone was fair game for mining. By this theory, Germany believed that it could fairly claim that ships entering the zone did so at their own risk, with exceptions allowed only under very strict circumstances.

The United States in World War I

This "unrestricted submarine warfare" was announced to passengers through newspaper advertisements such as the one that appeared the day the famous *Lusitania* departed for her final voyage from New York to Liverpool.

> NOTICE
>
> Travelers intending to embark on the Atlantic voyage are reminded that a state of war exists between Germany and her allies and Great Britain and her allies; that the zone of war includes waters adjacent to the British Isles; that in accordance with formal notice given by the Imperial German Government, vessels flying the flag of Great Britain or any of her allies are liable to destruction in those waters and that travelers sailing in the war zone on ships of Great Britain or her allies do so at their own risk.
>
> Imperial German Embassy
> Washington, D.C.

Notice posted on May 1, 1915, on the day *Lusitania* sailed on final voyage.[32]

The sinking of the *Lusitania* just south of the Irish coast resulted in the deaths of 1,201 people, including 94 children (35 babies). One hundred and twenty-eight Americans were counted in the lives lost. The submarine captain had referenced *Janes Fighting Ships* of 1914, which listed the ship as an auxiliary merchant cruiser that was used to transport troops. Furthermore, the ship carried 173 tons of American munitions (which may have contributed to the sinking), making it technically a legitimate target. Still, as the last ocean liner making the transatlantic run, its primary purpose was the transport of passengers. These civilian passengers, especially the children, had no control over the ship's cargo or its military designation. It was, in effect, the only accessible means to transit between continents. When the *U-20* torpedoed the ship without warning, it killed over a thousand people, most of whom were noncombatants.[33]

This attack created a furor around the world. Many Americans demanded war against Germany. American neutrality had been regularly infringed upon by both sides, creating fairly equal resentment against both. But the *Lusitania* attack demonstrated a complete imbalance in the scale of that infringement, i.e., the UK created inconvenience, while Germany created destruction and death. Any positive support for Germany began to dissolve. Still, President Woodrow Wilson resisted demands for immediate entry into the war.[34] Instead, he relied on diplomacy. Over the next two years, he continued using a diplomatic approach, encouraging peace even while his countrymen continued to supply matériel support to the British war effort. Disgusted by that support, resentment toward the United States festered in Germany.

On January 17, 1917, Germany sent its diplomat in Mexico a notice that unrestricted submarine warfare[35] would begin on February 1. This same notice directed that in the event that this action brought the United States into the war, the diplomat was to negotiate Mexican entry into the conflict. Upon victory, Mexico would be granted its old territories in Texas, New Mexico, and Arizona. Furthermore, Mexico was to negotiate Japan's entry into the war against the United States.[36]

This message, decoded by the British and released to President Wilson, infuriated the public. There remained only a direct, causal pretext to draw the United States into the war. That pretext came on March 12 and 18 when U-boats destroyed four American ships. On April 6, the United States declared war against Germany.[37]

Some of the earliest U.S. contributions to the war effort came in the form of anti-submarine warfare (ASW) support. Four weeks after war was declared, the first six U.S. destroyers arrived at the Irish port closest to where the *Lusitania* sank. Followed eventually by 73 more destroyers, these ships eased the burden on British ASW forces and made possible a convoy system that would stem the rising tide of U-boat victims.[38] In December of that year, the Royal Navy built the Dover barrage consisting of mines, nets, searchlights, and heavy patrols. Together these actions took a heavy toll on the U-boats, forcing them to transit around the northern end of Great Britain to continue their war against shipping.

In the interim, in October 1917, the British Admiralty approved plans for a joint United States/UK project to lay a mine barrage across the North Sea to block U-boat passage via this northern route. It was originally planned to include 100,000 mines, and the United States supplied about 56,000 moored mines at a price of $9,400 per mine in 2023 dollars.[39] For comparison, the minefield's total cost in present-day dollars was comparable to the average congressional cost cap for a single present-day littoral combat ship.[40] The mines provided by the United States were the Mk 6, a moored mine that was a sort of transition between traditional contact mines and the influence mines of later years. Instead of just the chemical horns or firing levers typically used, the Mk 6 used a new dielectric triggering device (the K-device). Proposed by a Massachusetts inventor, the K-device's construction involved a long copper wire attached to a float at one end and to a rotor winding inside the mine case at the other end.[41] If the steel hull of a submarine or ship passed within 140 feet above the mine, it would contact this wire. Due to the dielectric effect of the dissimilar metals, this contact produced a small electric charge sufficient to detonate the mine. This effectively created an area of "influence" that could control a vertical column that would otherwise require four traditional mines to control.[42]

While the United States and UK busied themselves manufacturing the mines needed for the North Sea Mine Barrage, an increasingly desperate Germany sent a series of seven submarines to the American coast. Deployed in spring and summer of 1918, these submarines were ordered to create chaos in hopes of drawing some of the now 36 U.S. destroyers in Ireland back to their home waters. Only five would reach their stations, but each was armed with torpedoes, deck guns, and mines. Together they destroyed 96 vessels. Their mines would damage six vessels off the eastern seaboard, including the armored cruiser USS *San Diego* off Long Island, New York (the only major U.S. warship sunk in the war). Ironically, the U-boat responsible for the USS *San Diego*'s sinking, *U-156*, was itself thought to have sunk in the North Sea Mine Barrage on its way home. The barrier's construction had begun several weeks before in the middle of the boat's deployment and may have come as a surprise.[43]

U.S. and UK minelayers laid the North Sea Mine Barrage between March and October 1918 from Scotland to Norway. Covering a region some 230 miles long and 30 miles wide, U.S. and UK minelayers deployed mines in 15 rows with 100-yard spacing. This minefield had mines at the depth of 45 feet, 160 feet, and 240 feet, and separated at 100-yard intervals. Featuring the MK 6 mine with its 140-foot antenna, these depth layers resulted in 240 feet of continuous vertical coverage.[44] However, the 100-yard spacing was perhaps too wide, for the beam of a U-boat was less than eight yards, meaning that a U-boat passing through a single mine line at a random point had roughly a 1 in 13 chance of accidentally striking a mine. Furthermore, it was reported that the antennae that made the Mk 6 such a revolutionary weapon at depths floated on the surface when the mine was near the surface, thus giving away the mine's position.[45] An obvious tactic, then, would be for a U-boat to transit the barrage on the surface and simply avoid all floats. Since the barrage was likely heavily patrolled, such a transit would probably be at night, probably using searchlights with red lenses to limit visibility of the light. While that approach is just the author's hypothesis, the fact is that the barrage's true effectiveness is unknown. Thirty to 40 U-boats per month continued to transit the area, often through unguarded and unmined Norwegian waters, while a mere four U-boats are known to have been sunk by the mine barrage.

With such a limited number of kills, was the North Sea Mine Barrage a waste of time and money? It depends on your definition. The total cost of the barrage was $40 million.[46] This, compared with the total cost of the war for the United States at $32 billion,[47] means that the barrage accounted for only 1/800th of the country's entire cost of the war. The four submarines were sunk at a minimum of risk to those laying the mines. Operationally, it changed the risk formula for U-boat commanders so that the risk of transit through the minefield, though far from 100 percent, would have been unacceptable in peacetime. The barrage might have been best categorized as a "defensive minefield," i.e., its primary function was to channel many (most?) U-boats to Norwegian waters. This in turn would do three things. First, it would make merchant shipping passing along the British east coast far less likely to encounter U-boats heading toward the British west coast. Second, it would increase the overall distance of the transit, thereby reducing on-station time. Third, it greatly narrowed the probable transit routes for U-boats departing from Germany or departing the Norwegian coast. This in turn allowed for more efficient use of Allied ASW vessels through increased concentration.

The other great value of the North Sea Mine Barrage (even if unintended) was the training value for U.S. mine warfare forces. Nothing helps build a collection of tactics, techniques, and procedures (TTPs) quite like repeating a thing over 50,000 times. The practical lessons from solving the problem of how to manufacture, ship, assemble, deploy, and later clear the 56,671 Mk 6 mines[48] would have filled volumes of postwar TTPs. This knowledge would go well beyond merely copying a small UK-written minesweeping volume and would continue to apply for the next 67 years that the mine stayed in U.S. service.[49]

Russian Mining During World War I

During the war, Russia concentrated its mining in the Baltic, the Gulf of Finland, and the Black Sea.[50] Considering its boxed-in position, such mining was primarily operational in nature, with the basic purpose of keeping Quadruple Alliance forces (Germany, Ottomans, Austria, Bulgaria) away from Russian ports and preferably bottled up within their own ports. Such efforts were all that she could do in the face of the effective blockades of both the Black Sea and Baltic approaches through which flowed most of the trade to and from the Russian heartland. Some 90 percent of its grain exports normally flowed through the Dardanelles, now blocked by the Ottoman Empire. It was upon these exports that depended much of the foreign cashflow into Russia. Furthermore, it severely reduced the flows of imports in both war matériel and commercial goods. This forced privation in goods led to skyrocketing inflation and a discontented populace. That in turn contributed to the Bolshevik Revolution that brought down the Russian Empire before the war's end.[51]

Outside probably defensive mining in the Black Sea, much of Russia's mining was focused against Germany. Early in the war, it laid mines across the Gulf of Finland and in the approaches to Kronstadt (an echo of the Crimean War). It further mined the Latvian coast, a move likely designed to limit the supply lines and amphibious landing options available to the German and Austrian forces considering advancing on the capital. Farther afield, it attempted to hinder the flow of metal ore from neutral Sweden into Germany. Germany, for its part, mined the straits passing into the Baltic from the North Sea to prevent supplies from the UK from reaching Russia. This infringed upon the neutrality of both Denmark and Sweden, between whose lands those straits flowed. They in turn laid their own mines to defend against such incursions.[52]

One point of intersection was in the British effort to deliver military supplies to the Russian port of Arkhangelsk on the White Sea. Being a long distance from the main German operating areas, it was among the very few somewhat safe options available for regional shipments. Germany, for its part, was not unaware of this end-run option and sent vessels to mine the Arkhangelsk approaches. To protect ships and mariners, Royal Navy minesweepers deployed to sweep these mines. Their officers would later complain that their Russian counterparts seemed to view these efforts indifferently and provided little material support.[53]

Postwar Mine Cleanup

If it takes months to drop tens of thousands of mines into the water, it takes about as long to find them and take them out of the water once more. Clearance of the U.S. Mk 6 mines took six months to complete over the course of seven operations between April and September 1919.[54] These operations involved some 88 vessels[55] and approximately 4,000 men working in danger at times 17 to 19 hours per day in the long hours of daylight of the North Sea in summertime. Many of these sailors were reservists held over well past the ends of their enlistments. Over the course of the clearance operation, one ship was sunk, 23 more were damaged, and personnel losses included nine enlisted and two officers killed.[56]

The British mines could be cleared using the techniques and equipment then in place. The U.S. Mk 6 mines were a different problem entirely. Their dielectric firing devices meant that any steel touching their antennae would cause them to explode, regardless of the lightness of contact. Furthermore, if one exploded, it might very well cause sympathetic detonations among its neighbors. The risk, then, was that if a steel-hulled minesweeper accidentally touched an antenna, all ships nearby could find themselves in the middle of a chain reaction of explosions. With a destructive radius of around 100 feet, any craft close to an exploding mine could be lost.[57]

Research and development of minesweeping techniques began after the barrage had been laid. The initial idea for avoiding detonating the Mk 6 mines was to use wooden-hulled ships (possibly the initial reason why later minesweepers were constructed of wood?). Experiments in wrapping ships in magnetized cables led to the discovery that these cables could produce a reverse-polarity current capable of neutralizing the dielectric current produced by the antennae. This technique, now referred to as "degaussing," would later be applied to MCM and non–MCM vessels alike to reduce mine risk.[58]

Lessons in sweeping were further learned during each of the seven phases of clearance that were subsequently applied to the next phase. In the early phases, sweepers cut the mine mooring cables, allowing the mines to float to the surface to be sunk by gunfire. This technique was the same as had been done throughout the war. Eventually, it was realized that the antennae that made the Mk 6 mines dangerous could also be exploited for clearance. Middle and deep layers of mines could be cleared by simply passing a steel wire between two ships. This caused them to explode at a depth less hazardous to surface vessels. The problems with mines fouling sweep gear and forcing resupply were solved by adding explosive cutters to the kites. Formation sweeping helped as well, with the lead ships using wires on antennae, followed by other ships sweeping deeper to cut mooring cables. Behind them, trailing ships destroyed any stray, surfaced mines and marked the cleared path with dan buoys.[59]

How World War I Shaped Naval Mine Warfare

The North Sea Mine Barrage helped shape mine warfare for the U.S. Navy as the core of all mine warfare doctrine to follow. Furthermore, it revealed the sheer level of resources required to clear mines. Despite the 4,000 men and 80-plus vessels working six months, only about 40 percent of the mines laid were found. The rest were assumed to have been set off by the North Sea weather or by sympathetic chain reactions, though mines would continue to be found on beaches and in fishing nets for years.[60]

World War I shaped mine warfare more generally in several ways. First, it highlighted the importance of maintaining a core MCM force capable of defeating the latest mining technology. It introduced the idea of using mines for anti-access and area denial (A2/AD) over wide geographic areas. By denying freedom of movement, mines could be used to neutralize the value of possessing Mahan's "fleet in being." Furthermore, it introduced new tactics like submarine mining, reseeding, mechanical minesweeping, and maintaining cleared channels. Finally, it cemented mine warfare as a weapon of war for which every navy must prepare.

Table 2. Overview of Losses* (Translated)[61]

	U-Boat (Subs)	Capital Ships	Destroyers	Torpedo Boats	Minelayers	Minesweepers, Mine Barrage Breakers	Gunboats, Picket Boats
Quadruple Alliance (total)	59	4	46	20	2	45	105
Germany	56 (19)	3	43 (1)	17 (1)	1	44	92
Austria	3	-	3	1	-	-	6
Turkey	-	1	-	1 (1)	1	1	6
Bulgaria	-	-	-	1	-	-	1
Entente (total)	30	13	31	13	3	34	294
France	2	2 (1)	2	4	2	-	29
Great Britain	21 (10)	8 (2)	20	5	-	13	242 (20)**
Japan	-	-	-	1	-	2	-
Poland	-	-	-	-	-	1	-
Romania	-	-	1	-	-	-	-
Russia	5 (3)	1	7	1	1	17 (2)	16
USA	-	1	-	-	-	-	3 (1)
Neutral	-	-	-	2	1	2	-
Netherlands	-	-	-	1	-	2	-
Sweden	-	-	-	1	1	-	-

**The figures given in parentheses are the part of the total number where the mine is either only assumed to be the cause of loss or was only a contributory cause, e.g., vehicles that suffered mine hits along with artillery or torpedo hits.*

***Of the 242, 193 were fishing vessels in service with the fleet.*

Chapter 7

World War II in the Atlantic

Mine warfare in World War II was a tale of two types of wars. In the Atlantic, it was a tale of mining and MCM technology. It featured many of the same strategies as World War I but executed with 20 years' worth of rapidly advancing technology on both sides. Consequently, to speak of the Atlantic War is to discuss this technology and its influence on the greater events of the war. Conversely, the Pacific war was a story of distance, timing, and maneuver. In the Atlantic, forces operated fairly close to their bases of operation and industrial centers. By contrast, in the Pacific, most of the forces involved operated hundreds or thousands of miles from base. This made mining more difficult to do in volume but more effective when employed against forces who were unsupported by appropriate MCM units. It was not until later that the mining fight was brought to the Japanese homeland, with devastating effect. Because the two versions of the war were so different, this book divides the war into two chapters by theater.

Germany versus the UK and Mine Warfare Technological Development

The German mining campaign against the United Kingdom began very much like that of World War I. Immediately upon the war commencing in 1939, German surface raiders (and even seaplanes) began laying contact minefields off the Thames Estuary and the British east coast. The Royal Navy, long expecting such an opening move, activated its own countermove by building up a reserve fleet of some 800 fishing trawlers and other vessels to mine sweep. Training proceeded as training crews and reserve vessels became available. By the end of the first year, the trained minesweeping fleet included nearly 300 vessels. Of these, 250 were organized into searching forces, while the remainder were clearing forces.[1]

What is interesting to note here is the preexistence of rapid expansion plans and the organization of the MCM forces. A nation at peace is never ready for war when it begins, especially when war is imposed upon them. In peace, a free society does not tolerate a massive dedication of resources to defense. As such, a peacetime military is little more than a bulwark to hold back adversaries until more resources can be diverted to the purpose. Then they act as the nucleus for building the wartime force. But to rapidly build forces, a practicable plan is needed, one that for efficiency's sake must be built on repurposing existing assets. For an island nation like the UK, the readiest available force of trained mariners for hauling gear attached to wire ropes was its massive fishing fleet.

The split of MCM forces into searching and clearing forces is especially interesting given the nature of newly activated reservists. Searching for mines takes a long time. Checking hundreds and thousands of miles of sea routes for mines can take countless hours. A force that only searches is generally not involved in the most dangerous work unless they find mines. After that moment of finding a mine, their job is to report and mark the mines' locations and then continue searching. This can be accomplished with limited training and equipment. Mine clearance is a more dangerous and specialized type of work requiring more training and equipment. With these factors in mind, when organizing rapidly expanding MCM forces, it is likely better to focus the experienced forces on the intermittent clearing of mines (with rest when not doing clearance). The inexperienced forces would be better concentrated on the more continuous but less specialized business of searching. As experience is gained and the need arises, these searching forces could graduate up to clearance.

Applied to more modern times, unmanned underwater vehicles (UUV) can be assigned to the searching mode. UUVs could be housed with local naval reserve units in peacetime to maintain a rapid-response force and a nucleus of trained operators. In wartime, more UUVs can be further assigned to fishing fleets with training coming from their local reserve centers. This would allow the active-duty forces to concentrate on the clearance function while also reducing the risk to commercial vessels from undiscovered minefields.

Aerial Mining—Although airplanes were available in World War I, they were still relatively slow, short-range craft with small payloads, not particularly suited to carrying heavy cargo over significant distances. By World War II, this state of affairs had changed. The world's military-industrial complexes continued developing airplane technology during the interwar period. When Germany began its aggressive European expansion, airplanes were more powerful in most every respect. Building on the country's successes with mining during the Great War, German engineers had experimented with using airplanes to deliver mines.

Aerial mining represented a huge leap forward in the story of mine warfare. First, it gave the ability to rapidly put weapons in the water at a distance in reasonable safety. Within a few hours of the decision to mine an area, airplanes could be dropping mines without the need to prepare a naval vessel and crew. Furthermore, airplanes could potentially reach places such as harbors and rivers that would be too hazardous for minelaying vessels to enter. Perhaps the most important consideration is that aircraft are not in the same medium as their mines. This allows the reduction of arming delay times, and it allows aircraft to lay additional mines in previously mined waters without regard for a mining hazard.

True, some airplanes of the time might have only been able to drop one or two mines apiece, but is that really such a problem? Consider: A ship with a crew of 100 people and carrying 200 mines could deploy to saturate an area with mines. But if that ship is sunk, it may lay zero mines and lose its entire crew of 100 people. Alternatively, 100 airplanes, each carrying two mines and operated by one person, could fly to the same area to drop mines. Even if 10 are shot down before they can accomplish their mission, there remain 180 mines that can be dropped by the 90 remaining pilots. Another advantage to small payloads is that they allow for less-predictable minefields. Due to operational efficiency, minelaying ships tend to lay mines in lines. This approach makes it more likely that they will sink the first ship that traverses the field, as it guarantees uniform coverage. Yet mine lines are predictable. Find one, and chances are you do not have to try hard to find the next one, and the next, and the next, all in a fairly predictable pattern. The danger is high near a detected mine laid by a ship but is quite low outside that line. With aerial mining, perfectly spaced lines are impossible. Instead, individual pilots lay their mines in approximate locations that may or may not be coordinated with the efforts of other pilots. This imprecision creates a much greater area of uncertainty (AOU) around any mines that are found. That, in turn, increases the search area in which MCM efforts must eventually be expended to remove the threat.

Yet another advantage of using aircraft is their small size and their ability to fly under the radar. This makes it difficult to estimate the number of mines to be found. For instance, if a submarine known to hold 24 mines is the only enemy craft known to have operated in an area, then it is reasonable to assume that there can be no more than 24 mines in that area, and more likely some factor of 24 (like six, eight, or 12). This would limit the perceived mining danger to places where that submarine could operate. On the other hand, if aircraft are known to have dropped some mines, but the number of aircraft operating in an area is unknown, the likely number of mines could be anywhere from one or two up to however many enemy aircraft are available. This factor adds a lot of uncertainty. Not knowing if it is a small nuisance

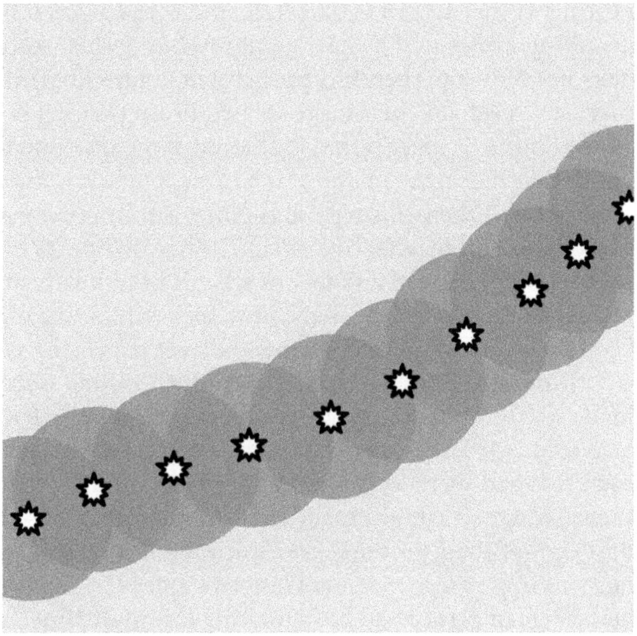

***Above and below:* Threat areas of 10 mines laid in a line versus randomly dropped by aircraft. Note that the true danger from each mine is the same in both, but the right side is less predictable (M. Mazurek, after original drawing by the author).**

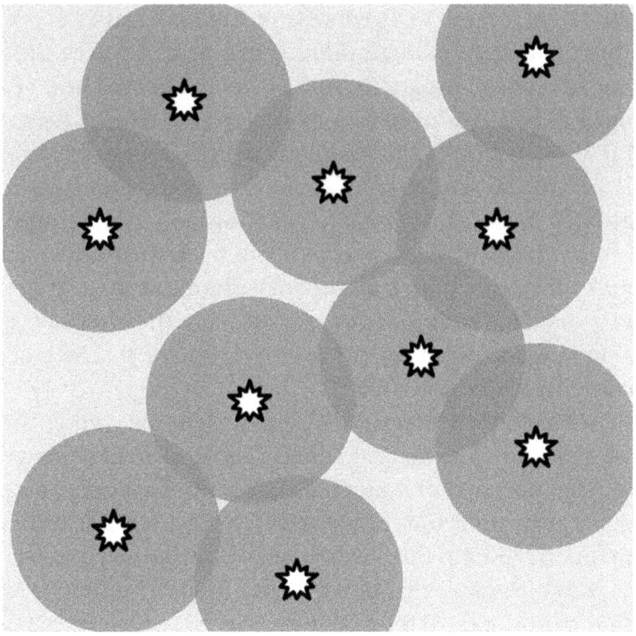

field or a major saturation would lead a prudent port commander to shut down a port entirely until MCM forces can better assess the area.

A final factor of note in aerial mining is that it is the least clandestine version of mining. If the aircraft is detected, and if the aircraft is observed, then the fact that it is dropping mines will be easy to figure out. When something falls from an aircraft and does not blow up, then it is probably not a bomb. If the object lands in the water, then there will probably be a large splash. From there, it is not hard to guess that it is either a naval mine or an external fuel tank. If a parachute deploys en route, then there is no doubt that the attached object is a mine. Once it is known that enemy aircraft are laying aerial mines, then all surprise is gone. But in mine warfare, that is again no bad thing. Minefields can be effective at shutting down ports even if no ships are damaged. The perception of danger, and fear of what is not known, are enough to bottle up a port until MCM efforts take place. That counts as a win for the miner, as it forces their adversary to keep their own ships in port and to divert transiting vessels to less-ideal ports.

Influence Mines—Up to this point in mine warfare history, influence mines had made no significant impact. While magnetic mines were first invented toward the end of World War I, they were not widely deployed operationally. By World War II, however, each side had secretly studied a variety of influence mines and possible ways to counter them.[2] Magnetic mines made their first appearance off the coast of the UK in 1939. The first sign of their presence was mine strikes in areas previously swept. These mysterious sinkings rose in number from two apiece in September and October to 13 in November. German planes had been making low nighttime passes over harbor approaches, but there was no confirmation as to their mission. On November 23, one of these planes was observed dropping something onto mudflats near Shoeburyness at the mouth of the Thames Estuary. Royal Navy mine warfare experts carefully disarmed the object and determined that it was a magnetically actuated mine.[3] This was when the Allies learned that mine warfare in World War II would be very different from that of World War I.

With the limited exception of anti-sweeping devices,[4] moored contact mines were relatively simple to sweep. Once mechanical sweep gear had passed through an area, ship captains could confidently traverse the area until reseeding made further sweeping necessary. Magnetic mines were different and could not be swept with the simple mechanical gear used for contact mines. Unlike contact mines, which normally float in the water volume, the magnetic mines (and other influence mines) used in World War II sat on the seafloor. As such, they could not be removed by simply groping around blindly underwater with a cable. Since navies could not yet see underwater, the only other option available was to try to trick the mines' sensors. One form of trickery was to make the mine blow itself up on a phantom ship, by generating a satisfying influence at a safe distance from the sweep platform. The alternative form of trickery was to mask the ship's signature so the mine could not see it. The challenge was designing devices that could pull off these tricks.

Magnetic Mines and Magnetic Minesweeping: In principle, a magnetic mine looks for changes in its local magnetic environment. A large metal object like a ship will tend to become a sort of magnet as it is built. During construction, as it sits in one place for a while, the molecules of the iron from which it is built align themselves with the earth's magnetic lines of flux. That magnetic alignment stays with the ship. As a sort of magnet, as the ship passes through an area, it will change the local total magnetic force, a change that mines can detect. The goal of magnetic minesweeping is to create an artificial increase in the local total magnetic force, at a safe distance from the sweeper, using equipment designed to survive mine explosions. That sweep equipment's magnetic signature is induced with some form of permanent magnet or electro-magnet. How that is done specifically has varied dramatically and is outside the scope of this book. The books *Allied Minesweeping in World War 2* by Peter Elliott[5] and *Weapons That Wait* by Gregory K. Hartmann[6] both give great exposition on the techniques used.

The British developed the first magnetic minesweeping techniques and equipment, one method involving a generator running a current through a cable in the water. An electrode streamed from the end of the electrode, providing an open path for the outbound electricity. A second cable of a dissimilar length provided the return path for the current. By running high voltage directly through the water, an outsized magnetic signature was formed. This signature could satisfy the mine sensors of the time. The Royal Navy shared their techniques freely with the U.S. Navy and even sent a sample of their mag-sweep cable in 1940. As a fun intersection with another aspect of naval history, the cable was given to Commander Hyman Rickover. Using that scrap of cable, he figured out the power requirements and designed an entire magnetic minesweeping system (including generators) to achieve the desired capability.[7] Rickover would later be known much more famously as four-star Admiral Rickover, the cantankerous "Father of the Nuclear Navy."

Making the Ship Invisible to Magnetic Mines: As described above, metal ships pick up magnetic signatures that make them susceptible to magnetic mines. During World War II, that risk was lowered by finding ways to reduce the magnetic signature. This was accomplished through at least one of three or more approaches. First, a ship could be built with as little magnetic material as possible. At a minimum, this usually meant a wooden hull (a technique that also helped with antenna mines in World War I). Another option was degaussing, a technique that involved running a current through cables wrapped around the ship to produce a different magnetic signature that lowered or negated the ship's permanent signature. An interesting alternative concept was to use that same degaussing equipment to achieve the opposite effect, i.e., creating an outsized signature to cause mines to actuate beyond their effective ranges.[8] A couple of additional methods were "deperming" or "wiping." Both techniques involved passing the ship through a strong induced magnetic field that would scramble the ship's permanent signature for a limited time. These techniques are again described in the same books listed in the previous paragraph.

Acoustic Mines and Acoustic Minesweeping: Within a year after Germany began heavy use of the magnetic mine, the British were conducting effective minesweeping to remove this threat. One day, however, in October 1940, ships began prematurely actuating mines in shipping channels after the now-routine magnetic minesweeping had been completed for the day. The implication was that a new threat had emerged.

Like with magnetic mines, British researchers had conducted basic research on acoustics before Germany introduced acoustic mines. Earlier in the year, they had recorded and analyzed the acoustic signatures of several classes of ships. By the time the acoustic mine came out, they already had an idea of what frequencies needed to be satisfied to sweep the new mines. After some trial and error, four different techniques were found to be effective.

1. *The Kango Sweep:* A Kango pneumatic road drill, banging against a diaphragm in a steel box and lowered over the side at the end of a boom.

2. *The Fessenden Oscillator:* Basically, the same concept as the Kango sweep, but fitted inside the hull of larger warships. Became less effective as mines received finer settings.

3. *Machine-Gun Fire:* A Lewis machine gun, fired in bursts, was found to be loud enough to actuate the anti-sweep devices in acoustic mines. While the approach would not actuate an acoustic mine, it would temporarily turn off the mine's firing logic, allowing the ship to pass while it was temporarily deactivated.

4. *Explosive Sweeps:* Grenades were dropped into the water through a tube at building-up intervals to simulate a large ship getting closer. Operators were protected by a blast screen. Some attention to detail was required.[9]

Pressure Mines and Pressure Minesweeping: As discussed, influence mines are actuated by various measurable effects of a ship passing through the water. The number of these effects is limited, and for MIW scientists (a niche job), predictable and worth investigating. One of these effects, water pressure, is caused by the fact that water pressure changes as the depth of water changes. The water pressure at the bottom of a body of deep water will be greater than it would be for shallower water. This pressure is localized, so that the pressure in a given location can be lower than deeper water only 100 meters away. And when a ship passes through the water, it displaces enough water to reduce the amount of water directly beneath it, which in turn reduces the water pressure beneath the ship. For the World War II MIW scientists of both sides, this was an influence signature worth studying. Germany rushed to develop pressure mines, while the UK rushed to find countermeasure methods in anticipation of pressure mines being used in the near future.[10]

Germany, for its part, had an operational pressure mine by 1943 but did not deploy it right away. It had learned its lesson from its mistake of releasing the magnetic mine too soon. In that instance, it deployed the mine early in the war and before enough of them were available to be decisive. This "preview" gave the British time to engineer countermeasures before the mine became a serious problem. Learning from this, the Germans held back from deploying the new pressure mines against the eventuality of the Allied invasion across the Channel. The idea was that by saturating the landing areas with this new weapon at a time when cross–Channel passage would be at its height, the Allies would be unable to adapt fast enough. Being at such a crucial time, when control was still being decided, German reinforcements could be pushed to the front faster than Allied reinforcements.[11]

Such hopes were soon dashed. In the intervening time, British scientists had run numerous experiments to find ways to affect the pressure signatures of ships and the sensors of then-hypothetical pressure mines. Accordingly, they had distributed speed tables out to the fleet within 48 hours of Germany deploying its mines during the Allied invasion at Normandy. They also had sensors and tactics that could be tried against this new weapon.[12]

The big weakness of pressure mines is that water depth and water pressure frequently change in nature. Tides can raise or lower the water depth just as much as can shipping. Engineers must design their sensors to account for the slow changes in water pressure due to the tide. With the German mine, termed the "Oyster" by the Allies, the problem was solved by a tube that could allow water to pass slowly from one side of the sensor diaphragm to the other. This meant that only quick changes in water pressure could actuate a pressure mine. Yet, quick changes also happen in nature. During storms, large waves can rapidly rise and fall even higher than the draft of the biggest ship. Knowing this, pressure mine designers would understand that an entire field of pressure mines could be actuated by a single storm. Therefore, they would design their mines to rely on at least two influences, such as magnetism and pressure. Theoretically, this would make the mine much harder for a small minesweeper to sweep. A magnetic sweep alone could not actuate the mine because the mine's logic would also require a sudden, sufficiently large change in pressure. The reality was that this provided opportunities for both active and passive MCM. And so it proved with the Oyster mine, which required the pressure signature before the magnetic or acoustic sensor could be engaged.[13]

The Allied passive countermeasure was for its ships to transit slowly in shallower water. By moving slowly, their pressure signatures would change too slowly to actuate a mine's pressure sensor. Without that sensor being actuated, the mine's other sensor was effectively disabled and, thus, presented no risk to the ship. The active countermeasure was to simply wait for a high-sea state before sweeping. The waves would constantly actuate the mines' pressure sensors, effectively turning them into simple magnetic or acoustic mines. They could then be swept with simple influence sweeps. Still other attempts were blunter, such as inducing artificial waves by hurtling small, fast ships toward the field and then turning them at the last moment to allow the large wake to wash over the field.[14]

Lend-Lease Program—Among the many challenges of a nation at war with a proximate enemy is how to acquire the matériel necessary to continue pressing the war. The

British divers recovering oyster mine at Bremen, 1945 (from the collection of Lieutenant Commander Rob Hoole, RN).

United Kingdom had this problem in spades during the war's early years. With Germany pressing her everywhere from the air and under the sea, the UK needed everything and quickly. Her difficulty was that her production facilities were vulnerable to attack, while the competition for their products was fierce. Engines were one type of item with many competing uses, which, combined with limited UK facilities for producing them, put them in short supply. That in turn affected the design and effectiveness of the many motorized units needed. Minesweepers were another product that could not be produced quickly enough. With German aircraft and submarines able to mine most anywhere in the islands, the Royal Navy was hard-pressed to deploy enough units to keep the channels clear. Minesweepers had to remain at sea most days either sweeping, pulling picket duty, or performing escort duties. Meanwhile, their diminutive size and the hazardous nature of their primary duties meant considerable losses to mines and enemy action. Much of the war matériel needed by the UK had to be filled from abroad.

Some of the need was filled from the British Commonwealth. Ships were provided by Canada, South Africa, Australia, and several smaller nations. But these countries together were insufficient to fill the need because they had their own defensive problems. This is where the Lend-Lease program with the United States held value. It was devised to skirt the sticky matter of U.S. neutrality. Technically, the United States had not declared war on the Axis powers, yet it considered it vital to do all that it could do short of war to prevent the UK from falling. Thus, in steps it provided that beleaguered nation with ships, aircraft, and other matériel under ever less neutral payment options. At first, it was "cash and carry," requiring the UK to pay cash for matériel and to transport goods on their own ships. Next there was the trading of 50 old destroyers for 99-year leases for bases in Newfoundland and the Caribbean (thus lowering the UK's defense workload). Finally, Congress passed what is known as the Lend-Lease Act. Under that act, war matériel could be sold to cash-poor belligerents on a "pay later" basis.[15]

UK Mining—Mining by the UK during World War II started on a similar pattern to that of World War I but changed as the Germans proved unexpectedly successful at invading unsuspecting countries. Upon declaration of war, the UK immediately activated its plan to build a mine barrage across the English Channel at Dover. Having expected the need for some months, it already had mine stocks and minelayers ready or in preparation and, in some cases, already loaded. The French, for their part, built further minefields at Dunkirk. After Germany suffered three U-boat casualties to this barrier, U-boat traffic slowed to a trickle as alternative routes were used instead.[16] Once France fell, German U-boats simply shifted their home ports to five new pens along the Bay of Biscay and later added a base at Toulon on the Mediterranean coast. From there, they enjoyed more direct lines of attack on Allied convoys.[17]

Other minefields were laid along the British east coast to protect shipping and between Cornwall and Ireland to protect against invasion. Another field like the North Sea Mine Barrage of World War I was planned and materials allocated, but the German invasion of Norway made such preparations pointless. After all, U-boats and surface raiders could simply be stationed in Bergen or just about any of the countless fjords north of the planned barrage. Instead, a system of minefields was laid in a line extending from Scotland to the Orkney Islands to the Faroe Islands to Iceland to Greenland. It was an imperfect barrier, as water depth and currents made a complete barrier impossible, but it helped to shape the battlespace and to concentrate German ships and U-boats where they could be more easily located. Other British minefields were laid in other parts of the world at the Bab el-Mandeb Strait, the Malacca Strait, Australia, and New Zealand.[18]

A more controversial set of minefields was planned for the ostensibly neutral waters of Norway (before the German invasion). Under coercion from the German

World War II minefields in European waters (from J.S. Cowie, *Mines, Minelayers, and Minelaying*, Oxford University Press, 1949, original substantially cleaned up and modified by the author).

government, Norway had been supplying iron ore to Germany and allowing Germany to transit through its territorial waters. Norway's coast is lined with islands that allow an inner passage for much of its length except for three points: Vestfjord, Bud, and Stadtlandet. The UK's plan was to mine these points, thus forcing German shipping out into open waters. When it relayed this plan to the Norwegian government, the news was met with violent outrage by that country. Despite Norway's objections, minelaying proceeded just as Germany began its occupation of Norway. Whether there was a causal relation between the mining and the invasion is unclear. Whatever the case, the Norwegian government joined the Allies and the mining quickly gained clear legitimacy. Soon thereafter, the British laid several more minefields along the Norwegian coast.[19]

Offensive mining became something of a UK pastime in the months ahead. British airplanes, submarines, minelayers, and destroyers blanketed the North Sea with mines. The Heligoland Bight, the Dutch and French coasts, and the Danish coast (to include the Kattegat and the Skagerrak) each became festooned with mines. Importantly, a 30-mile channel was left for the shipping of Sweden, who remained neutral throughout the war. Further minelaying occurred in the Mediterranean, though in smaller, more focused fields rather than the type of massive fields used farther north. It was in the waters of Oran, Algeria, that carrier-based planes first laid a minefield. In the northern Mediterranean, the Royal Navy disguised the fast minelayer HMS *Manxman* to look like the French cruiser *Leopard* using paint, canvas, and wood. In an exciting mission, this imposter succeeded in laying a minefield between the Corsican and Italian coasts.[20]

Cooperation Between MCM and Minelaying Staffs

A couple of incidents happened in the course of all of these mine warfare events that underline the value of having minelaying and MCM forces work together. The first event was when minelayers placed a series of minefields in the English Channel to stop an expected German sortie. Here, the MCM staff advised the minelaying staff about what route the enemy would most likely take if they were trying to avoid mines. This helped the miners to focus their minelaying efforts to defeat such plans. Later, off Madagascar, the minelaying staff helped the MCM staff to figure out where mines were most likely to be placed.[21]

The value of such cooperation cannot be understated. Both sides of the warfare area have their operational limits and their concerns, and both try to guess the potential actions of the enemy. To a minelayer, one patch of water may look as good as another for laying mines. Their MCM counterparts, on the other hand, may be more attuned to environmental factors that make some areas less susceptible to mine actuations and other areas nightmares to clear. In the other direction, an MCM expert may see an entire patch of water as requiring mine clearance. Here, a minelayer might have insight on the type of platform that would do the laying, its limitations, the navigational concerns of the platform's crew, and the type of pattern most likely to be employed. These factors could potentially narrow the area of probable mining. Both sides can provide valuable insight to the staffs of other components, such as routes where shipping might best be diverted to minimize mine risk or identifying opportunities for using mines to shape the battlespace. Such is the value of collocating battle staffs or, at a minimum, exchanging liaison officers (LNO) who have both expertise and the ability to think independently.[22]

Italian Mining—Though seldom described in detail, Italy laid the third greatest number of mines of the war after Great Britain and Germany.[23] Geographically, Italy's coastline makes it almost as vulnerable to amphibious assault as Great Britain or Japan. It is a long peninsula that is connected to mainland Europe only at its northern edge. That edge is separated from the rest of Europe by the Alps, a rugged mountain range that mostly isolates it from the rest of Europe except for a few vulnerable passes. At the beginning of the war, the country controlled present-day Libya, a huge country whose western and eastern edges are directly south of Rome and Athens, respectively. Furthermore, it had to worry about the coastlines of its large islands of Sicily and Sardinia. With all this coastline, Italy's problem as an aggressive expansionist was how to limit its exposure to counterattack. If it could limit the possible threat axis, it could concentrate its defenses.

Its answer was to build a mine barrier across the Strait of Sicily between Sicily and Tunisia using about 4,500 mines. This was part of a larger mining effort involving some 17,000 mines.[24] Since the primary threat came from the Royal Navy (and later, the U.S. Navy), this action helped screen the peninsula's entire east coast from easy attack, as well as the Libyan coast and Sicily's southern coast. The peninsula's west coast was screened by the large islands of Sardinia, Corsica, and Sicily. Together, the islands, the peninsula, and the north coast of Tunisia formed a mutually supporting triangle.[25] Any naval vessels in the area risked air attack from any direction.

A major side effect of the Sicilian Channel minefield was to isolate the British base at Malta. Being located on the wrong side of the channel, the island could not receive regular convoy resupply. In October 1941, a British battle group cruised into the area. This battle group succeeded in attacking Italian convoys but met with disaster when it attempted to bombard Tripoli in December 1941. Here, Italian mines laid a few months prior sank or damaged two-thirds of the battle group. Allied code breaking identified paths through the minefield, allowing a convoy to deliver 21,000 tons of supplies the following month. The Italian navy quickly closed these paths with more mines, thus cutting off the eastern half of the Mediterranean until fuel shortages precluded continued minefield maintenance. The garrison at Malta continued to defend the island until Italy finally surrendered in September 1943.[26]

Soviet Mine Warfare—Very little about Soviet naval mining is discussed in English-language books about World War II. With Russia's considerable success with naval mines during the Crimean War, Russo-Japanese War, and World War I, it is probable that this now-traditional weapon was also used extensively during World War II. As discussed in the next chapter, Soviet mining experts would go on to assist the North Koreans with the mining of Wonsan Harbor.[27] Expertise is not built by sitting idly by during a major war, and so credence must be given to what sources exist.

One source summarizes the USSR's use of mines during the war as being "extensive," consisting of 40,070 mines laid and accounting for the loss of over 200 enemy ships.[28] With all of the mines being laid by both sides, Soviet minesweepers were kept especially busy in the Gulf of Finland. After all, if the Soviets could use mines to keep the Germans out of the gulf, then Germans could use mines to keep the Soviets in the gulf. Minesweepers worked to prevent the latter effect. In the Black Sea, the Soviets used mines around Sevastopol and at the Kerch Strait (as they had attempted to do in the Crimean War). Though they claimed to have succeeded in destroying 200 enemy ships at Sevastopol, it appears that at the Kerch Strait, the German minesweepers managed to wrest sea control from its defenders.[29]

Youngblood (2006) explained that Soviet mining was largely defensive at first and only later extended to offensive mining. Once Germany declared war on the USSR in

June 1941, Soviet, German, and Finnish ships all began mining the Gulf of Finland. With its Baltic fleet bottled up in port, and Axis powers controlling all other approaches, resupply from abroad could come only via convoys taking the Murmansk run to the White Sea port of Archangel. To protect this single source of resupply, the Soviets mined the entrance to the White Sea. At first lacking influence mines or submarine-laid mines, they soon caught up technologically and began laying their own offensive minefields. This included submarine-laid fields in the Baltic. More interestingly, they conducted aerial mining in the Danube and Dnieper Rivers to cut off that source of resupply to invading German forces.[30] Both are huge rivers, major inland shipping routes, making both vulnerable to mines. Mining them harkened back to the riverine mining done in the Crimean War and the U.S. Civil War. Airborne riverine mining would make its appearance again during the Vietnam War.

Naval Mine Warfare in the Americas

Preparing for German Mining—The United States came into World War II much more aware of the German U-boat threat than it had at the beginning of the previous world war. At the beginning of World War I, the Germans had no submarines that could reach the United States, and the idea that it would try to mine American coastal waters seemed preposterous. By the end of that war, the Germans had begun to wreak havoc in those waters using mines and torpedoes alike. Consequently, when the United States finally entered World War II, it was well known that Germany could and would attempt to top its previous performance. Indeed, German U-boats laid a total of 338 mines in U.S. waters during the war, accounting for 10 ships sunk or damaged. To counter them, the United States expended a lot of matériel in mines, minesweepers, patrols, and port closures that far exceeded the cost of the German mines themselves.[31]

Research and planning into how to counter mines after the expected entry into the war began as early as 1939. It began with a member of the planning desk at the Washington Navy Yard being blessed by the chief of naval operations (CNO), who took a personal interest in planning for mine warfare. This officer surveyed available fishing vessels and other commercial ships as well as the inventory of stored minesweeping equipment. He studied the potential to build a scratch reserve minesweeping force, with at least 35 wooden fishing boats being eventually converted. Meanwhile, scientists in research and development were assigned to study German magnetic mines and British countermeasures equipment (discussed above), with emphasis on degaussing. The CNO directed that the whole fleet be degaussed. The few remaining legacy minesweepers from World War I were renovated and the keels were laid on a few new classes of minesweeper. Prominent among these new minesweepers were the small, wooden "yard motor minesweepers" (YMS). This class was intended to work in Stateside harbors but ultimately deployed to support clearance operations around the globe. A newly established Mine Warfare School began its first classes. By 1941, students were practicing sweeping in live minefields for the first time since the days of the previous war.[32] Through these efforts, the United States entered World War II in a state of near readiness to operate in the Atlantic theater, where mine warfare quietly played an even larger part than fleet-to-fleet actions.[33]

These preparations would prove useful when Germany began laying mines in U.S. waters (see Table 3). While some of these German minefields succeeded in closing individual U.S. ports for days or weeks at a stretch, sweeping forces were available and ready to minimize their impact. Had this been done as a sneak attack during the between-wars state of total unreadiness, these ports may have been paralyzed for months while solutions were sought.

Table 3. German Submarine-Laid Minefields Off the North American Coast[34]

1942	*Location*	*Boat*	*# Laid*	*# Detected*	*Found By*	*Ship Casualties*
May 14	St. John's, Newfoundland	*U-213*	15		Undiscovered	None
June 11	Boston	*U-87*	10		Undiscovered	None
June 11	Delaware	*U-373*	15		Sweeper	Tug *J.R. WILLIAMS*
June 12	Chesapeake	*U-701*	15		Ship	SS *ROBERT C. TUTTLE*, HM Trawler *KINGSTON CEYLONITE*, Coal Barge *SANTORE*, sunk; SS ESSQ *AUGUSTA*, USS *BAINBRIDGE* slightly damaged
July 20	Castries, St. Lucia	*U-66*	6		Ship	
July 24–25	Mississippi Passes	*U-166*	9		Undiscovered	None
end July	Charleston	*U-751*	12		Undiscovered	None
August 8	Jacksonville	*U-98*	12		Sweeper	None
September 10	Chesapeake	*U-69*	12		Sweeper	None
September 18	Charleston	*U-455*	12		Sweeper	None
November 10	New York	*U-608*	10		Sweeper	None
1943						
June 1	Halifax	*U-119*	66	55	Ship	SS *HALMA*
July 30	Chesapeake	*U-566*	12		Undiscovered	None
July 31	Chesapeake	*U-230*	8		Undiscovered	None
August 26–27	Charleston	*U-107*	12	1	Sweeper	None
October 8	Panama	*U-214*	15	10	Floater	None
October 9–10	St. John's	*U-220*	66	34	Floater	SS *DELISLE*, SS *PENOLVER*
October 27	Trinidad	*U-218*	14		Undiscovered	None
1944						
March 23	St. Lucia	*U-218*	2	1	Ship	None
April 1	San Juan	*U-218*	15	1	Premature	None

Fishing Boats as Scratch MCM Vessels

During both world wars, fishing vessels were pressed into service as minesweepers. This makes sense for a variety of reasons. First, they are often small vessels with nonmagnetic hulls designed to stream wire ropes used to haul heavy gear through the water. These traits are, in essence, those of coastal minesweepers. Second, their crews are trained for the purpose of hauling equipment and are conditioned to work in hazardous conditions. These are the prerequisite traits of minesweeper sailors. Third, they are very familiar with the underwater environments of the areas in which they normally operate and can recognize changes to those environments. This gives them an advantage when searching for mines in their home waters compared with navy sailors who are homeported in other areas. Finally, because mines threaten their own livelihoods, removing that threat is not just a duty, but personal.

American Protective Mining—One early U.S. response to the German mine threat was to mine North American waters from the Panama Canal[35] up to Portland, Maine. Of these, the Army laid 3,569 controlled contact mines and, later, 3,751 controlled magnetic mines. As in previous wars, these mines could be turned on or off from central control points, allowing friendly traffic to traverse these minefields unimpeded. All were part of the elaborate defenses recommended by the Endicott commission and built up over the previous 70-plus years. Room for debate exists on the effectiveness of these minefields, for not a single enemy ship is recorded as having been sunk or damaged by them.[36] This could be the result of no such craft being detected. Alternatively, and far more likely, the very presence of the defenses acted as a deterrent for U-boat captains considering entering American harbors. They could engage in depredations enough without risking mines.

While the U.S. Army had laid its controlled minefields in harbor entrances, the U.S. Navy laid several fields farther out to sea in hopes of sinking U-boats. Their actual effect, it seems, was to sink far more friendly shipping than unfriendly combatants. The problem was mostly human error and delays. Q-messages, meant to report the presence of minefields, would sometimes take weeks to reach affected shipping. At other times, merchant captains refused to heed the warnings and the advice of young naval and Coast Guard officers. In at least one case, a ship received conflicting reports as to which side of the marking buoy to pass. On the occasions when the presence of minefields (friendly or enemy) was unknown to convoys traversing them, confusion would reign as explosions were usually attributed to U-boats. In such cases, the tactic of scattering at high speed, appropriate for U-boats, was the exact opposite of tactics needed for traversing a minefield. All told, the U.S. Navy laid at least seven defensive minefields in U.S. waters. In so doing, their threat to friendly shipping was approximately the same as the threat from German mines in those waters.[37]

At times, naval vessels were tasked with patrolling the minefields. This was to serve at least two purposes. First, they could keep an eye out for submarines hoping to transit the harbor approaches through the minefields. If a submarine is alerted to the presence of minefields (a possibility that may be impossible to avoid), then their best hope would have been to observe the movements of shipping. With accurate observations, it may have been possible for them to discern the safe route. However, this would require extensive time lingering on or near the surface, making them susceptible to discovery by patrol vessels. More likely, instead of transiting, they would identify a point through which all transiting vessels seem to pass. That point would become the bottleneck in which the submarine would be guaranteed a hit. Patrol vessels would hopefully make that trick more difficult. The second purpose of patrolling was to interdict ships to prevent them from plowing into the minefield. This would commonly be done using signals, but if that failed, warning shots would help to bring merchant captains to heel.[38]

Why was there no record of enemy ships either successfully or unsuccessfully attempting to force U.S. minefields? There are several possible interrelated reasons for this: (1) Potential enemies knew about the mines and were deterred. (2) The ocean itself acted as a barrier, so that enemies saw little value in crossing an ocean to make a pinprick of an attack. (3) Because of the offensive actions of the United States and its allies, potential attackers of the U.S. homeland were tied up with actions closer to home. (4) There were enough targets closer to home that attacks farther afield were undesirable. (5) Following the attack on Pearl Harbor, coastal fortifications and controlled minefields were already deemed relics of the past. Why risk a capital ship to mines, guns, and attacking aircraft when a few bombers can achieve the same results?

Following World War II, the United States quietly discontinued its controlled mine program as being a relic of the past. In the new world order that followed the war, America's naval dominance was deemed sufficient to make U.S. coastal fortifications unnecessary. Instead of stopping ships from attacking U.S. cities, the new emphasis would be on strengthening anti-aircraft capabilities.[39] Indeed, this risk calculation is even more strongly supported today. Ships of today, if used at all, are most likely to attack points ashore using missiles or aircraft. It would be a waste of money to maintain permanent defensive minefields against the chance of enemy battleships forgoing all other means of attack and bypassing all other defenses in order to attack a U.S. city with direct gunfire. Other countries closer to adversaries may continue to find value in maintaining such systems, particularly those bordering major straits. With the advent of suicidal drone boats, descendants of the original mines, there may once more be a benefit to systems like nets; small, protective influence mines; and shore artillery able to track and destroy small boats.

Summary of All Minelaying—Overall, the two sides laid approximately the same number of mines during the war. According to a German source, the totals are as follows:

Table 4. Total Number of Mines Laid in World War II[40]

Axis Powers	*# of Mines*	*Allied Powers*	*# of Mines*
Germany	223,000	United Kingdom	263,376
Italy	54,457	United States	44,000
Japan	51,000	Soviet Union	40,000
Romania	6,500	Norway	5,000
Finland	5,500	Sweden	4,346
Bulgaria	1,000	Poland	110
		France	?
		Greece	?
	341,457		**357,200**

Minesweeping During Amphibious Assaults

One of the most prominent features of World War II was the number of amphibious landings made under fire. While most such landings were made in the Pacific, several happened in Europe as well (most famous of which was Normandy). Those landings are below. It is worth noting that none of these landings took place at sites previously mined by the Allies.

Table 5. Amphibious Landings in Europe in World War II[41]

Code Name	*Location*	*Date*	*Notes*
Operation Husky	Sicily, Italy	July 10, 1943	
Operation Avalanche	Salerno, Italy	September 9, 1943	near Naples
Operation Shingle	Anzio, Italy	January 22, 1944	near Rome
Operation Neptune	Normandy, France	June 6, 1944	
Operation Dragoon	Côte d'Azur, France	August 15, 1944	near Nice

Other landings occurred, though they were more traditional landings rather than the amphibious variety. They were conducted in Morocco, Algeria, Iran, and Tunisia. Though minesweepers searched, no mines were present to threaten these traditional landings.[42]

Naturally, mines can spoil any invasion, so the Allies took care to conduct mine countermeasures prior to every amphibious assault.[43] This took the form of the "assault sweep," a form of clearance done at speed and with the acceptance of perhaps a greater degree of residual risk than one might use in friendly waters. Assault sweeps took place in four phases: (1) clear a path to the landing area, (2) clear the assault lanes for the landing craft, (3) widen the beachhead area, and (4) keep sweeping in case of reseeding. The middle two phases involved much risk to the MCM units, with heavy attrition expected from shore fire, aircraft, and, of course, the naval mines they expected to clear. Consequently, it was not unusual for the operation to include dozens of such units. The implication is that when a contested amphibious landing is contemplated against a nation with a naval mining capability, enough MCM forces should be brought to be able to absorb heavy losses.

Postwar Mine Cleanup—The postwar mine clearance was a global effort taking years, with many units used in multiple theaters. It will be summarized in greater detail in the next chapter.

Summary—How World War II in the Atlantic Shaped Naval Mine Warfare

Mine warfare changed irrevocably because of the technology introduced by the Germans and British during World War II. Influence mines were the most important pieces of technology revealed in this period. These mines had several advantages, including their wider actuation range (no physical contact needed) and their variable settings. Because they did not need to be touched by their target, they could sit on the seafloor, where they were impervious to mechanical sweeping. Besides making the activities of minesweepers harder and less certain, it also meant that capital ships could no longer count on paravanes to keep them moderately safe from mines. Furthermore, because influence mines could sit on the seafloor, there was no need to design them with empty space for buoyancy, thereby allowing them to contain more explosives. With more explosives, they could damage ships farther away and cause greater damage. Combined with the wider actuation range, this meant that fewer mines were needed to provide coverage. Where moored contact mines were effectively single-point weapons, and antenna mines were vertical-line weapons, influence mines had a three-dimensional, hemispherical area of coverage.

To protect themselves from the threat of influence mines, navies had to develop two different families of technology. The first was tactics, techniques, and procedures (TTP) and equipment used to reduce ships' vulnerability to mines. These fundamentally changed how ships needed to be prepared for war and how they needed to behave in potentially mined waters. The other important family of technology was influence minesweeping. Different types of mines each required their own types of equipment, which in turn made it more difficult to rig platforms of opportunity for sweeping. The variable settings, dormant periods, and the ability to make mixed minefields complicated the sweeping task and greatly increased the number of sweep runs required to clear a minefield. That in turn suggested that greater command and control was needed to plan and track clearance efforts. No longer could a few boats with no prior information sweep a random patch of ocean in a few hours, drop a few buoys, and be strongly certain that their efforts were effective. Their efforts had to be closely tracked to ensure that all runs and techniques that needed to be done were being done. Furthermore, anti-minesweeper settings made it more likely that some minesweepers would suffer mine strikes, making it necessary to track efforts from outside the minefield. Finally, the sheer volume of effort required to clear a minefield, along with the greater risk of losing ships to mines, made it prudent to maintain a large core of minesweeping platforms. For a navy to do otherwise risked having their entire minesweeping force tied down in a single minefield for weeks, even as the size of that force steadily shrunk by attrition, loss of sweep gear, and wear and tear.

Aerial mining was another important technology introduced during the World War II campaign in the Atlantic. Made possible by the fully encased construction and wider coverage of influence mines, aerial mines could be rapidly deployed to locations unreachable by ships and submarines. Flying above the medium being mined, they could drop mines in previously mined waters, in enemy-held harbors, in places beyond enemy-patrolled chokepoints, and even in rivers well inside the enemy's territory. Dropping mines from the same platforms as bombs, there was little to distinguish a mine-carrying airplane from a factory-destroying airplane. Facing such a capability, coastal fortifications with their controlled minefields could prevent neither mines nor bombs from hitting the harbor area they were supposed to protect. Also, the small crew size of each bomber meant that the loss of any one unit did far less to damage the overall mining mission than did the loss of a minelaying ship crewed by a few hundred men. Just as the dispersed risk of bombers replaced battleships for threatening enemy ports, so too did they replace high-volume minelayers for offensive mining. *Note: Minelaying ships are still valuable for their ability to rapidly lay defensive and protective minefields.*

Thanks to the influence and aerial mining introduced in the Atlantic theater of World War II, the task of minesweeping has turned into a science that has forever altered the path of mine warfare.

Chapter 8

World War II in the Pacific

The Japanese Empire in the first year of World War II was a bit of an anomaly in world history. From Tokyo, its empire stretched as far north as the Aleutian Islands in Alaska and as far south as the island chains just north of Australia. To the west, it stretched deep into China, while to the east there was vast nothingness except for a few isolated atolls. With its rapid-fire actions in the first few months of the war, it effectively neutered the navies of the Western powers operating in the region. At Pearl Harbor, it caused massive damage to the U.S. Pacific fleet. This followed with lightning invasions of Thailand, the Philippines, Burma, Borneo, the Dutch East Indies, Singapore, and the Solomon Islands. Within short order, it had pushed its empire's outer perimeter some 3,000 miles.[1] The Western colonial powers would have to choose whether to accept this new order of a Japanese-dominated East Asia or to fight to return Japan to its traditional islands. They chose to fight.

Island Empires as Both Strengths and Weaknesses

At the beginning of the war, the Japanese Empire would have seemed nearly unassailable. Since the Russo-Japanese War, it had consolidated its gains in Korea and Manchuria (northeast China). Simultaneously, it built an enormous defense industrial base that churned out large numbers of some of the best ships and airplanes in the world. Its army dominated the government, while also dominating the emperor who held the loyalty of Japan's citizenry. The culture was a collectivist one based on honor and obedience to the emperor. It was by far the most powerful country in the region. It was the only non-Western great power, and its competitors had split interests that kept the bulk of their forces in other parts of the world. Hence, when Japan decided to launch its invasion, it could do so with numbers that overwhelmed the garrisons of the Western colonial outposts. Once the garrisons were defeated, any serious counteroffensives would have to come from thousands of miles away. The local people of the many invaded lands did not have the means to resist, perhaps in part because colonial rulers tended to discourage the establishment of strong indigenous military forces. Japan was the master of its empire and enjoyed relative freedom of navigation within its waters.

Island empires suffer both blessings and curses. Defensively, they have little to fear so long as they have control of the seas. For individual islands, the surrounding waters are like enormous moats. No land army can cross them without ships, the use of which requires sea control. Were this not the case, China would have long since invaded Taiwan, and England might have fallen to both Napoleon and Hitler. Furthermore, in

World War II, the islands of the Japanese Empire were also unsinkable aircraft carriers. Many of the islands hosted air bases as well as naval ports. If any island was under attack, aircraft and ships from surrounding islands could sweep in to harass the invaders or to replenish the defenders' resources. Should an attacker attempt to strike a base deep in the heart of the island empire, they would find themselves assailed from all sides. The attacker was isolated a thousand miles from their nearest base, while the defender operated from half a dozen bases within fighter aircraft range.

Even if one island is lost to enemy forces already on the island, the damage is isolated to that single island. So long as those forces cannot freely take to the sea, they are stuck on that island. Furthermore, if the surrounding waters are blockaded, their resources are limited to what can be produced on the island. The empire, enjoying sea control, can then choose to either retake the island with a concentration of forces, or starve it and use raids to deplete its defenders and industry through attrition.

Conversely, the situation reverses if the invading force can seize control of the sea and the sky or at least deny the defenders that control. Should that happen, any imperial forces on blockaded islands become neutralized from use anywhere else. They might be able to defend their present island so long as their will and supplies last, but they could not hope to help the defenders of the next island over. Because of these factors, the obvious first step in taking on an island empire is to deny its defenders control of the sea.

In World War II, as now, the Japanese home islands suffered from limited natural resources. With 73 percent of its terrain covered in mountains, only about a quarter of its land area is available for housing, industry, and farming.[2] Naturally, this limits the population that the land can support without bringing in food resources from the sea and abroad. Any rapid growth to industry or housing further reduces the land available for agriculture, which in turn increases the need for outside resources. At the same time, the land possesses little in the way of mineral and energy resources. What deposits that it has are normally small and hardly fit for the needs of heavy industry. The strategy for a resource-poor, highly populated country like Japan, then, is to use its large population to power a huge industrial base using resources from abroad. Those resources can be acquired either through trade (made possible by the products of industry) or through seizing control of territories blessed with the needed resources. Starting with the Sino-Japanese War, the country followed the invasion strategy. Only after the war did it switch to the trade strategy.

As discussed in previous chapters, waterborne shipping is by far the cheapest form of transportation. For a long, skinny, mountainous country like Japan, it made sense to build a robust maritime shipping industry. While mountains made land travel especially difficult, most of the Japanese population was and is only a short distance from the sea. Ships could easily travel between Japanese population centers or between those centers and foreign ports. That led to a sizable merchant marine. Between it and a massive fishing industry used to supply the population with food, the Japanese Empire enjoyed an immense reserve of trained sailors able to man its growing navy.

Because Japan had been trading for resources throughout East Asia, it would have been well-acquainted with the lands from which it acquired its resources. With the familiarity that one gains through frequent exposure, it likely possessed sailors with intimate knowledge of the many ports and water passages crisscrossing the area's many archipelagos. They would also be familiar with the defenses of such places, the location of important points in the supply chain, and what places could prove useful as naval bases. When Japan prepared to seize its island empire, the built-up intelligence provided by its merchant marine doubtlessly helped planners to devise the most efficient use of its military resources.

Once the islands and other new lands were under Japanese control, they exploited them for resources. In many cases, they simply took over the role of "colonial rulers" held by their predecessor. Instead of having to buy resources from the colonial rulers, they could instead take them directly from the people extracting them. Now suddenly they had access to all the resources they could need to feed the industry that was propelling the Japanese war machine. The merchant marine likely continued as it had done before, carrying resources from the same ports back to the home islands. Their country was at war, so there were new risks that required new precautions. However, with the empire enjoying air and sea superiority, the risk was manageable. Convoys could be assembled and escorts assigned to protect those convoys. Aircraft could provide coverage through much of the journey, and everywhere there were ports available to provide shelter. So long as that sea control belonged to the Japanese Empire, the flow of food and resources to the home islands could continue indefinitely.

It was the task of the Allied navies to take back that control.

Overview of the Island-Hopping Campaign

The strategy that the Allies used to wrest sea control away from the Japanese is commonly called the "island-hopping campaign." Its primary tools for denying Japan the use of its island colonies was blockade and/or direct amphibious assault. Weaker points were taken through amphibious assault, while stronger points were blockaded with mines. Japan was an empire heavily dependent on control of the sea and control of the skies, and its airstrips and ports were vital for mutual defense. The plan with the island-hopping campaign was to roll up this island empire from the outer edges, getting ever closer to the home islands. This would achieve several goals: (1) reduce Japanese access to natural resources; (2) reduce safe havens for Japanese ships and shipping; (3) deny Japan use of sea/air bases, lowering the threat to the Allied forces; (4) attrite Japanese shipping, navy, and airpower; and (5) capture launchpads for further Allied assaults.

Rather famously, the Japanese fought tenaciously for every island. Their defenses combined the fanatical loyalty of educated service members with some of the best weapon systems and platforms then available. Those defenses did not radiate solely from the island being attacked. Ships of the landing force could not simply steam outside artillery range and expect safety. Quite a few islands housed squadrons of aircraft and various sizes of warships, while still more could offer at least refuge and resupply for such ships. Consequently, attacks on Japanese-controlled islands had to be done in overwhelming force and with superior airpower.

The landings themselves were also no picnic. The best landing beaches were generally fortified with well-protected artillery and machine-gun nests. Defenders had the advantage of prepared beaches, tunnel networks, and concrete pillboxes, i.e., camouflaged guard posts impervious to small-arms fire and grenades. Surf zones were littered with obstacles that could tear the bottom out of approaching landing craft, drowning heavily laden marines. Artillery was sighted and measured ahead of time to ensure that a wave of destruction would rain down on these same approach lanes. Land mines awaited them ashore. Prior to an attack, the Allies would flatten the area with gunfire and bombs, only to find these pillboxes manned and ready for action. Often the only way to take them out of action was to sneak up on them and insert grenades directly into the loopholes through which guns were fired. The tunnel networks included entire barracks blocks and command posts, as well as multiple egress points. This allowed an area to be devoid of soldiers one minute and flooded with them the next. These soldiers, though often physically smaller, fought with the vigor that comes from the conviction that the only acceptable outcomes were victory or death. Where a Western soldier would surrender if they no longer had the means to fight, the Japanese ethos met surrender with scorn. Thus, victory over a Japanese-held island came only after most of the defenders were dead.

For the Allies, the strategy was to peel back the overlapping Japanese defenses one island at a time. Starting in the southeasternmost islands, far from the homeland, they captured or neutralized each island chain in turn. With the United States leading the overall campaign, the effort was divided between two fleets. The Seventh Fleet, under the operational control of Army General Douglas MacArthur, concentrated on the chains of large islands that extended from the Solomon Islands to the Philippines. With many of these islands being the size of European countries, they featured extensive ground fighting. The Third Fleet, under the operational control of Navy Admiral Chester Nimitz, attacked the chains of tiny, remote islands that made up the mid-ocean defenses threatening supply lines. Since many islands could host squadrons of aircraft, they each represented an interlocking threat against any thrust directed at the heart of the Japanese Southeast Asian empire. By attacking the island chains, the Allies could tie up forces that might otherwise be used for further expansion and exhaust the experienced military resources at the end of a long supply chain that was under continual attack. With each island captured, Japan lost a source of supply while the Allies gained a launchpad for further attacks.

Invasions of island chains followed a basic template. First, a massive fleet would arrive in the area, with ideally enough overwhelming force to defeat any possible assault by ships or aircraft. Carrier aircraft attacked the various airfields in the chain to limit the enemy's capacity for counterstrikes. Next, ships' guns and aircraft softened up the targeted island's defenses, striving to silence all sources of counterfire. For their part, Japanese aircraft and ships conducted frequent raids in hopes of sinking the fleet. While the better-armed warships and aircraft fought their fights, minesweepers swept for mines. Working under cover of darkness, underwater demolition team divers surreptitiously destroyed obstacles, conducted reconnaissance, and mapped out lanes for landing craft. When it appeared that the residual risk to the landing force was acceptable, landing craft would be launched under cover of more gunfire and close-support aircraft. Once ashore (always under fire), the marines began the dangerous process of digging out the enemy. When a beachhead was established, combat-loaded supplies[3] were sent ashore and Seabees began the process of building airstrips, field hospitals, etc.

The difference in distances involved between the Atlantic theater and the Indian and Pacific theaters had serious impacts on the ability of the Allies to manage and maintain their minesweeping fleets. While the sweepers of the Atlantic theater often operated from their home port or as part of large invasion forces, Pacific sweepers worked more remotely. They did their work in contested waters among far-flung islands often

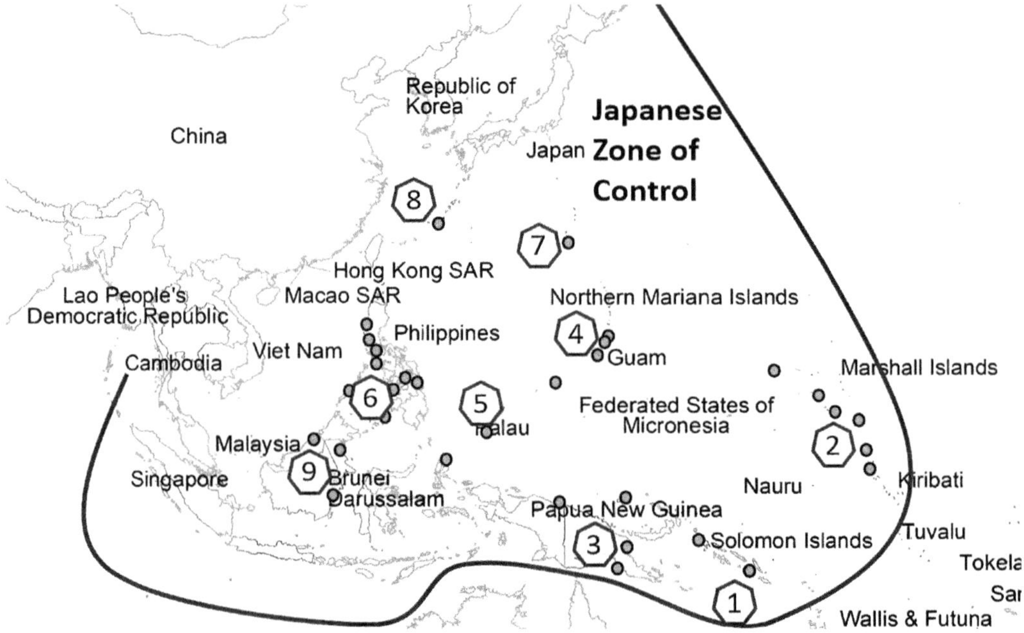

Locations and order of Allied amphibious landings in island-hopping campaign (from P. Elliott, *Allied Minesweeping in World War II*, Patrick Stephens, Ltd., 1979, 154–155, map created by author to illustrate Elliott's table).

thousands of miles from any permanent bases of operation. Sweeping proceeded as Allied troops fought entrenched Japanese troops ashore only a few miles away. Once cleared, the fleet moved to its next battleground hundreds of miles closer to Japan, with minimal facilities remaining at previous fleet anchorages. The nature of such fighting meant that new bases of operation were needed frequently, lest ships in need of supply or repairs be forced to travel thousands of miles to established bases. With major, permanent facilities making little sense, the Allied fleets adopted what may best be described as "floating naval bases" capable of providing most of the services of land naval bases. To that end, both the UK and the USA built "headquarters" ships (or "command" ships) and minesweeper tenders. At times built by simply modifying existing mine countermeasures ships, these were fitted for the specific needs of their respective missions.[4]

Headquarters ships are used to house the minesweeping commodore and his/her staff. The commodore declares their intent; the staff turns that into different courses of action (COAs); the commodore chooses a COA; and then the staff turns that COA into detailed plans and orders. Upon receipt of those orders, the ships execute the orders to the best of their abilities while dealing with all the exigencies that come from operating at sea and in a theater of war. For the commodore and staff to be able to do their work, they needed spaces to plan and to track the mission progress, and they needed robust communications equipment. They also needed additional berthing to accommodate the extra people. Therefore, a headquarters ship made from an available asset, which was often an MCM vessel, had to be modified for those needs. Tiny minesweepers would be stripped of at least some of their minesweeping equipment and personnel to make room for the extra equipment and staff. Likely they would participate little in sweeping efforts,[5] as doing so would both interfere with the staff's ability to track operations and put them at risk. Though deck-plate sailors will often see little use for a commodore's staff, that staff is vital for tracking and adjusting the overall effort and for keeping higher headquarters informed. To lose them would force another ship out of the fight to assume those duties in earnest, without the benefit of the records lost with the previous headquarters.[6]

Minesweeping tenders take on the role of floating intermediate maintenance activities. They carry spare parts beyond what sweepers can carry, as well as precious repair specialists with more skills than required by individual sweeper crews. A minesweeper tied to a tender could have broken engines, generators, and cables repaired or replaced. They could expect mail, and maybe some fresh stores. Tenders also played the important role of offering a place to relax away from the far-too-familiar settings of a ship that crew members may have not left for weeks or months on end. Furthermore, they could act as staging points for new crew members and for crew members departing due to illness, injury, or simple transfer. To perform these roles, the UK repurposed some already-defunct mine destructor ships (themselves converted coastal colliers).[7] The United States converted the cruiser/minelayer USS *Terror* for the purpose, as well as used it as the minesweeping flagship. Both U.S. and UK versions were fitted out with derricks, repair spaces, foundries, large sick bays, and crew rest spaces.[8] One advantage of the use of a cruiser/minelayer would have been its increased firepower. The crew of any minesweeper tied to it would have been able to do so under the belief that the *Terror* could protect their ship far better than they could do so themselves.

Allied Mining in the Island-Hopping Campaign—Mining played an important part in the island-hopping campaign. Its main function was to deny Japanese forces

MCM command ship USS *Terror* (*CM-5*), 1942 (National Archives photo).

full freedom of the sea. At the beginning, when Japan was at its strongest and the Allies at their weakest, mines were barriers to prevent direct assault on Allied forces. Farther afield, mines could disrupt reinforcement and resupply routes, thereby preventing the enemy from gaining strength at critical times. Mines sub-laid or air-laid in Japanese-held ports could temporarily stop the flow of men, ships, and materials at that port. This in turn could have downstream effects on frontline reinforcements and on war matériel production in the Japanese homeland. If a Japanese navy base was especially strong, possessing a sizable fleet, a heavy infusion of mines could bottle up the fleet in port while the Allies hit easier targets. These were all goals of the Outer Zone Mining Campaign.

It took time for the Allies to ramp up for war in the western Pacific. Japan, for its part, had fought in the region almost continuously for decades. It had expanded its territory to include all of Korea, Taiwan, and the most populous parts of China. Through participation in World War I, she had gained former German island colonies in the South Pacific. She had even fought significant battles with the Soviets. When World War II raged in Europe, diverting the attention and resources of the European powers, Japan took the opportunity to seize coveted possessions in Southeast Asia. In 1940, she took control and occupied the northern portion of French Indochina after France surrendered to Germany. The following year, Japan claimed joint protectorate status over the whole colony. This directly led to the United States sanctioning Japan to include an oil embargo. When diplomatic negotiations broke down, Japan conducted its surprise attack on Pearl Harbor. Almost immediately thereafter, the country used its regional military superiority to seize possession of all Western colonial possessions in the region. The colonial governments, being manned and armed only for limited engagements, had no hope of opposing this well-armed, militant, veteran force. They were swept away, giving Japan control of this massive territory.[9]

The Allies were not prepared to fight this sudden, full-scale war against Japan that spanned thousands of miles of islands. They had the technology and the nucleus of equipment, but everything was located thousands of miles away. What preparations they had made had assumed limited, sequential incursions by the Japanese, rather than strikes in all places at once. But, all at once, the huge fleet at Pearl Harbor was crippled and the colonial forces at Singapore, Hong Kong, the Philippines, Guam, Wake Island, and the Dutch and British East Indies all fell to overwhelming forces. Where only a few months prior they had enjoyed control of a patchwork of islands throughout the South Pacific, now the Allies were left with the fringes.

The fight back began with submarines within hours of the Pearl Harbor attack. Submarines could sneak into the strongly defended Japanese territory without being seen and wreak havoc on shipping. In 1942, the first full year of the war, these submarines were the primary method used for Allied mining. A boat could go out armed with about 18 mines along with torpedoes. If it failed to find shipping to attack, it could still contribute to the war effort by laying a small minefield. Such missions accounted for 36 sorties to lay 658 mines in 33 minefields stretching from Indochina to the Japanese home islands.[10]

By April 1942, the three American aircraft carriers in the Pacific began to fight back. Outnumbered at first, they achieved success first with the Doolittle Raid, when B-25B bombers launched from the USS *Hornet* raided Tokyo. A few weeks later, two aircraft carriers achieved a strategic victory at the Battle of the Coral Sea by stopping the Japanese advance on Port Moresby, Papua New Guinea. A month after that, in early June, three American carriers near Midway Island surprised a Japanese fleet that included four carriers, sinking all four. From that point onward, the Allies had the naval dominance to launch their counteroffensive, beginning in the Solomon Islands in August.

Allied surface minelaying began on February 1, 1943, starting with a field off Guadalcanal. Placed at high speed to avoid the expected arrival of the nighttime raiders known as the Tokyo Express, the minelayers of MineDiv1 laid their field in 14 minutes, then retired at high speed just as a battle broke out behind them. They had their first victim that very night, the destroyer *Makigumo*. Eight days later, the six-month-long battle for Guadalcanal was over, the first successful campaign of the counteroffensive.[11] Over the next 15 months, surface minelayers would lay a total of 2,819 mines in 17 fields throughout the region.[12]

Both navy- and land-based bombers were involved in aerial mining. Carrier-based aircraft had neither the capacity nor the endurance to lay large minefields deep in enemy territory. Instead, they laid tactical minefields in direct support of naval operations. In one example, naval aircraft dropped mines in the anchorages of four islands in a chain. Denied the use of the refuges, Japanese shipping mostly disappeared from the area. In a twist, dummy mines were laid at Kwajalein, thereby tricking the Japanese into ruling that port out as a target needing to be defended. The ruse made for an easier landing than normal. Later, when submarines had driven shipping into hugging the coast, naval aircraft dropped mines among the islands of Korea's southern coast. With those myriad paths closed to them, cargo ships were forced out into the open water where submarines could reach them.[13] When one considers that submarines sank 54 percent of all Japanese shipping during the war,[14] it's clear that minefields helped submarines find victims by forcing ships to stay at sea longer or to take less-sheltered routes.

Long-range, land-based aircraft flew deep into Japanese territory to mine along the mainland and the large islands of present-day Indonesia, Malaysia, and Papua New Guinea. Flying from China, India, Sri Lanka, and Australia at first, and later from the Philippines and the Marianas, they could deliver mines in volume at points unreachable by naval assets. The results were like those of other means of mining, i.e., they accomplished the anti-access/area denial (A2/AD) mission to interrupt enemy supply chains and reinforcement efforts.[15]

Minesweepers in the Island-Hopping Campaign—Minesweepers played an important role during island invasions by removing the underwater menaces that may have otherwise prevented other craft from getting close to their target. So long as they remained active, they would eventually clear paths for their larger cousins. Without them, all else would grind to a halt until replacements arrived. This made minesweepers targets for defenders. Combined with their limited defensive abilities compared with larger craft, it would have made them tempting targets for Japanese pilots. As such, they had as much to fear from kamikazes (suicide) planes as their larger cousins. However, even a direct hit was not always fatal. For example, the four-stacker (converted World War I destroyer) minesweeper USS *Dorsey* was hit by a kamikaze during the Okinawa campaign. She was struck by a plane whose bomb she had shot away, though the collision was plumb in her icebox. Doubtlessly bereft of fresh stores, she continued supporting the minesweeping efforts by acting as a mine disposal vessel. She further acted as an observation vessel for the minesweeping admiral embarking her from his flagship, USS *Terror*.[16]

The USS *Aaron Ward* (*DM-34*) survived an even bigger attack. While on radar picket duty off Okinawa at dusk on May 3, 1945, she encountered a formation of Japanese aircraft bound for the fleet. Soon she found herself the target of one kamikaze pilot after another. After succeeding in knocking down several attackers, her luck eventually ran out. Six kamikazes struck her in rapid succession, killing several crew members, including the first four U.S. Navy minemen casualties.[17] The *Aaron Ward* survived the war,

Kamikaze damage to destroyer-minelayer USS *Aaron Ward*. Mine visible on left (U.S. Navy photo, Naval History and Heritage Command).

and its crew was awarded the Presidential Unit Citation.[18] The ship and its story were the subject of the classic book on resilience in war, *Brave Ship, Brave Men*.[19]

Other minesweepers were not so lucky. On July 21, 1945, while clearing mines off Phuket Island, Thailand, a British mine clearance force found itself under attack by three kamikazes. The low-flying planes were thought to be coming in on a strafing run. Then one hurtled itself directly into the midships of RMS *Vestal*. Only 78 meters long, the *Vestal* did not fare well. A large crater, ablaze in flames, was gouged out under the quarterdeck, killing 20 crew members, destroying the radar, cracking the deck in two, and setting the ship on fire. Communications between the fore and aft sections of the ship were severed, so messages were passed by swimmer. With the ship beyond repair and lacking all ability to contribute to the mission, the commanding officer ordered the ship to be abandoned. She was scuttled by gunfire, receiving 30-plus shells before accepting her new home in Davy Jones's embrace. The crew was taken to the battleship HMS *Nelson* to be mothered back to health along with the crew of HMS *Squirrel*, another minesweeper that was sunk by a mine the day before.[20] Little did anyone involved realize that the war would abruptly end just weeks later.

Unluckiest of all were the minesweepers tasked with clearing for the planned amphibious landing at what is today Balikpapan, Indonesia, in June 1945. It was here that, for the first time, the American minesweepers met with American and Allied influence mines. Up to that point, their experience had been with the relatively easy-to-sweep Japanese moored contact mines. It was at Balikpapan that they discovered that they had neither the training nor the equipment to clear all that they had laid. With influence mines having been set in ever-more-sensitive combinations, they soon discovered that their own ships were alone sufficient to set off these mines. Their force largely used wooden-hull YMSs, whose engineering plants gave off enough of a magnetic signature to detonate the more sensitive mines. In the ensuing fiasco, these mines sank seven minesweepers.[21]

Note: If there is only one lesson to be learned from this entire book, it is to make sure that no new mines are fielded without first verifying that a technique exists to clear them at a safe distance. The failure to do so guarantees death on your own side. If a mine is made undetectable by sonar, it should have some other system for ensuring that it can be found or neutralized. Examples might include a known-good sweep setting that will detonate them, a function that will turn them off permanently after a delay, a reliable self-sterilizing function, an acoustic transponder that actuates after a set date, etc. If no such techniques are available, then the old adage remains: "Just because you can, doesn't mean that you should!"

Japan's Use of Mines—Japan did not come into World War II with the same emphasis on mine warfare as did the West. To an offensively minded military, mines were a defensive weapon unnecessary to a maritime empire enjoying full sea control. While she had large, respectable mines, they were little changed in basic form from what was available during the Russo-Japanese War. The two most common types were the Type 93 moored contact mine (deployable to 1,000 meters) and the Type 92 moored RECO mine.[22] She had other models of moored contact mines, but all were of the basic Hertz horn type seen in the previous world war.[23] Japan used them primarily to build anti-submarine minefields ranging from the La Perouse Strait at the northern tip of Hokkaido down to the Dutch East Indies (modern Indonesia). These fields were laid deep enough for surface vessels to pass over them, and in multiple layers extending down to below submarine crush depth. On the surface, patrolling navy vessels deterred submarines from passing over the barrier. The purpose was to limit the ability of submarines to access the vital mainland shipping lanes bringing raw materials to Japan.[24]

Although the majority of Japan's mining was defensive in nature, she also laid some offensive fields near Allied strongholds early in the war. Affected ports ranged from Darwin, Australia, up to Kiska, Alaska. During the fights for Singapore and the Philippines, her mines helped to shape the battlespace and limit the options of the defenders.[25] Except for the minefield at Kiska, Japan made no attempt to mine the U.S. coast. Oddly enough, she still managed to do so due to the vagaries of the wind and currents. Beginning in 1947 and for a few years afterward, Japanese mines (mostly Type 93) in their hundreds kept appearing off the U.S. West Coast. These became the problem of local U.S. Coast Guard, Army, or Navy units to deal with as they were found.[26] It is both a good thing that the requisite skill set was in plentiful supply and sad that it took a traumatic event like World War II to bring about those skills.

For their part, Allied submarines enjoyed a few of their own advantages. These included the capture of Notice to Mariners charts marking the location of these minefields, as well as the earliest form of mine-detecting FM sonar. Though the rudimentary sonar only returned the echoed sound of a sonar beam bouncing off an underwater object, it proved useful for painfully picking one's way through a minefield.[27]

Japan did not make extensive use of minefields in its surface defenses until late in the war. Though Allied minesweepers swept for mines ahead of every assault, they usually found few mines, if any. For the first two years of amphibious landings, no naval mines were found in the landing areas. It was not until the assault on the Marianas Islands that the first major minefields were discovered. Peleliu was where the first mines were encountered, with nearly 100 mines to bother the 45-ship MCM force. Ulithi Atoll, which would become a major Allied forward operating base, was protected by over 300 Japanese mines. This island chain was perhaps the most critical in the fight against Japan. It became the staging point for the invasion of the Philippines and the heavy bombing of the Japanese home islands. Within the Philippines themselves, the number of mines varied, with some places guarded by hundreds of mines and other places having no mines. No mines were found at Iwo Jima, while at Okinawa, kamikaze aircraft offered more danger than the 510 moored mines found in the surrounding 2,500 square nautical miles of water. In all places, only moored contact mines and controlled mines made significant appearances[28]

Why did mines not play a greater part in Japanese island defenses during the war? Obviously, they had them and had them since at least the Russo-Japanese War. Most likely it came from the complicated calculations of a maritime empire dominated by its army. An army-centric leadership will generally think of the navy as having a mere supporting role while its main assault arm was aircraft. Mines would have been thought of as having little consequence. After all, Japan had captured half of the Pacific without needing to concern itself with mine clearance except for perhaps an Allied base or two in the Philippines and Indochina. They attacked suddenly and at many different points. Likely, from their perspective, they had no real peer competitor in the region the way that Germany and the United Kingdom had each other. They prepared their defenses to repel the only type of attackers who had been successful in the region, which is to say, themselves. Mines, if they were to be considered at all, were likely seen as a greater danger to their own supply lines than to those of their enemies. The early Allied attacks in the Solomons, Marshalls, and Papua New Guinea were annoying but seemingly unimportant. It was not until the Allies had several successes under their belts that Japan would have seen them as a real threat that had made amphibious landings into a science. Even then, mines might have been used only if specifically requested by individual base commanders.

An alternative explanation, of course, is that mine production was limited due to competition with ships, aircraft, and ground weapons. With such a huge empire, the few available mines were probably prioritized for use at the more important bases and for protecting the core of the empire. After all, mines were used to create a series of huge barrier fields to deter submarines from attacking the inner supply lines, though these mines were initially held in reserve for defense of the home islands.[29] If it was a simple case of maximizing use of limited assets, then the outer islands like those of the Marshalls were treated as expendable.

Operation Starvation

Most naval mines used in war are for tactical or operational reasons. They are used to block ships from leaving a contested port or to prevent the resupply and reinforcement of that port, as was the case at Port Arthur during the Russo-Japanese War. Mines are often used to help shape the battlespace, to put rails up to limit the movement of enemy combatants. Only a very few mining campaigns have ever been strategic campaigns designed specifically to destroy the enemy's will to continue the fight. Arguably the German mining blockade of British waters during World War I was the first attempt at strategic mining, though it failed for a variety of reasons. Operation Starvation, on the other hand, was the first successful strategic mining campaign in history. It proved wildly effective at achieving its intended aims, the culmination of which was cut short only due to the first atomic bomb attacks on Hiroshima and Nagasaki.

Throughout the war, the Allies had been steadily rolling back the extensive Japanese Empire. But it was an island campaign that primarily affected newly acquired island colonies. Japan still retained control of its continental possessions in China, Korea, and Indochina. Though the Allies had destroyed most of Japan's sea power and airpower, it still had the industrial capacity to replace those assets.[30] Materials and supplies could still flow in from those continental territories. Although plagued by submarines in open water, they could still route their shipping along coastal routes before making the final dash across the Tsushima Strait. From there, shipping could either spread east and west to the many ports along the Sea of Japan, or it could transit the Shimonoseki Straits between the islands of Kyushu and Honshu. Once through, ships enjoyed relatively safe passage through the sheltered Inland Sea to Japan's industrial heartland.

Operation Starvation was the mining portion of a much larger operation by the U.S. Army Air Force's 21st Bomber Command. It represented only 5.7 percent of the larger operation's total effort, which otherwise concentrated on the destruction of Japan's industrial base. The object of the mining effort was to deny Japan the use of its waterborne transportation routes. Such a denial would mean very little to a country like Germany, where a robust road network connected everything and where the ocean was merely peripheral. But for a long, thin, mountainous, resource-poor country like Japan, the ocean was the primary transportation network. Most of her oil and iron, as well as a significant portion of her coal and food, arrived via waterborne imports and three-quarters of her intra-island transport was waterborne. Mining the ports would not only bottle warships up in port, but it would severely hamper the internal movement of troops and matériel as ground-based transportation networks were pushed to capacity.[31]

The Five Phases of Operation Starvation

The mining campaign took place in five phases beginning in April 1945. Those phases are graphically depicted at the end of the chapter and are summarized as follows:

Phase 1: Mining intended to prevent a Japanese counteroffensive against the Allied expeditionary forces invading Okinawa. In this phase, aircraft mined the major southern Japan naval bases at Sasebo and Kure to bottle up fleet units that were in port. They also mined the internal transportation route of Suo Nada and Shimonoseki Straits to prevent the masked movement of any naval ships not in port. With this route mined, any remaining ships desiring a fast transit south would be forced to an open-ocean transit where they risked discovery by the patrolling Allied submarines and aircraft. Of note, this was the only phase that included Sasebo.[32]

Phase 2: Mining intended to blockade the seaborne flow of goods and materials to and between the major Japanese east coast industrial cities. This was the only phase that included Tokyo Bay or Nagoya. While this author does not have further details, it is possible that this mining, combined with follow-on saturation bombing of these cities, was sufficient to deny Japan its use of major industrial ports. It may also be that later mining was avoided to give Japanese forces an opportunity to clear the port in time for the expected surrender a few months later.

Phase 3: Now that the major east coast cities were blockaded, attention shifted to mining the west coast cities that offered the safest alternative harbor for ships unable to transit the straits. This would prevent Japan from being able to set up efficient land/sea hybrid routes. To underline their importance, the Shimonoseki Straits continued to be mined through all phases.

Phase 4: With most of the major ports blocked (except for Aamori in the far north), the miners expanded their focus to include more intermediate-sized ports as well. They also re-seeded the fields around Osaka, and of course, Shimonoseki Straits.

Phase 5: By this point, the industrial capacity of Japan's home islands was severely damaged. The miners further expanded their efforts to include the major industrial ports of Korea, preventing the shifting of production overseas. They also hit even smaller ports in Japan and continued reseeding previous fields. It was during this final phase that atomic bombs were dropped on Hiroshima and Nagasaki on August 6 and 9, respectively. Even after the atomic attacks, mining continued. Just hours before Japan formally surrendered on August 14, 1945, the allies laid yet more mines at three ports and in Shimonoseki Straits. This author notes that no atomic bombs were dropped in Japan during those final hours leading up to its decision to surrender … but mines certainly were.[33]

Japan's Response to Operation Starvation

Because Japanese waters had been mostly ignored by miners except for occasional small-scale submarine mining, the Japanese home islands did not have a robust MCM force as the operation began. However, once it became apparent that mining would be ongoing, she proved quite adept at developing effective techniques to maintain some level of access to shipping lanes. She swept for magnetic and acoustic mines and developed tactics to deal with pressure mines. In places where mines were suspected but could not be swept, divers recovered or disposed of mines. Upon inspecting mines, experts could devise the best means to counter them (American mine modification units frequently changed settings to make it difficult to use only one MCM approach).[34] She reorganized her forces so that instead of spreading her sweepers across all ports, she concentrated them to allow the rapid clearing of only certain ports. For example, Miyazu, Maizuru, Ohama, and Tsuruga all share a wide bay, and all were mined. Instead of clearing all routes, local MCM forces focused on the route to Maizuru. Fushiki, with its deep natural harbor, proved especially easy to clear. Despite being mined seven times between May 24 and July 25, it remained in near continuous use. Its narrow mineable harbor area edged by a rapidly deepening continental slope was both hard for mining aircraft to hit and easy for MCM forces to clear. As such, Fushiki never stayed closed for long. Overall, through the use of intelligent, nationwide action against the ongoing mining assault, Japan was able to maintain some level of continuous access to at least some ports.[35]

In addition to dropping mines, mining aircraft also dropped leaflets into the population centers describing the mining and the prediction that the mines would lead to starvation. A form of psychological warfare, the goal was to ensure that the public knew that the mining was taking place. The leaflets also blamed the mining on the government's insistence on continuing to ship in the raw materials of war. It was hoped that the message would cause shipping to stay in port in a sort of self-initiated blockade.[36]

Results

What were the results of Operation Starvation? Japan's smart and heroic efforts prevented a complete blockade, yet it could not prevent its shipping from coming to harm. The plethora of mines yielded much attrition in places where ships chose to sail despite the blockade attempts. The effect was to reduce Japan's afloat shipping by almost two-thirds from 1.8 million tons to just 0.66 million tons.[37] The flow of ships in and out of the ports of Kyushu and most of Honshu south of Aamori slowed to a trickle. Shipments were forced to be offloaded at locations not optimized for the rapid movement of goods to and from industrial centers. As such, transportation and storage infrastructure became clogged in these areas, which in turn would have created shortages everywhere.[38]

The significance of Operation Starvation is hard to state with absolute certainty, though the opinions of some Japanese officials give clues. Prince Konoye suggested that the mining was just as effective as the bombing of industry that made up the other 94 percent of the 21st Bombing Command's efforts.[39] One steel industry leader reported, "It was not only the bombing of factories that defeated us, it was the blockade which deprived us of essential raw materials."[40] A Japanese mine warfare expert was even more specific: "The B-29 mining was so effective against the shipping that it eventually starved the country. I think you probably could have shortened the war by beginning earlier."[41] Whatever the case may be, the atomic bomb attacks did happen, and the war ended shortly thereafter. Unknown is if the Japanese surrender was because of the atomic bomb attacks and Russian declaration of war alone. Would it have done so had the mining and industry bombing not devastated the country first? Equally unknown is if Japan would have soon surrendered due to the mining and bombing even without the atomic bomb attacks.

Lessons Learned from the Mining Campaign

The Naval Ordnance Laboratory in its report *Mines Against Japan*[42] provided several lessons yielded from the mining campaign. That report is enlightening and, at times, entertaining and well worth the read. Those lessons learned are paraphrased below.

1. It is possible to blockade an entire fleet with aerial mining. (*Note: It is not guaranteed. Germany tried it against the United Kingdom but failed to achieve a complete blockade.)*
2. The *byproduct* of a heavy mining blockade is heavy attrition. The *goal* of the blockade is to stop enemy ships from sailing in the first place.
3. The heavy mining must be everywhere, or the enemy will shift to alternate ports.
4. Mining must be everywhere, every night, and the shipping channel needs to be hit. (*Note: This makes sense. Minefields that sink no ships are not effective threats.)*
5. Area mining is enough to blockade. Precision mining isn't needed.
6. Long-term blockades require unsweepable mines.
7. If the mineable area is small, clearance will be quick even for hard-to-sweep mines.
8. If the mineable area is large, clearance will be long and difficult if mining is heavy.
9. The best way to clear a mine is to stop it from going in the water in the first place.

Postwar Cleanup

On August 28, 1945, a Japanese minesweeper led the American battle force into Tokyo Bay. Behind her sailed "23 carriers, 12 battleships, 20 cruisers, 116 destroyers and escorts, 12 submarines, 157 other ships, and 28 mine-craft."[43] For most of these ships, the war was over. The hazards of combat were behind them. They were going home. Not so for the 28 minecraft and their almost 2,000 sister ships of all affected nations spread across the globe. They still had the war's 698,657 mines with which they had to contend.

Like any war where mines are deployed, mine clearance operations began before war's end and extended long after hostilities had concluded. One British mine-sweep sailor of the time recalled starting clearance operations in Malaysia and Indonesia before shifting to Greece, then Norway (where they were helped by German mining charts), and then finally to Scotland.[44] Such will always be the life of wartime MCM sailors. What made World War II special over previous wars was the new influence mines, some of which had no corresponding clearance technique. Clearing a minefield has always been hazardous, but when there is no official means to clear them, often it is the minesweepers themselves rather than their sweeps that cause the mines to explode.

In Japan and Korea, the Japanese minesweepers were made to take the greatest risk. The first sweeping, and much of the last sweeping, was done by Japanese crews. Perhaps the theory was that the vanquished should take the greatest risk, when the minefield was fullest. And it was their people who would be most affected if anything was missed at the end. They were not the only ones to risk greatly. While they labored interminably in the American-laid minefields, the U.S. minesweepers labored to clear the great Japanese-laid barrier minefield between Taiwan and Japan.[45] Others worked in places as far-flung as Singapore, the Hainan Channel off southern China, the Kattegat, the British Isles, the Mediterranean the Panama Canal approaches, and the rest of the dozens upon dozens of places targeted by mines. All told, it took 1,900 minesweepers two years to clear over 125,000 square nautical miles of water. Even with that effort, not all mines were found. Many moored mines broke loose from their moorings to drift along with the currents for the next decade, with some appearing for years along a beach or in a fisherman's net. Influence mines not actuated by sweeping would simply remain until they self-sterilized, ran out of battery power, or became useless due to sea growth or burial.[46]

Mine clearance operations in Japanese home waters following Operation Starvation would continue for years after the war. American MCM forces helped the 350 small vessels of the Japanese MCM force in sweeping 22,000 square nautical miles to clear 12,000 mines. When they turned over the effort to the Japanese fleet eight months later, tons of work remained. Japan would continue clearance, yet after 26 years, it was estimated that at least 2,000 mines remained.[47] This gave them both an expertise in mine warfare and a deep-seated understanding of the need to maintain a professional MCM force. Fully aware of the impact that mining can have on their economy, they retain robust mine countermeasures capability and a highly professional MCM training program.

Over the course of the war in the Pacific, as well as the Korean War a few years later, the belligerents planted naval mines in more than 200 locations. As late as 1967, the U.S. Naval Oceanographic Office listed around 235 danger areas, of which mines made up the lion's share of the residual threat.[48]

An important lesson was learned from this complex cleanup of mixed minefields and variably set mines. This lesson was that a new technology was needed that bypassed the need to sweep the same area dozens upon dozens of times. Mine-detecting sonar, having already been introduced on submarines, appeared to be the technology of the future.[49] Eventually, countries of the world began developing ever more sensitive and specialized mine-hunting sonar and remote-operated vehicles (ROVs). These led through successive generations, many of which the author worked with in his own career: AN/SQQ-30 and AN/SQQ 32 v1, v2, v3, and v4, as well as the AN/SLQ-48. The newest generation includes side-scan sonars run from helicopters and drones—all because of the problems that came from clearing hundreds of thousands of mines from World War II.

Supplement: Phases of Operation Starvation[50]

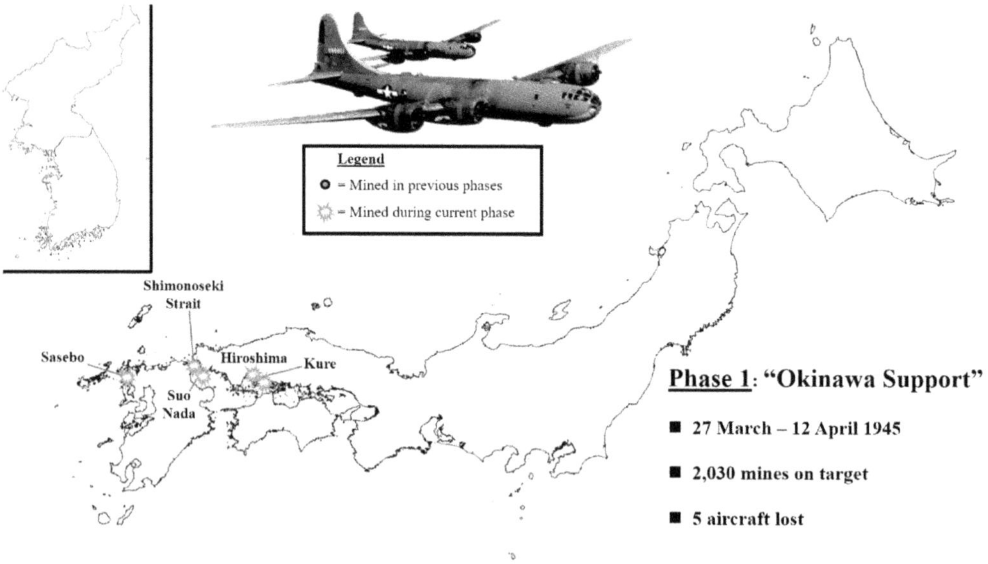

Operation Starvation Phase 1. Results: Traffic in Shimonoseki Strait reduced to 25 percent of normal, with about 35 ships sunk (created by the author using USAAF data from *Starvation: Phase Analysis of the Twentieth Air Force Strategic Mining Blockade of the Japanese Empire*, U.S. Army Air Forces, 1945, 1-35).

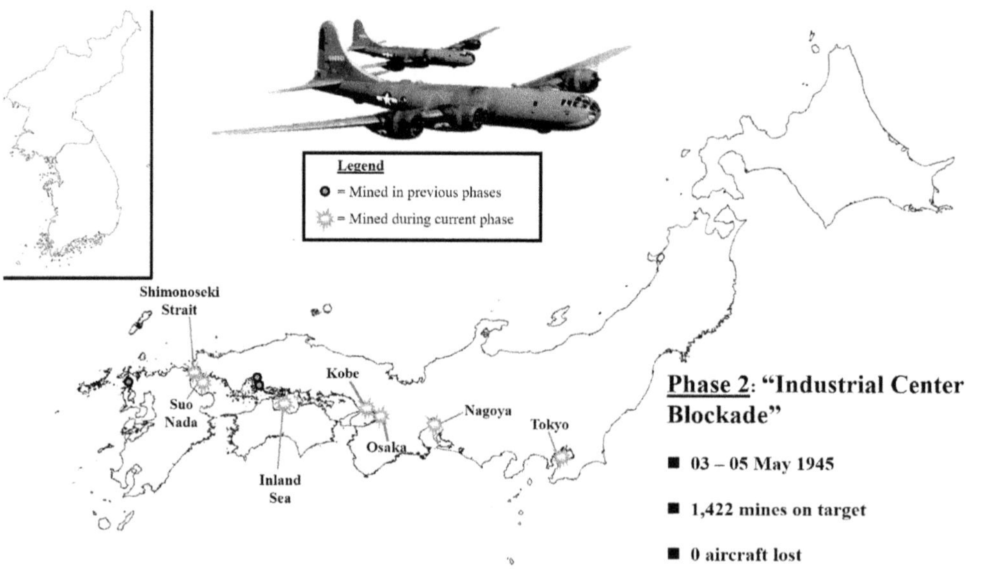

Operation Starvation Phase 2. Results: Strait traffic only two to four ships. Only sailboats and wooden boats used in Inland Sea (created by the author using USAAF data from *Starvation: Phase Analysis of the Twentieth Air Force Strategic Mining Blockade of the Japanese Empire*, U.S. Army Air Forces, 1945, 1-35).

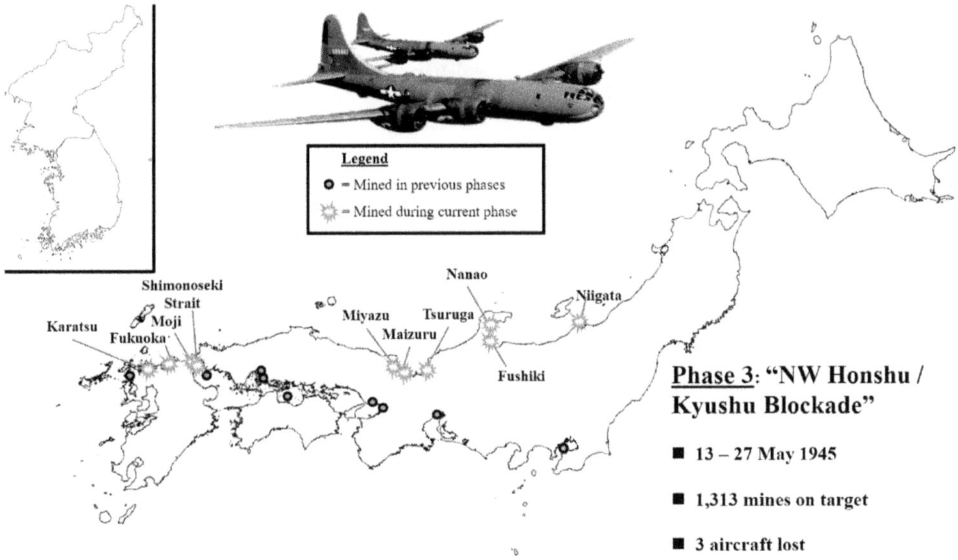

Operation Starvation Phase 3. Results: Strait traffic 1/10th of normal. Three hundred thousand tons of shipping sunk. Port traffic somewhat reduced (created by the author using USAAF data from *Starvation: Phase Analysis of the Twentieth Air Force Strategic Mining Blockade of the Japanese Empire*, U.S. Army Air Forces, 1945, 1-35).

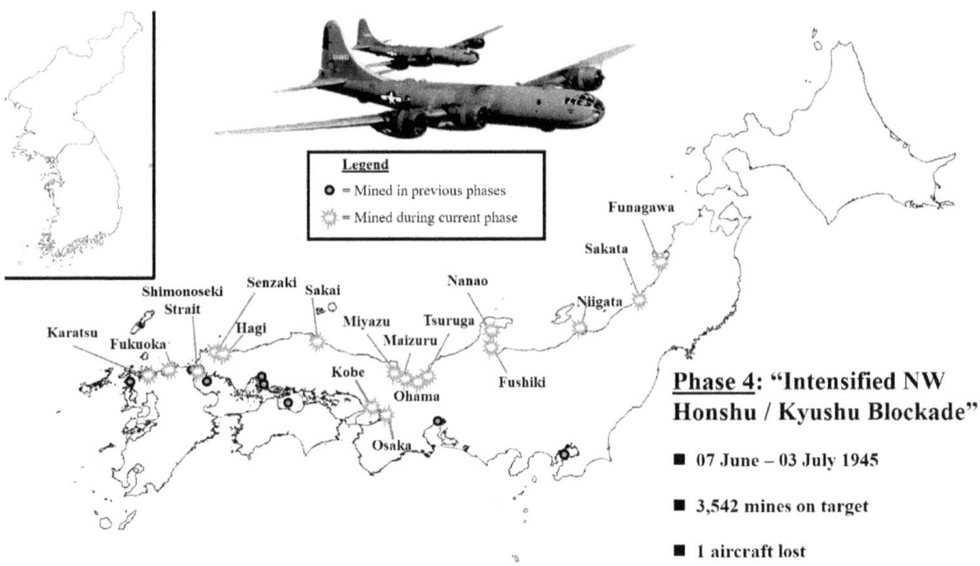

Operation Starvation Phase 4. Results: Strait traffic 1/10th of normal. Three hundred thousand tons of shipping sunk. Port traffic drastically reduced (created by the author using USAAF data from *Starvation: Phase Analysis of the Twentieth Air Force Strategic Mining Blockade of the Japanese Empire*, U.S. Army Air Forces, 1945, 1-35).

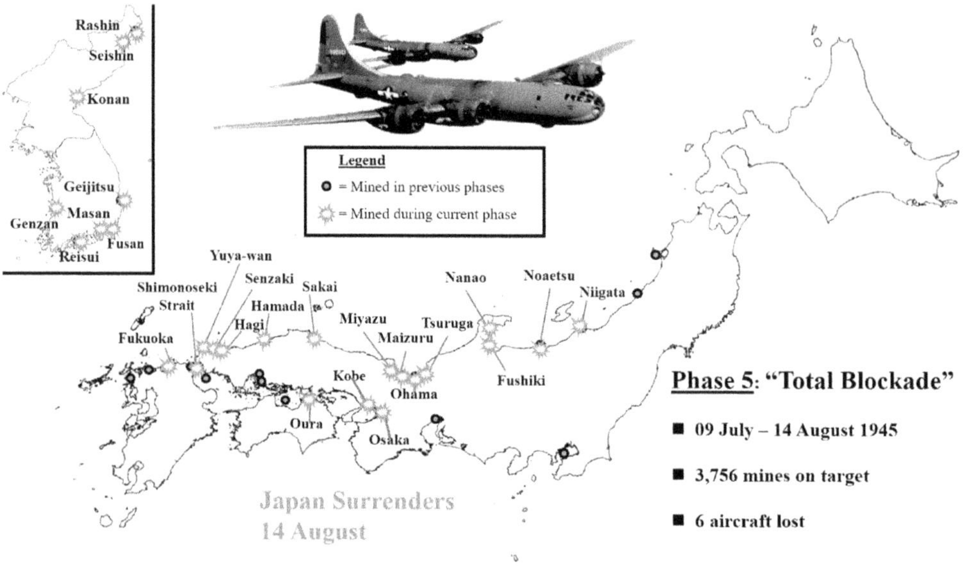

Operation Starvation Phase 5. Results: Total Japanese shipping reduced from two million gross tons to less than 400,000 gross tons. All raw material shipments ceased. Food shipments are a fraction of pre-operation levels (created by the author using USAAF data from *Starvation: Phase Analysis of the Twentieth Air Force Strategic Mining Blockade of the Japanese Empire*, U.S. Army Air Forces, 1945, 1-35).

CHAPTER 9

Korean War

Introduction

World War II ended with the United States having learned the wrong lessons from the war. In the minds of the decision-makers sitting in Washington, D.C., the atomic bomb suddenly became the center of military planning. After all, wasn't it the atomic bomb that had driven the once-mighty Japan to surrender? Who would possibly be willing to challenge the United States against the threat of such a terrible weapon?

Perhaps they forgot that wars are not fought with just atomic bombs?

Mines and MCM Situation in 1950

United States—The United States ended World War II with a massive Pacific minesweeper force of over 500 ships crewed by 33,000 men.[1] By 1947, the Navy maintained a global force of 37 active minesweepers, with an additional 143 in mothballs. Even this force dwindled in 1949 when President Truman transferred minesweepers to France, which was then fighting communist guerrillas in Indochina. Foreshadowing the events of the next chapter, these guerrillas were mining rivers and harbors in hopes of driving away the French. By the time the Korean War began, the Pacific fleet, responsible for nearly half the globe, had just 12 MCM vessels ready for immediate service.[2]

Planning was underway for the MCM force of future wars. Based on experience in World War II, these plans included destroyer-minesweepers (DMS)[3] for fast clearance under fire, wooden-hulled auxiliary minesweepers (AM) for basic minesweeping work, and a combination of minesweeping boats (MSB)[4] and divers for clearing harbors. Plans also included research and development (R&D) into improved influence sweeps, sonars and other mine-locating technology, improved navigational equipment, a system for conducting aerial mine watches, and underwater mapping. The theory was that R&D done in peacetime would speed the ramp-up into wartime production should the need arise. Thus were the plans set that would mostly map out mine warfare for the next 50 years.[5]

Enemies never wait for plans to come to fruition. The Korean War came at a time when the U.S., Japanese, and Korean MCM forces were all suffering this sort of identity crisis. They would meet it with a motley collection of technologies and a severe capacity problem that left little room for losses and no room for flexibility.

North Korea and the Soviet Union—North Korea began the Korean War with no navy of its own, no domestic mine warfare program, and few vessels larger than fishing boats. What they did have was a powerful ally in the form of the Soviet navy.

In the West, it is common knowledge that when the Soviet Union captured German lands, they also captured German nuclear and rocket scientists. These scientists made up the core of the future Soviet nuclear and space programs. Far less known is that the Soviets also captured the remaining German mine stockpiles, mine materials, and mine experts. These captures became the root of the Soviet Cold War mine program. Historically, Russia and the later USSR had maintained a strong interest in the defensive possibilities of mine warfare ever since the days of Nobel in the mid–19th century. Their R&D efforts had atrophied between the world wars, but now the technological boost of their lucky captures put them on the cutting edge of the technology. With this boost, they built a true, professional mine warfare and MCM force with both a huge mine stockpile and the largest MCM fleet ready for all environments.[6]

Soon after the Korean War began, the USSR sent mines and mine experts to North Korea to help it compensate for the lack of a navy. The types of mines and training provided were ideal for a country with no existing expertise: simple, rugged, and tied into a larger defensive picture. Only four mine types were provided: the contact mines MKB, M-26, and Myam (respectively, large, medium, and small moored mines) and the magnetic bottom mine R1A. Most minelaying would be done with moored mines planted in small lines of all the same types, planted by boats possessing weight-handling equipment (like commercial fishing boats). Mine plant locations were fixed visually using readily identifiable points of land, yet they were positioned to complicate any attempts at shore bombardment. Magnetic mines, though few, were assembled and set under direct Soviet supervision and planted where they could cause the most chaos. Myams came only later in the war and thus played little part in the war's great MCM events. Unlike the other moored mines, they were small enough that they could be manhandled by two or three men. This gave them the flexibility to be deployed from the smaller boats most likely to be remaining after years of larger craft being lost through attrition.[7]

Background to the Invasion—The roots of the Korean War began well before 1950. Like many conflicts, it started with the decisions of a freedom fighter turned national leader. Prior to the Russo-Japanese War, the Korean Peninsula had gone through a long period of turmoil, with periods of isolationism followed by openness, all amid shifting political control. At the turn of the century, Korea was ruled by its first emperor, Gojong, who ruled from 1864 to 1907. In 1884 during his reign, China and Japan agreed to share hegemony over the peninsula. By the Russo-Japanese War, Japan had seized that hegemony solely for itself. The Treaty of Portsmouth, which ended that war, formally recognized that power. In 1905, Japan pressured Gojong to agree to Japanese protectorate status over Korea. Two years later, it forced him to pass his throne to his son and to disband the Korean army.[8]

In 1910, with its position as a regional power cemented, Japan annexed Korea outright. Two years later, a boy named Kim Sung-ju (later named Kim Il Sung) was born. As the Japanese rule became ever more severe, in 1925 the Kim family fled the country when he was just 13 years old. Over the following 15 years, he lived in Manchuria. It was while there that he joined a communist youth organization and later became a guerrilla fighter against the Japanese along the Yalu River (which forms much of the Chinese/Korean border). This movement, in which the now "Kim Il Sung" played an important role, was joined in the fight by the Chinese communist movement.[9] In 1940, he fled to Russia as Japan spread its influence across China. It was there that he received formal military and political training and was later promoted to the rank of major in the Soviet army.[10]

After the defeat of Japan in 1945, the United States and the Soviets divided Korea along the 38th parallel into two spheres of influence. Among the Soviet troops to accept the Japanese surrender in the northern portion was Kim Il Sung. After three years of occupation, the UN held elections to determine the first leader of a reunited country of Korea. The Soviet-dominated northern portion of the peninsula declared that the UN had no such authority and abstained from these elections. This was effectively a declaration that North Korea would hold itself aloof as a separate country. Later that year, as the Soviets prepared to leave, they held elections of their own for the first North Korean premier. Kim Il Sung was USSR's preferred candidate because, as one of the few Koreans to have served in their army, he was a known entity. The North Koreans supported him because he had lived in both the USSR and China and thus would not likely be completely dominated by either country. He won the election and became the first premier of the new nation of the Democratic People's Republic of Korea (DPRK).[11]

Two years later, in 1950, Kim began giving speeches about reunifying the two Koreas. Possessing a well-trained military armed with Soviet weapons and believing that 50,000 South Koreans were ready to rise up and join him, he thought that he could take the entire south in just three weeks. After convincing Stalin that the United States would not intervene, Kim gained Stalin's blessing for an invasion.[12]

The Korean War

On June 25, 1950, North Korean troops breached the 38th parallel and flooded south. Three days later, they captured Seoul and continued pushing south and east. In response, the newly created United Nations (UN) authorized the United States to lead a unified command to push these forces back into North Korea. By August 4, the North Koreans had chased their southern neighbors and the UN forces all the way to the southeast corner of the country.[13] It was there that UN forces established the defensive line known as the Pusan Perimeter. It started on the south coast near Changwon, then followed the Nakdong River north to the vicinity of Daegu before turning east to finish along the Hyeong San River in Pohang. Behind these natural barriers, U.S. and Republic of Korea (ROK) forces were able to hold despite repeated North Korean probes. This small area, totaling less than 10 percent of the country's original area, was all that remained until reinforcements and relief could arrive. That relief came in the form of a daring amphibious landing, ordered by General Douglas MacArthur, at Inchon on the Seoul outskirts. If they could retake Seoul, the heretofore victorious North Koreans would suddenly find themselves cut off from their own sources of reinforcement and resupply.[14]

Inchon Clearance—For the first couple of months of the war, there had been no indications of North Korean mining in South Korean waters. True, minesweepers swept a path into Pohang on July 15 to enable U.S. troops to land there and reinforce the northern edge of the Pusan Perimeter. But they had found no mines. By early September, however, there were clear signs that the North had gained that capability. Mines were spotted near Chinnampo as well as near Inchon where a ROK Navy (ROKN) patrol boat had destroyed a small boat engaged in minelaying. All signs indicated that mines would threaten the Inchon landing scheduled for September 15.[15]

The approaches to Inchon were like much of the Korean Peninsula's west coast: shallow mud flats going miles out to sea; a massive 33-foot tidal range; five-knot currents as that tide rushed in or out; and an extremely limited number of torturously twisting navigable channels. In the case of Inchon, the main channel was the "Flying Fish Channel." For the landing troops, the tidal range was a huge problem. If they landed at the right time, their landing craft could push all the way to good beaches by Inchon. If they arrived at the wrong time, they could find themselves miles from those beaches, slogging slowly and painfully through thick mud, fully exposed to the defenders who could be leisurely reinforced, with survivors being drowned by the rapidly rising tide. The "right time" happened only once every few weeks.[16]

That huge tidal range also presented a problem for the defenders. When contact mines are being used, they must make physical contact with a vessel's hull to actuate. Ideally, they would be placed shallow enough to always make that contact but deep enough to be not easily spotted by semi-alert observers. But where is that sweet spot when dealing with a 33-foot tidal range? If placed, say, seven feet below the surface at low tide, then any ship with a draft of less than 40 feet would pass safely over it at high tide. If placed, say, 15 feet below the surface at high tide, then it would be floating on the surface for perhaps half of each day. At Inchon, the defenders leaned toward the latter option.

What resulted would be what one might expect in such a situation. On September 13, surface combatants arrived ahead of the minesweepers to soften up the shoreward defenses with gunfire. Arriving near low tide, they easily spotted mines bobbing on the surface in Flying Fish Channel. To their crews, the next step was obvious. Target practice commenced, with the prize for a good shot being a big explosion! Before long, the mines were all destroyed. The minesweepers went to work shortly thereafter but had nothing left to sweep. They found no further mines, either in the channel or in the inner harbor. On September 15, right on time, the troops landed and stormed over the seawall to begin their battle for Inchon. By the 25th, MacArthur and his troops had retaken Seoul.[17]

Had they stopped there and merely chased the North Korean troops back to the 38th parallel, subsequent history in the region would probably have turned out quite differently. Without the counterinvasion that followed, the North Korean leadership would have had little ammunition with which to build an entire society around anti–American paranoia.

Lessons Learned: (1) If there is a large tidal range, look for mines at low tide and land at high tide. (2) It does not take a specialist to destroy a mine that is floating on the surface.

Every mine destroyed by gunfire as soon as it is spotted results in that much less time that needs to be spent reacquiring it and equipping/deploying an EOD tech or a piece of high technology. By preserving the limited number of specialists for jobs that only specialists can do, a navy reduces the risk of losing them entirely and wears them out less quickly. It also allows the clearance force to work more quickly at times when speed matters. Finally, it preserves the specialized equipment used by those specialists. Specialized equipment is usually far less available than bullets. If it is routinely wasted, especially early in a conflict thousands of miles from the factory, then it may take weeks or months to replace.

As an example, consider the next section, Wonsan, where there were 3,000 mines.

Map of the Korean Peninsula (modified by the author from creativefabrica.com).

Thirteen mines were cut in just four minutes. Suppose that the policy was to send EOD in a small boat to each mine as it was mechanically cut. After spending, say, 10 minutes looking for and transiting to each mine, EOD spends, say, 20 minutes rendering it safe and setting charges. Fine if it was one ship cutting one mine every 30 minutes. EOD would need to destroy only two mines per hour. But that policy would become a real problem in a thick minefield if there were three ships that managed to cut 13 mines in four minutes. Had that rate continued, EOD would have to clear 195 mines every hour, all while mines drifted all throughout the area. Or, a few people could be assigned to shooting the mines as they surfaced. There would still be a risk of the mines remaining dangerous somewhere beneath the surface. But that risk would be less than having that many mines drifting around waiting for EOD to get to them.

First Ships Hit Mines—When the UN landed at Inchon, the North Korean forces found themselves caught in the claws of a classic pincher move. With enemies to their north and to their east, the northern troops withdrew back toward their original lines. However, they did not leave things as they were prior to their arrival. In their wake, they left harassment minefields in the water off the southern coast. Farther north, Soviet mining experts trained North Koreans in mine warfare and supervised mine assembly and planting. Anticipating the UN counterattack, the communists began laying further minefields and mine lines along the North Korean coast. Though the first sighting of mines off the coast was on September 4, mines did not become a significant problem until later that same month.[18]

The ROKN and the U.S. Navy ships learned the extent of North Korean mine efforts beginning on September 26. Over the next week, five different ships struck mines. The destroyer USS *Bush* was the first of these. Working with another destroyer to find enemy gun emplacements near the North Korean town of Tanch'ŏn, she struck a mine that killed 13 and wounded 343. By the time she got back to Sasebo, Japan, more ships had fallen victim to mines. ROKN ship *YMS-509* struck a mine on September 28 somewhere off the south coast but still managed to limp back to the main ROKN base at Chinhae. The following day, the destroyer USS *Mansfield* struck a mine while searching for a downed pilot near Changjon, suffering 28 men injured.[19]

The worst day for UN mine strikes prior to the Wonsan clearance operations fell on October 1. On that day, two ships struck mines. The first was the former fishing boat turned minesweeper USS *Magpie*. Sweeping a channel into the small South Korean port of Chuksan, she struck a mine that blew her to splinters. The explosion killed 21 of her men and left the 12 survivors injured and floating in the water. That same day, while pulling into Mokpo in the southwest of the country, ROKN *YMS-504* struck a moored mine. The explosion set off sympathetic detonations from two other mines, leaving her with a ruined engine room and five injured men.[20] As damaging as these mines were, just 2.5 weeks later, mines would present the UN forces with even greater challenges.

As often happens in wars where sea control is contested, the admirals detailing ships to the task force had underestimated the potential of mines to screw up operations. Once mines started damaging ships throughout the Korean theater of operations, Chief of Naval Operations Admiral Forrest P. Sherman gave the permission needed to reinforce the theater's MCM force. Accepting the risk to other parts of the Pacific, he directed the redeployment of several auxiliary minesweepers and destroyer minesweepers to the conflict zone. Arriving too late for the Wonsan operation, many arrived in time for the Chinnampo operation. More locally, Rear Admiral Arleigh Burke, deputy

USS *Mansfield* (*DD-728*) with mine damage to port bow (U.S. Navy photo by Lieutenant Commander Edward L. Rollins, Naval History and Heritage Command).

chief of staff for COMNAVFE,[21] requested and acquired the support of a large Japanese MCM force.[22] The latter force proved instrumental at Wonsan when casualties decimated the small U.S. minesweeper force.

Wonsan Clearance

Wonsan was the big event of the war that the naval mine warfare community remembers. The Wonsan landing did not need to happen. U.S. and ROK troops were rolling up the peninsula, keeping their North Korean counterparts on the run at a clip of 14 miles per day. At such a fast pace, the Army needed Wonsan's large east coast port facilities to resupply the UN troops. To the Navy, used to seeing land as a flat brown area on a map, the obvious solution was for the Army to simply drive the hundred or so miles to Wonsan. From the Army's perspective, the Korean Peninsula's mountainous terrain seemed formidable. They assessed that such a drive would result in the loss of half of their heavy equipment. They felt that an amphibious landing there would likely get there faster and intact.[23] And why not? The Navy had just landed them at Inchon, on time and with little problem. Besides, why should the Army have to do all the work and take all the risks?

The Navy objected to such a plan. First of all, it would tie up a huge number of ships that were needed to support other missions such as gunfire support, anti-submarine warfare, interdiction operations, and close ground support.[24] Furthermore, it was a logistics nightmare of loading out troops in a heavily used port currently being used to onload supplies. This would be followed by an 830-mile journey around the peninsula during typhoon season, taken while coordinating the simultaneous arrival of 250 ships from throughout the region. Upon arrival, they would need to perform mine clearance to enter an enemy's port where the garrison had enjoyed plenty of time to prepare defenses, a port whose value was obvious to planners on both sides. What's more, the Navy did not have enough minesweepers for such a big clearance job.[25]

During World War II, more than 100 minesweepers cleared ahead of the fleet for each of the landings at Balikpapan, Okinawa, and Leyte. For Wonsan, of unknown mining levels, the fleet had only six minesweepers assigned: two fleet minesweepers (AM) and four coastal minesweepers (AMS...formally YMS).[26] The MCM force had been lucky at Inchon with its high tidal range. But at Wonsan, if mines were present in large numbers, the limited ships could clear a path either quickly or sufficiently, but not both. The loss of any one of those minesweepers would reduce their MCM capacity by nearly 17 percent. The Navy recommended the overland route as being quicker and easier. The Army won the argument in favor of an amphibious landing.[27]

Wonsan sits on a bay at the back of a large concave bend in the North Korean coastline. From that point, the continental shelf extends well out to sea before transitioning to the continental slope. Because of that, its approaches are flatter and gentler than most of the rest of the country's east coast, which quickly plunges into the depths of the Sea of Japan in a manner similar to the California coastline. Two islands, Ung-do and Yo-do, guard the harbor entrance. Once past those islands, one encounters a second line of islands in the middle of the bay. The port of Wonsan lies beyond the second island chain, while the planned landing beach was on a spit of land anchoring the southern end of that chain.[28]

Mine Clearance Begins—When the U.S. task force arrived at Wonsan, they had little intelligence about the mining situation. A helicopter flying from a cruiser had spotted mines in the area a few days prior, but little else was known. It was known that the Soviets had transited between the two guardian islands, but it was decided that the best approach was directly toward the harbor, south of Yo-do Island.[29] Sweeping began efficiently. The 10-mile approach lane to the 30-fathom line was swept in a matter of hours with 18 mines getting neutralized. It was near the 30-fathom curve that a scouting helicopter reported five mine lines in the planned approach lane.[30] The minesweepers confirmed the presence of large numbers of underwater objects using their crude mine location sonars (brand-new MCM technology at that point).[31] The mine problem, it appeared, was more difficult than originally anticipated.

The best advice in dealing with a minefield when in a hurry is always to go around. Sweep efforts halted while aircraft scouted the area. Finding the channel between the two islands (through which the Soviets had passed) to have fewer mines so far out to sea, the MCM planners shifted the approach lane north to that channel.[32] This made sense. With the threat axis being from the south, a more northerly approach would probably have fewer mines. Subsequent analysis indicates that the mine-lay was meant to hamper amphibious landings by cluttering the most likely staging areas and preventing unimpeded direct gunfire support. That conclusion was indicated by the fact that the moored mine lines extended out to the visual horizon of 12 miles and followed the coastline east and south.[33]

The following day, work started over again from the 100-fathom curve. More assets joined the force (including one more AM and two more AMSs) and the task group reorganized itself to sweep mine contact locations previously reported and plotted. Riding in their shallow-draft landing craft personnel ramp (LCPR), underwater demolition team (UDT) divers ranged ahead of the ships searching for and clearing mines that might threaten the following vessels.[34] Sweeping proceeded without incident, approaching within four miles of the first island line at day's end.[35] At midnight, the command team met and decided that a determined push the next day could very well get them all the way ashore, where they expected little resistance, for the ROK Army First Corps had captured the city earlier in the day.[36]

Disaster Strikes—All positivity died on October 12. The minesweeping team passed the final four miles to Ung-do and Yo-do Islands in a formation sweep of three AMs, with USS *Pirate* in the lead, followed by USS *Pledge* and USS *Incredible*. Two AMSs followed to drop marker buoys and to shoot floating mines cut by the minesweepers. Behind them, the DMSs USS *Doyle* and USS *Endicott* followed to provide gunfire support.[37]

As the fleet of minesweepers passed between Ung-do and Yo-do Islands, they encountered a line of mines spaced only 50 yards apart.[38] For context, the modern aircraft carrier USS *Gerald R. Ford* has a beam of just under 45 yards.[39] Once past these two islands, they found themselves in a tangle of mine lines in the small area between those islands and two islands of the second island chain. Within four minutes, the three leading minesweepers together cut 13 mines (no doubt a neutralization nightmare!). The scouting helicopter reported three more mine lines close aboard[40]; *Pirate*'s sonar reported a mine only 100 yards ahead, and her lookout reported a mine close off the bow. With nothing to do but move forward, *Pirate* struck a mine at 1209 and capsized in four minutes. When its mast fell, it missed the commanding officer and several crew members by the smallest of margins.[41] Her crew suffered six killed and 43 wounded.[42] The survivors found themselves adrift with help near at hand but surrounded by hazards still closer. Just as *Pledge* stopped to put her ship's boat in the water, artillery and small-arms fire erupted from two nearby islands. Now, with gunfire and loose mines all around, the sunken *Pirate* and more mines somewhere below, and *Pirate*'s sweep gear drifting around like the tentacles of a kraken, *Pledge* cut her gear and attempted to rescue survivors. Still under fire, she attempted to maneuver back into swept waters when she, too, struck a mine only 11 minutes after *Pirate*'s mine strike. She suffered seven dead or missing and 36 wounded. At the same time, *Incredible* reported the loss of her engines. After USS *Endicott* rescued the survivors of both ships, the MCM task group retired to regroup and plan.[43]

Clearing the Way into Port—The Seventh Fleet commander, Vice Admiral Struble, arrived later the next day. With all three of their AMs down, the minesweeping force now included only the six small, wooden AMSs and the oversize DMSs. Priority shifted to "safe sweeping," meaning that no minesweepers would operate in waters not already cleared of mines.[44] Clearance efforts that day took the form of organic MCM, with helicopters spotting and numerous small boats combing the area looking for mines. Several North Korean fishermen from Ung-do Island used their local knowledge to assist in exchange for either cigarettes or rations (depending on the era of the description). UDT divers neutralized some mines, while aircraft neutralized other mines with strafing. By the 14th, this patchwork force had cleared the channel's top layer of moored mines. Magnetic minesweeping could begin.[45] In those efforts, they would be supported by the arrival of eight Japanese minesweepers (JMS). These ships, which had swept almost continuously since World War II to clear the remaining influence mines from Operation Starvation, were probably the most experienced influence sweepers in the world.[46]

With no information on what, if any, magnetic mines lay in the inner harbor, or their setting, the minesweepers had to guess at how to sweep them. For the next six days, a mixed force of U.S., Japanese, and South Korean minesweepers swept back and forth with every magnetic sweep configuration available. In the process, JMS-19 struck a contact mine while operating in the unswept waters just south of Yo-do Island.[47] On October 18 (D-2), the minesweepers made their last turns between the final two islands, a mere two miles from the landing beach. Suddenly, explosions started going off all around. First, two explosions shot from the water a few hundred yards astern of USS *Redhead*, followed by a third explosion that tore apart ROKN YMS-516. At long last, influence mining was confirmed with the associated implication that another week was needed to clear the inner harbor.[48]

The following morning, it was revealed that Lieutenant Commander Don DeForest had gone ashore over the previous few days to look for intelligence about the mines and the mine-lay. Picking his way through the city, avoiding snipers and asking questions, he found the information he sought. From the train station master, he learned about the numbers and types of mines that had passed through that station. From perhaps the same informant (available descriptions are contradictory), he learned of a Soviet magnetic mine's search coil hidden under a stack of straw.[49] Elsewhere, he found workers who had helped lay the mines. From them, he learned that 3,000 mines had been laid at night using 32 small boats, often with powered sampans being used to tow nonpowered sampans crammed with some 10 mines each loaded on wooden rails. Minelaying was conducted at night using the most basic of navigational equipment to guide their positioning.[50]

Armed with information about how to sweep the mines and where to find them, the UN minesweeping force finished clearing the harbor on July 25 with no further casualties. The experienced JMS crews led the efforts, while small boats armed with makeshift equipment swept the shallows. When at last the landing force reached the shore after three weeks of turning laps at sea, they found a USO troupe featuring Bob Hope waiting for them.[51] Only 225 of the 3,000 mines had been cleared. Sweeping continued long after the landing, and long after the port itself was abandoned, against the chance of landing there again.

Chinnampo Clearance

Chinnampo, or simply Nampo, is the main shipping port supporting the North Korean capital of Pyongyang. Geographically, it is on the northern bank of the wide Taedong River estuary at the point where it narrows to a mere two-kilometer width. In 1950, it had islands in the estuary and was extremely shallow on either side of the main shipping channel.[52] It suffers from a huge 30-foot tidal range and a main shipping channel a mere 500 yards wide. With currents over 3.5 knots, it was a nightmare to clear, yet it was the only port available on North Korea's west coast.[53] Today, clearance would be even worse, as it is protected by a railroad-bearing barrier known as the "West Sea Barrage."[54] However, no such barrier existed during the Korean War.

What did exist was an outer mine barrage of 217 moored mines and 25 magnetic mines arrayed in 12 mine lines starting 12 miles from shore. They could get by with so few mines because the harbor approaches were no more than channels cut into an otherwise very shallow (three to seven feet) shelf. Such natural defenses meant that the limited mining resources could be concentrated at the few places where shipping could pass. Many of these mine lines were placed near islands that might otherwise be used for cover and concealment. All mine lines were placed near points that the North Koreans working from small boats could readily identify by sight. With this intelligent placement of mines, the North Koreans were ready to repel the expected Allied amphibious assault.[55]

The U.S. Army desperately wanted to access Chinnampo port to resupply its forces that had charged up the peninsula. To support that endeavor, the order to clear a path into Chinnampo came three days before MCM forces had finished doing the same to Wonsan. When one considers that UN commanders had rejected an alternative plan to land at two different beaches north of Wonsan because of a lack of minesweepers, this was an ambitious order. Sweep operations at Wonsan could be expected to last for months, yet there was now the order to sweep another harbor on the other side of the peninsula.[56]

The task to clear mines at Chinnampo fell to two officers, Commanders[57] Archer and Clay, only recently arrived in Sasebo. With flag backing, Commander Archer assembled a "scratch" team from whatever somewhat-useful units came close. Their team included a couple of DMSs and three AMSs just arrived from overseas, four ROKN YMSs, 14 MSBs, a UDT, a helicopter, a salvage ship, a destroyer, and a few amphibious ships. These were the units that were either unready or unavailable a few weeks previously for use at Wonsan. Before Chinnampo clearance was complete, Japan contributed 13 minesweepers, a guinea pig ship, and an MCM command ship.[58] Topping off the force was Lieutenant Commander[59] DeForest, the man whose intelligence gathering in Wonsan had proved invaluable to clearance operations at that port.

Commander Clay, meanwhile, went to Chinnampo to collect information about the environment and the mine threat. Here he found a North Korean whose knowledge of the minelaying operations helped pinpoint the mine line positions.[60] He also acquired the assistance of U.S. and UK aircraft, whose pilots searched for and found 340 mines, many of which they destroyed by strafing. Ship-based clearance in the increasingly icy waters was straightforward. Starting on October 29, the two DMSs swept an approach lane beginning 39 miles farther out than the intelligence gathered by Commander Clay suggested. With the approach lane cleared, the rest of the MCM force began work in the suspected mined areas on November 2. As they swept, the helicopter and UDT ranged ahead conducting aerial spotting. With the minefields' edges defined, sweeping operations began outside these fields and ran in concentric circles parallel to their shrinking boundaries. By sweeping around minefields and not through them, the danger to the sweeps was reduced and casualties were prevented. When the force reached Chinnampo harbor, they found 212 enemy mines and one blocked approach channel. These, too, were cleared. When all was complete, Commander Clay proved the effectiveness of the clearance operations by riding a North Korean tug from dockside to the open sea.[61]

Mine clearance operations were finally finished on December 3, 1950.[62] By that point, however, the mission and purpose had become entirely different. A week before, on November 27, China sent 100,000 troops flooding over the Yalu River, where they began pushing back the UN forces. What had been a task to clear a resupply point near Pyongyang suddenly became an evacuation mission. Two days later, on December 5, the

PBM aircraft destroys mine with gunfire near Chinnampo, North Korea, 1953 (U.S. Navy photo. Naval History and Heritage Command).

UN forces evacuated Chinnampo to escape the Chinese onslaught.[63] Despite their evacuation, minesweepers continued to sweep the harbor approaches off and on in hopes of tying down communist defense forces against the possibility of an amphibious landing that never happened.[64]

MCM Operations 1951–1953

Starting with the Chinese counter-counterinvasion across the Yalu River, the Korean War became a different conflict. Instead of the lightning amphibious strikes up the coast, the war's pattern shifted to primarily land forces fighting for hard-won ground. The Navy's role shifted to one of support for these ground forces. Surface combatants conducted gunfire support where they could, while aircraft supported beyond that range. The primary MCM mission changed to clearing fire control areas and routes to those areas. Another mission was tactical deception. They frequently cleared routes to potential landing beaches, including Wonsan, Songjin (today's Kimchaek), Hungnam, Chinnampo, and perhaps others. This helped to tie up enemy forces to defend against a landing, while also keeping the path clear in case leadership desired a future landing.[65]

Tactics changed on both sides. On the UN's side, a landing ship tank (LST) joined the MCM force as a helicopter carrier. Paying attention to the lessons learned at Wonsan, the helicopter preceded the sweepers, doing aerial spotting and other forms of customized, rapid aerial reconnaissance. It also helped to find safer paths out of minefields that proved too thick for the sweepers.[66] On the North Korean side, mining technology saw changes. One of these, the minuscule Myam mine, sported a mere 44 pounds of TNT. It was usually anchored only a few feet below the surface to target minesweepers and other small boats.[67] The other was a change in mine-laying technology. Some mines were hung from floating trash such as barrels and logs using pelican hooks with soluble plugs. These mines were cast adrift into rivers to be pushed out to sea by the current. Once the plug dissolved, the mine would stop drifting and anchor itself. Despite these tactics, no other ships were sunk by mines during the war.[68]

On July 27, 1953, the conflict ended when China, the United States, and North Korea signed an armistice. Both North and South Koreas retained the lands held at the time of the armistice's signing. South Korea refused to sign the armistice because it did not agree to the principle of a divided Korean Peninsula. For that reason, the war is considered to have never technically ended.[69] The peninsula remains divided, with both sides ever preparing for the other side to invade. Should the two countries one day reunite, through either diplomacy or combat, it will be an even more dramatic event than that of the German Reunification of the late 1980s.

Postwar MCM Renaissance

Although mines appeared throughout the Korean War, it was the lack of sufficient MCM capabilities at Wonsan that gave the United States and its allies their biggest lesson. A mere five years after World War II, and even fewer years into the Cold War, the USSR humbled the West with humble mines.

The United States responded to that embarrassment by updating and expanding its MCM fleet and those of its allies. In reality, this rebuilding program was merely the execution of the planning done prior to the war. But without the Korean War, that program may have never been built in a military then trying to strike a balance between budget cuts, increased operational commitments, and a major shift to an atomic weapons-centered defense policy. Doubtlessly it was the visceral picture painted by the commander of the amphibious task force for the Wonsan landing, Rear Admiral Allan E. "Hoke" Smith, in his report to Admiral Sherman that helped motivate Washington, D.C., to fund such a building program: "We have lost control of the seas to a nation without a Navy, using pre–World War I weapons, laid by vessels that were utilized at the time of the birth of Christ."[70] Several other such statements followed (see the end of this chapter).

The message was clear. The most powerful navy and amphibious force in the world is useless if simple technology can deny it sea control. Just because atomic weapons could now destroy cities did not mean that mines could not still destroy fleets and economies. Both had their uses, but as the poor man's weapon, mines were far more likely to be encountered.

Ultimately, the U.S. Navy began in the 1950s to build a huge MCM force that included 139 vessels plus other supporting units:

- 65 oceangoing minesweepers (MSO)—Had the new USQ-1 mine-hunting sonar.
- 47 minesweeping boats (MSB), two inshore minesweepers (MSI).
- 24 coastal minesweepers (MSC) (159 built, most transferred to other navies).
- 1 coastal minehunter (MHC)—Had a towed-array mine-hunting sonar but no countermeasures/sweep capability.
- Four ships converted to mine countermeasures command ships (MCS). Three of these each carried eight minesweeping launches (MSL).
- River minesweepers (MSM) and patrol minesweepers (MSR).
- R&D on helicopter minesweeping beginning with an HRP-1 helicopter pulling an Oropesa sweep. The intended purpose was to target contact mines moored shallow enough to threaten the already shallow drafted minesweepers.

With such a large number and variety of minecraft, MCM began to look like a feasible officer career path for a short period of time. After all, the larger vessels required division officers and department heads, and all vessels required either commanding officers or officers-in-charge (or chief-petty-officers-in-charge). These would further be grouped into numerous flag staffs, squadrons, divisions, and flotillas, maintenance groups, training groups, etc. Such a large grouping made it possible for a young officer to progress from ensign to admiral, building an ever-deeper level of expertise. They gained the experience of early command and the breadth of technical skills that come from being part of a small crew with all the programs required of a ship with 10 times the crew size. Unfortunately, within a decade that viability came to an end as the Navy became stripped of assets as the military refocused on other concerns, including the impending Vietnam War.[71]

It was not until the mid–'90s, post–Desert Storm, that the U.S. Navy began to rebuild this professional mine warfare expertise using enlisted personnel instead of officers. The new concept was to expand the mineman rate to start with traditional shore-based mine shop roles and to add to it the MCM-specific roles from minesweepers. This included former sonar technicians, operations specialists, gunner's mates, and boatswain's mates. Aboard ship, minemen could specialize in one skill set or gain experience in all. Ashore, they could work at a mine shop, teach, inspect, work at intermediate maintenance activities, or work in a variety of roles for whichever squadron or flag staff iterations came in vogue. With this core of expertise, officers temporarily working in the warfare area had experienced professionals available to help fill in the knowledge gaps that resulted from frequent shifts between unrelated warfare areas.[72]

Quotes from Leadership Discussing Wonsan

> "The Navy able to sink an enemy fleet, to defeat aircraft and submarines, to do precision bombing, rocket attack, and gunnery, to support troops ashore and blockade, met a massive 3,000 mine field laid off Wonsan by the Soviet naval experts.... The strongest Navy in the world had to remain in the Sea of Japan while a few minesweepers struggled to clear Wonsan."
>
> —*Rear Admiral Smith's After-Action-Report*[73]

> "The main lesson of the Wonsan operation is that no so-called subsidiary branch of the naval service, such as mine warfare, should ever be neglected or relegated to a minor role in the future. Wonsan also taught us that we can be denied freedom of movement to an enemy objective through the intelligent use of mines by an alert foe."
>
> —*Vice Admiral C. Turner Joy,*
> *Commander Naval Forces, Far East*[74]

Chapter 10

Vietnam War

Over all of mine warfare history, the state of the warfare area during the Vietnam War most closely resembled that of the U.S. Civil War. On the one side, a richer, more technologically advanced force operating far from home. On the other, vigorous fighters operating in familiar territory, using every force-multiplying trick available to sting their enemy. Like in the U.S. Civil War, guerrilla fighters attacked armored gunboats plowing through rivers and swamps by using improvised explosive device (IED)-style mines, limpet mines, and ambush. To counter this threat for which they were ill-prepared, the United States needed to develop new technology, tactics, and organizational structures. Yet, at the same time, they did little mining along the coasts where all U.S. mine countermeasures (MCM) doctrine was then directed. With the exception of minesweeping at the very end, the massive MCM fleet built up following the Korean War was wasted patrolling mine-free waters.

On the other hand, U.S. offensive mining capabilities still found a part to play in the war. To attack North Vietnamese logistics lines, they dropped the innovative Destructor (DST) mines on land and on rivers to target truck traffic. Toward the war's conclusion, when a North Vietnamese offensive caused peace talks to break down, the United States replicated Operation Starvation by mining their major harbors. After fits and starts, this strategic mining campaign became the linchpin for the peace agreement that followed. The war concluded with the first combined surface and airborne MCM operation in history. That MCM force worked for five months ensuring that no active mines remained to threaten North Vietnam or its trading partners.

Roots of the Vietnam War

It is impossible to pin down an exact start date for the war in Vietnam. Americans think of it in terms of their own involvement, yet it really began much earlier. Perhaps it was as early as the 1880s when a defeat to France ended with the country declaring itself a French protectorate. Various nationalist movements and other unrest often strained that protectorate relationship. French control disappeared in 1940 when the Japanese seized the country, leading some Vietnamese groups to support these new Asian masters. But the Japanese only lightly held the country, giving communist guerrillas the opportunity to build mountain fastnesses. When World War II ended in August 1945, the communists were quick to evict the defeated Japanese. In the temporary power vacuum, the northern communists established the Democratic People's Republic of Vietnam (DRV), otherwise known as North Vietnam. In the south, the communists met with competition from other nationalist groups and so had to share power. The following month, the French arrived to reclaim their lost territory and so drove the DRV and the local communist supporters (Vietcong, or "VC") away from the urban areas. Forced into hiding, the VC became guerrillas, fighting against the French.[1]

Over the next four years, the French and the communists fought an ongoing series of skirmishes and battles, with sporadic peace talks thrown in as seasoning. Losing ground, the French first set up Vietnam as a "free state," followed in 1950 by establishing it as an autonomous country under an appointed government. The communists, viewing the new government as illegitimate, continued fighting while running a propaganda campaign in France to foster resentment toward the war.[2] Fighting formally concluded in July 1954 with a peace treaty that officially ended French rule[3] and established the 17th parallel as the border dividing the areas controlled by the two rival Vietnamese governments. Elections for a unified government were scheduled for 1956. They did not happen. The premier declared the southern half to be the "Republic of Vietnam" following a national referendum and declared himself to be its leader. His autocratic rule lasted until 1964, when it fell to a military coup. When the new government proved incapable of handling increasing resentment and bedlam, the new Johnson administration in the United States feared the country would fall to the communist north. In 1966, Johnson sent U.S. combat forces to awe the VC into submission. It did not work. The northern government had big friends of its own in China and the Union of Soviet Socialist Republics (USSR). By 1968, the yearslong series of guerrilla fighting turned into a full-scale conflict with the Tet Offensive. At that point, there was no question that the United States was part of an undeclared war in Vietnam.[4]

Vietcong Riverine Mining—Vietnam is a long, thin country with rugged terrain and often dense foliage. It is 1,650 kilometers long but only 50 kilometers wide at its narrowest point. Around a quarter of the country's land area is river deltas covering over 70,000 square kilometers, with over half of that area being the Mekong River Delta in the south.[5] During the Vietnam War, the United States, UN, and South Vietnamese ("Allies" or "Allied") enjoyed relatively good control over the cities and the roads. The countryside, however, was a different story. There, the separatist National Liberation Front held entrenched footholds in the rainforests, where they could be reached only by troops slogging through unfamiliar terrain. Unable to use the roads to transport weapons, military supplies, and personnel without subterfuge, the guerrillas used the innumerable waterways as highways from the coast. Small, 100-ton trawlers, hugging the coast to avoid Allied patrolling ships, would hide in any of the countless concealed inlets and offload their supplies to awaiting guerrillas. From there, small boats transported the supplies the remaining distance to their jungle outposts. Roads were not needed when footpaths and waterways met their logistics needs.[6]

One area of particular concern for MCM forces in South Vietnam was the river delta area that formed the entire shoreline of the country's heavily populated southern end. Within that area, innumerable channels dump the waters of the great Mekong River (12th longest in the world) and its tributaries into the ocean. Immediately adjacent to this delta, the Soài River forms a smaller delta, fed by those parts of the South Vietnamese interior that somehow missed the Mekong River. Though the Mekong River starts in the Himalayas, the tropical monsoon rains experienced by the region each year feed the torrent of water inundating this highly fertile region. It is a massive alluvial fan where water, countless islands, swamps, flood lands, and vegetation dominate most of the rural countryside.[7] It was within this wild, waterlogged landscape, especially in the Rung Sat special zone (swamp) of the Soài River Delta, that many VC found refuge.[8] From there, they could directly threaten shipping headed up the Long Tau River destined for the capital.[9]

Just as waterways made excellent transportation networks for the VC, so too did those same waterways provide transportation routes for Allied forces. In places of deep cover, often they were the only means for moving small numbers of personnel and supplies in and out of remote locations. If operating near Allied troops, they could provide extra gunfire support or provide timely transportation. More often, Allied watercraft patrolled the waterways to interdict and cut the VC's logistics. Every weapon stopped was one less armed guerrilla. Every bullet stopped was one less shot taken at an Allied troop. Every trainer stopped was a multitude of VC left without a skill set useful in killing Allied troops. And so Allied boats patrolled the many channels of South Vietnam as part of the larger strategy of strangling the VC's campaign to take the South.

Echoing the American Confederacy's method for countering ironclads, mines proved the answer to the VC's problem of countering Allied ships and patrol boats. Where small-arms fire might hole a boat's sides a thousand times and yet leave it afloat, an underwater explosion could crack the hull and/or knock the engineering plant out of alignment. Or tear the boat apart into a thousand pieces. Any of these outcomes would result in the boat being knocked out of service. However, the other part of the equation was the reason why the VC defended these waterways. In terms that they might not have used themselves, it was to retain freedom of navigation for themselves while denying it to their enemy. Autonomous contact or influence-actuated mines deny freedom of navigation to friend and foe alike. While the VC could have littered their rivers with simple contact mines, they would be left with the problem of having to navigate through minefields every time they went to pick up supplies. Alternatively, they could use remote-controlled (RECO) mines. As in other wars going back to the Crimean War, RECO mines could be left switched off when approached by friendly boats and turned on or actuated when approached by enemy boats. RECO mines could be used to deny freedom of navigation to their enemy, but not themselves.

Although the VC acquired and deployed a small number of advanced Soviet mines,[10] they otherwise used RECO mines almost exclusively in the form of improvised explosive devices (IEDs). Actuated either electrically or by pulling a string (but usually electrically), the VC's mines were set somewhere midstream in the targeted channel. Often placed at likely ambush sites where vessels were forced to slow, they could be anchored to the bottom, to a log or other midchannel obstruction, or to the opposite riverbank. In some instances, the operator would rig the mine to be able to be pulled directly into an oncoming vessel's path. If a boat passed over the correct position, the operator would actuate the mine—followed by an explosion, a splash, screams, and then one less boat to challenge the VC's assertion of authority.[11] In places closer to Allied controlled ports and anchorages, VC divers often attempted to use improvised limpet mines to target ships at rest.[12]

Allied Riverine Warfare Craft—Because of the ongoing threat from VC along the southern rivers and swamplands, U.S. policy in the early 1960s was to transfer combat craft to the South Vietnamese. In 1962, the Navy planned to provide 212 minesweeping launches (MSL) but instead sent a dozen larger motor gunboats (PGM). Other craft provided in the early years were frequently excess landing craft left over from World War II. Built to traverse

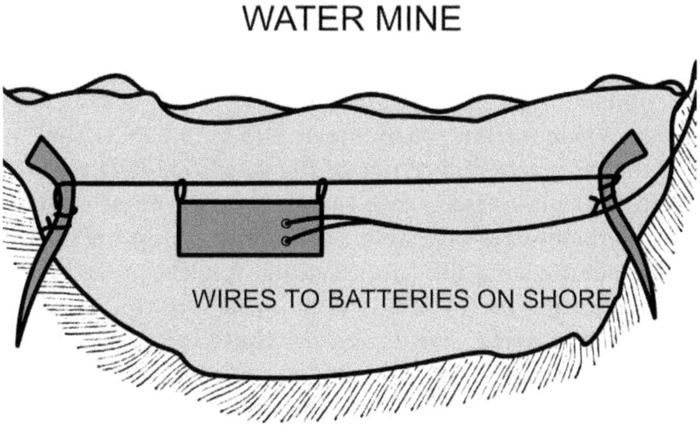

Above and below: **Two types of VC underwater IED mines from a U.S. Army field manual, 1965 (*Mines and Booby Traps Used by the VC in South Vietnam [Min Bay Do VC Xu Dung Tai Mien Nam Viet Nam]*, Department of the Army, 21-25, original drawings recreated by Milena Mazurek).**

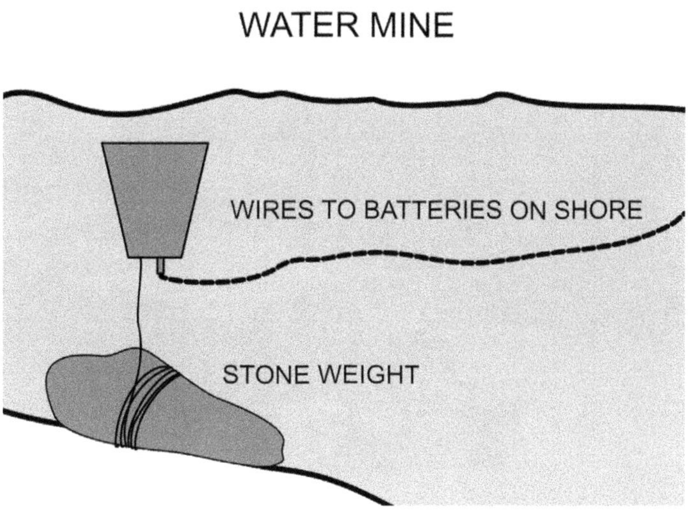

shallow water, these craft did everything from carry troops to minesweeping to acting as floating forts. A simple mechanism of heavy chains rigged with welded-on cutters accomplished the basic service of cutting RECO mine wires ahead of patrols.[13] This local solution was improved in 1964 when the technical center at Panama City, Florida, identified that the previous system could miss mines by skipping along the bottom.[14] Soon the shop-made chain sweeps were replaced with Panama City's improved chain sweeping system and tactics.[15]

In 1965, with no sign of the North giving up trying to take South Vietnam, the United States drastically increased its scope of in-country operations in hopes of forcing the issue. It started a yearslong operation of bombing and blockading meant to destroy the DRV's ability to support the ongoing conflict.[16] At sea, ocean minesweepers (MSO) and coastal minesweepers (MSC) were primarily used for OPERATION MARKET TIME patrolling, though they were not designed for that purpose. Optimized for low magnetic and acoustic signatures while running for short periods at slow speed, they suffered significant wear and tear from the stresses of continuous patrolling. At the time, many MSOs were scheduled for refitting with the "Shadowgraph" side-scan sonar and the "Sea Nettle," a wire-guided mini-torpedo for mine neutralization (like the "Sea Fox" of the 21st century). Refitting never took place, as most available funds were diverted to supporting operations. Thirteen of 65 MSOs did receive upgrades to Waukesha engines and the SQQ-14 sonar.[17]

Note: The author's first sonar to maintain was an SQQ-30, a modified version of the SQQ-14. It featured vacuum tubes, synchro-and-servos, and some truly scary capacitors.

Closer to shore, 56-foot-long minesweeping boats (MSB) arrived in the same year to conduct ongoing harbor clearance operations. Quickly their mission shifted to support riverine warfare, particularly OPERATION GAME WARDEN, meant to deny the VC the resources and refuge of the southern delta regions. This followed the successful mining of 77 vessels over the previous 26 months. Starting in 1966, U.S. forces concerned themselves primarily with maintaining freedom of navigation on the 45-mile section of the Long Tau River leading from the sea to Saigon.[18]

For years, the VC used the movement of traditional river folk to hide their movements while using threats against these same people to extort supplies. The GAME WARDEN patrols sought to win their hearts and minds into accepting curfews and providing cooperation. In response, the VC spread rumors of U.S. atrocities that, in turn, the United States countered with kindness, messaging, and direct civil assistance. At the same time, the VC increased mining and terrorizing the population in hopes of coercing compliance.[19]

To counter this new reality, two Allied task units were formed. The U.S. Navy modified its MSBs with non-magnetic armor, a brace of weapons, and chain-drag sweep gear.[20] Task Force (TF) 116 combined MSBs with river patrol boats (PBR) and

Assault support patrol boat rigged to sweep for controlled mines passes armored troop carrier, 1967 (National Archives photo by CDR D.K. Dagle).

helicopters to continue patrolling the river and keeping it clear of mines.[21] The combined Army/Navy Task Force (TF) 117 was the assault force assembled to take the fight to the entrenched VC. Its Navy component used as its workhorses a fleet of 52 armored troop carriers (ATC), 32 assault support patrol boats (ASPB) that combined gunfire with minesweeping, and 10 landing craft, mechanical (LCM) monitors. This last served a similar purpose to the monitors of the Civil War ... think of a tank that floats.[22] Often under heavy fire,[23] TF 116 and TF 117 were tasked with keeping the river clear of mines and the shoreline clear of attackers. They did this through minesweeping, patrolling by air, sea, and land forces, combined with rapid convergence on hot spots. In these efforts, they were further supported by shoreline defoliation (possibly "Agent Orange"?).[24]

The fight on the river was far different from that experienced by blue-water forces operating at sea. U.S. Navy minemen[25] celebrate their most decorated member, MN1 Cecil Martin, for his actions in the brown-water fight in Vietnam. On November 21, 1968, while he was acting as the senior boat captain on a two-boat patrol, his partner boat was hit by two rockets. With every crew member aboard injured, the boat ran hard aground[26] under enemy fire. Seeing his teammates in this perilous situation, MN1 Martin charged his boat back into the fray. Directing fire and often firing a machine gun himself to beat back the attackers, he shielded the sister craft with his own boat. He then gave first aid to the other boat's crew, reported the situation to higher headquarters, and led the evacuation of wounded personnel to his own boat. For his leadership and valor under fire, MN1 Cecil Martin received the Navy Cross, the U.S. Navy's second-highest decoration.[27] This made him the highest-awarded mineman in history.

TF 116 and 117's base of operation was the Mekong River and the Nha Be anchorage nine miles south of Saigon[28] where the river forked just before making a sharp turn from south-southeast to west-northwest.[29] Needing to be able to rapidly deploy in this flood-prone environment under constant threat from VC sappers, divers, artillery, and snipers, the river force operated from floating bases. Using the APBs to which were tied other support vessels, these bases provided housing, workshops, recreation facilities, and even helicopter pads.[30] They also proved to be highly visible and tempting VC targets. The base was heavily armed and equipped with mine nets. Topside rovers shot at floating garbage that frequently hid IEDs. Security periodically tossed grenades into the water to discourage attacks by VC sappers who were often as skilled as any special forces.[31]

Other MCM craft used elsewhere in the war included patrol minesweepers (MSR), the large MCM support ship USS *Epping Forest* (MSC-7) carrying 16 minesweeping launches (MSL), and modified motor launches (MLM). Like their French counterparts before them, they also used landing craft, mechanical (LCM) modified for minesweeping. These operated in the delta regions as well as places farther north, such as Da Nang, Cam Ranh Bay, the Perfume River, and the Cua Viet River.[32]

One type of craft that would be surprising to modern MCM sailors were the minesweeping drones (MSD). Introduced in 1967, two versions of these were made, with the 23-foot Hydrodyne boat from Panama City being deemed better suited. They were essentially remote-controlled (RC) boats; operators riding in following boats could control them from half a nautical mile away. Eventually the United States would give these to the South Vietnamese, deeming them insufficient for their needs.[33] By 1971, the Chinese invented their own minesweeping drones, the Type 312 Futi class drone minesweeper that was used in North Vietnamese mine clearance efforts.[34]

American Mining—During the Vietnam War, the use of naval mines by the United States centered on one main goal: interrupting the supply lines bringing weapons and supplies to the DRV troops fighting in South Vietnam. Because the strategy for much of the war was to prevent escalation that might draw in more direct Chinese and Russian support, most early efforts were south of the 20th parallel. Over time, these expanded to bring the fight ever closer to the northern capital of Hanoi. The first U.S. naval mining was in northern rivers to try to interdict supplies being brought in through the cost- and resource-efficient waterways. Later, the Americans laid magnetic influence mines near river crossings and on land along the Ho Chi Minh Trail in hopes of slowing the trucks bringing supplies via that famous supply line. Finally, mines were laid in the harbors of the main North Vietnamese ports to directly blockade the country to encourage them to negotiate.

In the early years of the war, approximately half of the supplies going to VC and the North Vietnamese regular army troops came via water transportation up the river systems. In the South, patrolling boats helped to stop this traffic while in DRV rivers south of the 20th parallel: Cua Sot, Kieng Giang, Song Ca, Song Ma, and South Giang. These efforts helped to push the VC's supply lines farther inland.[35]

All mines laid in the effort were air-laid acoustic and magnetic mines. After an initial batch of 143 acoustic/magnetic mines laid south of the 20th parallel between February 26 and May 21, 1967, the effort shifted to purely magnetic mines dropped farther north. For the most part, mines had to be laid in the wider, deeper river sections near their estuaries, though later efforts attempted to cut off Hanoi from its port at Haiphong. In these parts of the rivers, the water was deep enough for larger watercraft to operate, though it is unknown if the water was deep enough to satisfy the mines' arming depths. What is known was that the mining did not have nearly the anticipated effect, not least because mines were laid at levels far below the concentration believed to be required. Where Operation Starvation involved huge quantities of mines, reseeded frequently, in North Vietnam they were laid sporadically and in small numbers. With only handfuls dropped occasionally, the North Vietnamese military had time to adapt.[36]

In fact, the mining was assessed as being ineffective, having done very little damage to shipping. Still, it represented a threat that forced North Vietnam to expend far more resources and manpower than it would have had the mines never dropped. The government set up a mine watch system and dragged the rivers for mines using nets slung between sampans. Any mines found were actuated using simple methods rather than disarmed. Mined rivers were marked to warn off shipping during countermeasures. River traffic in mined areas hardly changed. In areas where mines had been dropped near bridges (a tactic meant to target ships and trucks alike), trucks were simply diverted to other bridges farther inland.[37] Unlike in World War II Japan, where Operation Starvation caused clogging of the country's inadequate rail infrastructure, in the DRV the roads were already running at far below capacity. Furthermore, the amount of war matériel needed by troops fighting in the south was relatively small. Even with the increased road traffic, the road network was still used at only 5 percent of capacity. As such, they provided a readily available alternative to using waterways whose availability was made unreliable by the infrequent mining.[38]

If any lesson were to be gained by the U.S. riverine mining efforts, it would be that such mining should be done by mines that are effective in shallow waters. Moreover, these mines should be effective against wooden boats, and they should be laid in such concentrations as to make them unavoidable. If acoustic mines are used, they should be designed for the acoustic environments of rivers. Another lesson is that riverine mines should be reseeded frequently to enforce the impression of being an enduring threat that makes any water travel seem hazardous. Finally, aerial riverine mining would likely have a greater effect if part of a larger, concentrated operation. Instead of spending a day laying 10 mines in each of five rivers, why not lay 50 mines in one river while also destroying every bridge passing over that river? Isolated minefields can be avoided. It is much harder and more expensive to compensate for entire rivers made impassable.

Accidental Mining in the Gulf of Tonkin—On January 27, 1968, when the riverine mining campaign was well underway and the Ho Chi Minh Trail mining was still going through growing pains, a mishap sent patrolling MSOs into action sweeping in the Gulf of Tonkin. Shortly before, U.S. aircraft dropped a load of DST mines somewhere north of the 17th parallel under unknown circumstances. Ordered to sweep a danger area at that location, the MSOs would later receive tasking to sweep a danger area at about 16°N 110°E. This second location was the closest point from which aircraft carriers could launch aircraft to attack the DRV.[39]

Mine Division 91 received the task to sweep these danger areas. Its ships, only two days into a transit from Subic Bay, made all preparations for minesweeping. While in transit, they performed their prechecks to ensure their equipment was in operational order. At Da Nang, the port closest to the action area, they made their final mission preparation.[40] Once in the designated mine danger areas, they first attempted to locate the mines using sonar but were unsuccessful.[41] They followed those efforts with 30 hours of magnetic minesweeping in combat conditions while suffering through cold and rainy weather. Upon completion, with no mine explosions reported, they declared the areas safe. With that done, they returned to the interminable patrolling of OPERATION MARKET TIME, having completed the first wartime minesweeping at sea against potentially active influence mines since the Korean War. For this accomplishment, they received a BZ message from commander, naval forces (COMNAVFOR).[42] *Note: Given the nature of how and where the mines were dropped, they were most likely to have been jettisoned for safety reasons. Jettisoned DST mines were not likely to arm but would still need to be treated as active until sweeping confirmed that they were not in an active condition.*[43]

Mining of the Ho Chi Minh Trail

The Ho Chi Minh Trail was a DRV logistics route that snaked through the adjacent countries of Laos and Cambodia. Originally created to avoid Allied patrols, it was less a single trail so much as a web of approximately 9,000 miles of trails and (later) roads. Troops and supplies traveled along this route for anywhere from six weeks to nine months on their way to South Vietnam's battlefields. To support this journey, they built several underground facilities in which troops could rest, resupply, and recuperate. The Ho Chi Minh Trail was vital to the DRV's ability to continue pressing the war in the South.[44]

Not surprisingly, the United States and South Vietnamese considered it to be of prime importance to cut this path. Initially, the approach was to use air-laid munitions to attempt to cut the trail and/or provide indications of movement along the trail. Later, they used ground troops in hopes of disrupting traffic flow.[45]

The airborne campaign began with a grab bag of different munitions dropped at various chokepoints. Magnetic DST mines were laid along rivers near bridges under theory that they could be actuated by passing heavy vehicles or boats. However, local military had little trouble finding and neutralizing them (i.e., have troops patrol both sides of the road, and plant explosives on any DSTs found). Acoustic sensors designed to snag in the trees provided another tool, transmitting a signal whenever sound levels increased above a specified volume. Devices known as "gravel mines" helped actuate these signals with firecracker-sized explosions when weight was applied to them by feet or vehicle wheels. "Dragontooth" devices, essentially caltrops, had the ability to puncture tires and cause delays. Air-laid anti-personnel mines caused casualties by deploying several trip wires over the wide area encompassed by their explosive damage radius. All these devices were individually scattered along the trail with only limited effect. DRV troops learned how to counter them as they were discovered, or to simply go around.[46]

In March/April 1969, planners identified a synergistic combination of air-laid weapons that proved successful in temporarily stopping traffic. First, a laser-guided bomb was dropped onto the road at a chokepoint, creating a crater impassable to trucks. Next, magnetic DST mines were dropped in the area to destroy any construction vehicles sent to fix the crater. Finally, anti-personnel, wide-area mines were scattered over the area to prevent the DST mines from being easily moved or detonated. By using this combination, traffic could be stopped in these sections for as many as four days while engineers worked to clear the tangle of mutually supporting ordnance. As effective as this combination was on a temporary basis, it did little more than slow the traffic flow as it was diverted along the many alternative paths.[47]

Airborne MCM Comes into Vogue

On July 1, 1970, a new U.S. chief of naval operations (CNO) stepped into office. This CNO, Admiral Elmo Zumwalt, had previously served on and/or commanded destroyers, cruisers, and Cruiser-Destroyer Flotilla 7, followed by service as commander, naval forces Vietnam.[48] One of his goals, based on his past experience, was to rebuild mine warfare.[49]

Every person anywhere builds their perception of how different programs work based on their interfaces with those programs. And how had Admiral Zumwalt experienced mine warfare? In short, he experienced its units operating in conditions for which they were not designed. In November 1945, as the prize master of a Japanese river gunboat operating in China, he observed minesweepers struggling to clear unexplored minefields while fighting heavy currents in the mouth of the great Yangtze River.[50] What would he have seen? Worn-out ships creeping along and probably snagging every piece of flotsam and jettison careening along in their path. His next interactions with MCM ships were in Vietnam, where the mine war was entirely riverine during his tenure from 1968 to 1970.[51] What would he have seen? MSOs endlessly patrolling, a condition for which their non-magnetic engines were not designed. The strain wore out those engines and the wooden hulls, which were increasingly repaired with locally available ferrous-metal parts.[52] He would have seen them as being unable to do their primary mission. His third interaction with MCM was the small boats combating the VC mines in the rivers of South Vietnam. What would he have seen? MSOs were useless in the shallows, while smaller vessels were vulnerable to enemies and mines alike. Small boats and drones plodded slowly through the rivers, under frequent attack by ambushers and by swimmers armed with limpet mines. Often, they needed to be rescued by helicopter gunships swooping in to provide fire from above. Helicopters seemed to be the lords of the battlefield, while the surface force was a liability.

These would have been his impressions as he stepped into his role as the CNO. He recognized the importance of MCM, but he had doubts as to the effectiveness of the surface-based MCM force. He believed in the need to improve MCM technology, yet the Department of Defense (DoD) was suffering from budget cutbacks. His challenge was to procure units that delivered more capability for fewer resources. Helicopter-based MCM, then under development, offered a potential solution. Conceptually, the original idea was to use helicopters to pull sweep gear ahead of the surface mine countermeasures (SMCM) ships and to work in shallower waters. The first task was important. One of the lessons learned from the Korean War was that SMCM units were safer if a helicopter flew ahead of them looking for shallow-moored mines. The next logical step was to have those helicopters clear the mines that they spotted. By sweeping a path ahead of the leading SMCM unit, that unit had less to fear from mines set to specifically target SMCM units. After the helicopter finished, the SMCM could operate in relatively safe waters as they cleared a wider, deeper path. Furthermore, a helicopter would be able to operate closer to shore in shallower waters than an SMCM could hope to operate. The effective speed could be a few times faster than an SMCM, with much less risk to the much smaller helicopter crew. On paper, it reads as fewer people incurring less risk doing the same job but at three to four times the speed, with far lower maintenance costs. Like the littoral combat ship (LCS) concept of the early 21st century, to a non–MCM admiral looking from the outside, it would seem like the perfect solution to the problems that they perceived.

Zumwalt reshaped the Navy to get rid of the plodding, high-maintenance, brown-water minesweeping craft. He also accelerated the decommissioning of blue-water minesweeping ships. Those craft purpose-built for sweeping Vietnamese rivers were turned over to the South Vietnamese. The MCM command ships (MCS) with their loads of 16 MSLs apiece were decommissioned as being no longer needed. The MSBs were returned to the States, perhaps because of their value as carcasses for parts. By 1976, the Navy decommissioned the last of the coastal minesweepers (MSC).[53] All that remained were the MSOs, worn out from constant patrolling. Between 1970 and 1973, over 30 MSOs were sold, placed in mothballs, or scrapped. Of the remainder, 24 were transferred to the Naval Reserve where they were crewed by a caretaker crew of 39 active-duty sailors, with 31 reservists brought aboard for their "one weekend per month, two weeks per year" training. There they wallowed from a lack of practice, few experienced personnel, and few monetary resources.[54] Zumwalt had greatly downsized the surface force built as a lesson learned from fighting an enemy competent in ocean mining. In its place, he championed airborne mine countermeasures (AMCM), which were then only in the development phase. Like with the 21st-century transition to LCS, he was making an educated guess that the new concept would be ready before the world once more had to deal with mined waters.[55]

The minesweeping gear itself would have been relatively simple to develop. The navies of the developed world had spent decades sweeping for mines from the backs of destroyers, boats, launches, and even drones. They had the basic sweeping technology solved and needed only to make adjustments to allow it to be deployed and pulled from a higher angle. Minehunting technology also had reached a state that it could conceivably be modified for deployment from a helicopter. After all, the Shadowgraph side-scan sonar had been available since the mid–'60s and would have been used to upgrade the MSOs had all available funding not been needed for the ongoing conflict.[56]

What was needed next was to identify an acceptable AMCM platform. It would need to be strong enough to pull heavy equipment against water resistance, spacious enough to store (most) equipment when not being towed, and robust enough to remain in the air should a rotor engine stop functioning. It needed to be deployable within 72 hours and be heavy-lifted by aircraft to anywhere in the world. The Marine Corps' CH-53A helicopter, modified for towing, offered the desired features. Often described as the size of a school bus, it had a seven-bladed propeller powered by three engines that offered a combination of both lift and redundancy. It had the power needed to pull the magnetic-minesweeping hydrofoil sled recently engineered for the purpose by the research and development facility at Panama City. Perhaps of equal interest to an efficiency-minded CNO, its heavy-lift capability meant that it could be used by the Navy for other missions. Moreover, the Marine Corps already maintained a fleet of the helicopters to perform heavy-lift missions. Those helicopters could be used as a pool of additional aircraft and pilots to scale up the MCM mission when needed. Tactics and training would have come in part by intentional design (probably by modifying existing SMCM practices) and from the lessons learned of exercises.[57] The first AMCM unit, HM-12, was stood up on April 1, 1971.[58] By all appearances, AMCM would be ready when called upon to support mine clearance operations at Haiphong Harbor in 1972.

The Mining of Haiphong Harbor

By early spring 1972, the United States had mostly finished withdrawing its forces from South Vietnam. Ground forces and ground-based aircraft had all departed the country, leaving the original U.S. commitment of military advisers. The Navy continued patrolling the seas, with carrier aircraft and Air Force bombers flying from abroad continuing to provide ground support and supply route bombing. In Paris, representatives from each of the countries conducted ongoing negotiations aimed at achieving a final peace. Then, on Good Friday, the DRV threw the idea of peace out the window. Conventional forces poured over the demilitarized zone (DMZ) while other forces, having been spirited down the Ho Chi Minh Trail during peace talks, attacked out of Cambodia. With well over 100,000 troops and hundreds of tanks battling throughout the northern half of the country, the two on-scene American carriers launched attack after attack to help slow what was known as the "Easter Offensive." After a week, two more carriers arrived, followed a week later by yet another. Six weeks after the attack began, a sixth carrier arrived in mid–May, having traveled from Norfolk nearly half a world away. Now, with the largest concentration of carriers since World War II, the United States took off its gloves and prepared to deliver a combination of blows that would inspire North Vietnam to "give peace a chance."[59]

Up until that point, the United States had maintained a policy of restraint in Vietnam. Painfully aware of the Chinese intervention in the Korean War, policymakers accepted escalations only in stages. They had mined DRV rivers and infrastructure and bombed some industry to impede that country's ability to supply its forces operating in the south. But they had not yet used one of the more powerful tactics in their arsenal, one that helped encourage Japan to consider peace. They had not yet used naval mines to blockade the country's ports. The danger was that they might mine one of the Chinese or Soviet ships that supplied the country with war matériel. Doing so could potentially lead to a much larger conflict extending well beyond the proxy war then taking place in Vietnam. The advantage, however, was the potential to cripple the DRV's economy. The capital's port city of Haiphong received 85 percent of the country's imports and all of its fuel. If that port could be mined, as well as every other major port, and the overland routes seriously damaged, then the DRV would have a tangible reason to stop fighting.[60] Furthermore, such a blockade would help deescalate the conflict by constricting the supply of war matériel. When there is an unlimited supply of bullets, there is nothing material to prevent unlimited shooting. Conversely, when that bullet supply suddenly becomes finite, one must be more selective in the choice of targets.[61] Thus was born OPERATION POCKET MONEY, only the second successful strategic mining campaign in history.

The minefield was designed to target large merchant vessels and smaller craft alike using a mix of 1,000-pound MK-52 and 500-pound DST mines. MK-52s targeted large ships in deeper water, while the much more numerous DSTs targeted them in shallow water. DSTs also targeted the smaller vessels that might be used to attempt to bypass or break through the blockade. Of the total mine count, MK-52s represented less than 1 percent. Part of this was because the mines were dropped entirely in territorial waters where the water was mostly shallow. Another part was probably simple supply chain practicalities. MK-52s had to be assembled at the closest mine shop (Subic Bay, Philippines). DSTs, on the other hand, could be assembled aboard aircraft carriers by attaching components from kits onto the general-purpose bombs already available on aircraft carriers. This was not unlike the concept of the Civil War keg mines where crates of fuses could be brought to an area and attached to existing powder kegs (see chapter 3). The mines were given a variety of sensitivity settings to target varying sizes of target, very low ship counts, and sterilization periods of only a few months. The first mines were also given a 72-hour arming delay to allow neutral ships to evacuate following the mining announcement.[62]

Minefield planners created the POCKET MONEY mining plan under the constraint that it must be easily sweepable for U.S. MCM forces yet convincing as a still-active field. And the easiest-to-clear minefield is a mostly self-clearing one where the mine locations and settings are known. Therefore, the POCKET MONEY minefields were all set with relatively short self-sterilization or self-destruct times. Miners used only bottom influence mines. With no contact mines, there was no need to worry about mechanical sweeping, the danger of shallow-moored mines, or managing large numbers of surfaced mines. Self-destruct settings would result in mines blowing up at random times (keeping the risk of transit palpable).[63] Conversely, short self-sterilization times meant that one could be reasonably safe when sweeping after the waiting period. If a mine is already sterilized, it will not blow up regardless of the magnetic signature being generated.

POCKET MONEY's timeline, along with that of the preparations for the follow-on MCM effort, OPERATION END SWEEP, meshed with the DRV's reluctance to negotiate in good faith. The first batch of 36 mines splashed into Haiphong Harbor on May 8, 1972 … less than five weeks after the Easter Offensive began and before the sixth aircraft carrier arrived.[64] President Nixon then announced the mining and asserted that the minefields would stay until the American prisoners of war (POWs) had been returned. Upon this announcement, 27 foreign-flagged vessels chose to remain in port rather than risk the 72-hour time-delayed mines.[65] Over the next eight months and at six different ports as well as Haiphong,[66] carrier planes flew 1,149 sorties to seed or reseed roughly 12,000 mines. During the same time, U.S. Air Force planes dropped an additional 2,000 or so mines inland to interdict ground transportation.[67] When negotiations broke down in mid-December, U.S. forces cranked up the pressure even further by reseeding the minefields and reintroducing bombing in the north, to include the bombing of Hanoi. The last mines dropped on January 14, 1973.[68] A peace agreement was reached just nine days later.[69]

1972 Solar Storm and Keeping the Minefields Operational—A section of chapter 2 described a spontaneous detonation of mines at Kronstadt, thought to be the work of an electrical storm. The incident was substantially repeated as the result of a solar storm from August 2–4, 1972. During that period, mines in the hundreds were estimated to have detonated in a matter of minutes. At least part of those detonations was witnessed by American pilots who saw about 24 mine explosions in a 30-second period in a part of a minefield where no ships were passing. During that same time, people across the globe were experiencing sudden problems with their electronic systems. The cause of this widespread disruption was thought to be the massive solar flares that had erupted from a sunspot. Solar flares send huge amounts of radiation through space. Earth's magnetic field protects the planet from the worst effects of the radiation (i.e., everybody getting microwaved), but it also gets distorted by the radiation. Many of the DST mines used in Vietnam, sensitive to changes in the local magnetic environment, reacted to this sudden distortion by exploding.[70]

The solar storm happened only a couple of weeks after the minefield reached its peak strength. The widespread explosions demonstrably reduced that strength well

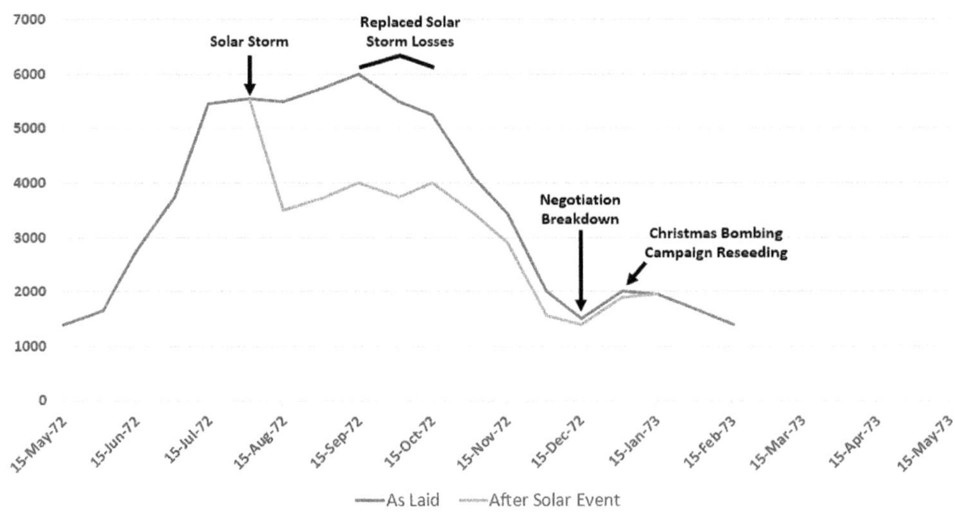

Estimated number of active mines remaining in coastal fields (derived by author from U.S. Navy chart, "The mining of the DRV 08 May 1972 to 14 January 1973," Mine Warfare Project Office).

ahead of schedule. It was always expected that the minefield would reduce in strength over time because of the self-sterilizing and self-destruct settings.[71] Self-destructing mines had a few effects beyond reinforcing the minefield's credibility, such as convincing the DRV that the United States had the ability to neutralize its mines remotely.[72] A less desirable effect was that every mine self-destruction gave a visual indication that the minefield had weakened by one mine. Although that outcome may be acceptable when it is one mine at a time, *hundreds* of mines exploding all at once harms the minefields' credibility. At a local level, individuals might conclude that the minefields simply disintegrated and were suddenly worth risking. Importantly, the minefields' damaged credibility likely also lowered their power as leverage in negotiations. *(Conjecture: Possibly this accounts for some of the DRV's actions during negotiations.)* Reseeding the minefields became important. After a short period while the Navy figured out how to resolve the mines' weakness to solar storms[73] and to get the changes to the fleet, the prematurely detonated mines were replaced. Although self-sterilizing times for many earlier mines were expiring by that point, the display of reseeding may have helped shore up the lost credibility from the mass detonations.

Mine Clearance Preparations—Preparations for clearing the POCKET MONEY mines began almost as soon as the main fields were finished being laid. Officially they began a mere three hours after the first mines hit the water, when the chairman of the Joint Chiefs of Staff sent a message advising the Pacific fleet commander that mine clearance would probably prove necessary.[74] Two days later, commander Mine Warfare Force (COMINEWARFOR, or "CMWF," a role not unlike COMINEWARCOM of the 1990s and 2000s[75]) held a planning conference to figure out what would be needed to clear the planned minefields. This probably included an assessment of what mines and mine settings could be countered by which techniques. That, in turn, likely drove which mines and mine settings would be used in the mining. From the meeting, a Mine Warfare Project Office (PM-19) was established to oversee the acquisition and logistics of all necessary matériel. With basic planning complete, President Nixon had a reasonable reassurance that a path forward existed to make his May 16 announcement that the mine danger would be removed once the American POWs were released.

Over the next few months, the MCM force received important equipment like a production MK 105 magnetic minesweeping sled and the precision navigation system Raydist (radar distance ... it acted like a land-based version of GPS). They exercised with this equipment in Charleston and Panama City, which in turn yielded the types of practical lessons learned needed to build realistic plans for MCM operations.[76]

Meanwhile, the Navy began moving its remaining active-duty MSOs to Guam to be available for the expected clearance orders. On Halloween, USS *Inflict* was engaged in environmental survey operations off the coast of the DRV[77] when the Joint Chiefs of Staff order came out alerting the fleet about the expected MCM operations.[78] Over the next several weeks, all available AMCM and SMCM units converged on Subic Bay, where they fell under CTF 78 headed by the COMINEWARFOR, Rear Admiral Brian McCauley. There, they exercised in preparation for the upcoming clearance.[79]

Some of the ship movements were disrupted in December when talks broke down.[80] One reason for the breakdown was the South Vietnamese desire for security guarantees following the U.S. withdrawal. Another was U.S. objections to the DRV demands that the United States clear inland waterways and that all mines be "removed" (a major safety hazard). With an impasse reached, on December 18 the United States resumed mining and bombing, which had been suspended as an act of good faith since October 23.[81]

The final mine clearance protocol, signed on January 27, 1973, still required U.S. responsibility for clearing inland waterways, but with full DRV participation. It still included the term *removal* of mines, but only as an option along with more realistic means of rendering the mines harmless.[82] In reality, later on-scene negotiations resulted in the United States providing mine clearance equipment and training the DRV personnel in their use. Meanwhile, nearly all mines were already rendered safe by their self-destruct and self-sterilization settings.

As one might expect, the bulk of the negotiations about the MCM protocol happened without advice from a U.S. MCM expert. When one was finally assigned just nine days before the agreement was signed, his first task was to brief the negotiation team on U.S. mine countermeasures capabilities.[83]

Operation End Sweep—Operation End Sweep was perhaps the least-typical harbor clearance operation in MCM history, one that more closely resembled a play than a serious war necessity. As mentioned above, the United States had set its mines with U.S. sweeping in mind. There was no mystery as to what types of mines were in the water, or their settings. It knew which batches with which settings lay in the different parts of the harbors. Furthermore, toward the end it enjoyed the luxury of being able to sweep in areas where it knew that the mines had long passed their sterilization dates. Finally, it knew well in advance that it would sweep these waters, and no others. With this overwhelming collection of available information, the planners could create the perfect plan that was guaranteed to leave zero mine risk with the loss of zero ships due to mines. Furthermore, with a long, unknown lead time, the sweeping assets were able to rehearse and re-rehearse the plan.[84] With all the odds stacked in their favor, the MCM task force simply could not fail.

U.S. mine clearance in Haiphong Harbor began on February 6 with four MSOs sweeping the deep-water anchorage for the Rear Admiral McCauley's flagship and MCM support ship USS *Worden* (CG-73). This action gave early evidence that the United States was pursuing its treaty obligations while also giving the newly equipped AMCM units more time to assemble and train. A few weeks later, the AMCM force of four amphibious ships carrying CH-53A helicopters joined the MSOs. On February 27, 1972, HM-12's executive officer (XO) flew the first AMCM mission, with Rear Admiral McCauley aboard, against a live minefield. The following day, the United States pulled the entire MCM force out of the area in protest over the DRV delays in releasing American POWs. Returning a week later, after the POW dispute had been settled, the force resumed mine clearance. On March 9, an AMCM unit successfully detonated its first mine, the only mine detonated by any unit involved in END SWEEP.[85]

Note: Three other mine detonations were witnessed during the operation, but all too far from any MCM unit to be attributable.[86] The lack of frequent explosions was cause for at least one reporter on March 8 to question if the whole situation was a ruse used to fool the DRV.[87] Indeed, it was a fair question. When mines do not explode during influence sweeping, planners should question their sweeps' effectiveness. A lack of explosions may mean: (1) the sweep setting being used is ineffective against the mines in the field, (2) more sweeping is needed, or (3) there are no mines in the area being swept. In the very special case of END SWEEP, the cause was already known.

Throughout END SWEEP, local negotiators from both sides met on a frequent basis to negotiate every operational aspect, from what areas to sweep to when, how, and with what method of monitoring. One important change came in early March when the DRV announced that it alone, and not the U.S. forces (as originally demanded), would sweep the inland waters. U.S. responsibility was limited to providing advice, training, and equipment. One such piece of equipment, the magnetic orange pipe (MOP), was little more than a long, floating, metal cylinder with an increased magnetic signature. The relatively lightweight device could be towed by helicopters or even by groups of people or draft animals dragging it through canals and rivers. Classroom training and a limited demonstration continued until April 16, when the DRV unilaterally declared training to be completed. From there, it assumed all responsibility for inland clearance, though it continued to ask for additional equipment.[88]

Meanwhile, minesweeping by MSOs and by Navy and Marine helicopters continued for the next several weeks. The fleet suffered noncombat casualties such as the sinking of USS *Force* after detaching, an engine room fire aboard the USS *Enhance*, a 17-knot collision between an MK 105 sled and a fishing boat, and two helicopter crashes. The helicopter crashes led to a stand-down of all aircraft to allow for maintenance, while the collision led to the DRV negotiators agreeing that U.S. sweeping could be limited to the main channels. As operations continued, various ships and helicopters rotated through the force, relieving or replacing units worn out by the continuous strain of operating in a minefield for weeks on end.[89] Before the operation had ended, this force would see 10 MSOs, nine amphibs, 31 CH-53A helicopters, and 28 other supporting ships rotate through END SWEEP duties.[90] On April 14, special minesweeper MSS-2, flying the American flag, commenced check runs of the Haiphong main channel.[91] A former tank landing ship (LST), filled with foam and padded for the protection of her volunteer skeleton crew, her job was to prove that full-sized steel ships could pass through the channel without detonating mines.[92]

CH-53 helicopter sweeping in Hạ Long Bay, Vietnam, during Operation End Sweep, 1973 (U.S. Navy photo by PH1 George Norris, Navy History and Heritage Command).

On April 17, in protest over DRV failure to comply

with the cease-fire in neighboring countries, the MCM force once more departed the area. Following the signing of a further agreement, the MCM force resumed operations on June 13. Its task was to complete sweeping within a month. Since all mine timers had expired long before, this posed no problem since the minefields had already cleared themselves. Sweeping resumed in the country's southern ports. Efforts were consistently hampered by DRV unwillingness to provide necessary information or to agree to U.S. forces operating in multiple areas at once. On July 5, with all reasonable efforts accomplished, Rear Admiral McCauley notified his counterparts that clearance operations were completed in those areas. Despite ongoing DRV complaints (believed to be for propaganda purposes), no more U.S. clearance efforts took place after that date. On July 18, after a final bilateral meeting, Task Force 78 departed the area.[93]

Results and Lessons Learned—The operation's total cost was around $21 million,[94] nearly double the $12.5 million or so spent on the mining operations.[95] For that $33.5 million spent, the United States had secured a cease-fire, the release of all U.S. POWs, DRV recognition of the South Vietnamese government (temporarily), and domestic support for the new AMCM force. As an instrument of strategy, the two complementary operations proved a remarkable success.

Despite the operation's feat in achieving diplomatic goals, it cannot be seen as an example of a typical mine clearance operation. It was done under the perfect conditions of being a high national priority, having all available support at its disposal, and with mine types and settings being simple and known. In his final report, Rear Admiral McCauley appropriately offered this caution:

> It would be a mistake to attempt to devise general, long-standing mine warfare conclusions from the specific operational and political arena in which End Sweep was conducted. End Sweep was a unique solution to a unique problem and did not present a challenge of nearly the magnitude that can be expected in the future. The location, type and settings of all mines were known. The vast majority of mines were the DST-36, a very sensitive magnetic or acoustic fuse placed on a 500-pound aircraft bomb. The magnetized pipe (MOP) was effective against this mine. It will not counter properly designed sophisticated mines.... The objective of the sweeping was largely accomplished prior to laying the mines when the self-destruct time was set into the fuse.[96]

Conclusion

The U.S. experience with mine warfare in Vietnam was nothing like anything that it had experienced in the previous 100 years. It was a guerrilla war against an enemy with powerful friends, yet one that did not practice the more classical forms of mine warfare. Where traditional U.S. minesweeping units came into their own only when the United States started laying mines in coastal waters, the nontraditional brown-water Navy had a far different experience. Suddenly, most of the scientific technology of the previous half century gave way to small crews on small boats using bladed chains to cut RECO mine detonating wires. Half sailor, half soldier, they were among the tiny fraction of the Navy to ever see the whites of the eyes of the enemy trying to kill them. But in the end, it was their own mines that ended the war, cleared by traditional MCM ships and the newfangled minesweeping helicopters that seemed destined to replace them.

Riverine mine warfare had little use in the desert wars that followed over the next half century, but it does not mean that it would never return. It takes certain conditions for riverine mines to have value, conditions where it is easier to travel by boat than vehicle. The Southern United States during the Civil War was one such place. Vietnam was another. It is impossible to say where such warfare might return in the future. The Crimean Peninsula? Finland? The Netherlands? The rainforests of South America, Africa, or Southeast Asia once again? When and if it does return, the military leaders involved will be best served by reviewing the history books to unlock the lessons learned in the Vietnam War.

Chapter 11

Middle Eastern Mining (1984–2021)

Introduction

It was the 1980s. Cold War hostilities continued around the globe. Both the United States and the USSR provided funds and arms to battling factions in the Third World. Military dictatorships and guerrilla revolutionaries became increasingly more common. In some parts of the Muslim world, terrorism became a tool of guerrilla warfare, pushing outsized security costs on adversaries. As power vacuums formed, military dictators fought rivals in ways not countenanced by the Geneva Conventions that were signed before their countries were formed. One of those ways was unannounced minefields.

Over the next quarter century, mines appeared throughout the Middle East in places as far-flung as Tripoli Harbor, the Gulf of Suez, the Gulf of Oman, and the Arabian Gulf.[1] These mines both helped and hurt the mine countermeasures (MCM) forces of Western navies. They helped by demonstrating to the West their vulnerability to mines, which in turn provided visceral examples that normally lead to the rebuilding of capabilities. At the same time, they hurt by giving a whole generation of leaders an inaccurate picture of mine warfare. With only one major mining event, it became easy to forget the lessons brought about by the realities of working in the more complex mine environments of past wars. Two decades of mine warfare on "easy mode" does little to prepare navies for fighting mines laid by a competent peer-competitor navy. What it saw instead was poorly assembled mines, poorly set and laid in senseless ways that often allowed their targets to get away with less-than-perfect countermeasures. Despite the miners' incompetence, they still managed to damage three U.S. warships and about 20 merchant ships.

Background

Iran—Unlike most of the Muslim world, Iran had been a part of neither the Ottoman Empire nor any other European power since Alexander the Great. As Persia, it had a long and storied history of conquering civilizations and being conquered by other civilizations. Through much of its history, its people had lived as one with Islamic values and sensibilities under various internal and external empires. That harmony changed in the early 20th century when factions began pushing for a more secular society. A series of monarchs took power in the country, aided by foreign assistance from Russia, the United States, and the UK. In 1951, the country nationalized its oil industry at the expense of the foreign companies then operating in the country. Two years later, a CIA-funded coup led to Mohammad Reza Shah seizing the reins of power. Over the next quarter century, he implemented enormous reforms that increased the country's gross domestic product while destroying many traditional ways of life and ideals of morality. His actions helped foreign oil interests even as they improved quality of life and women's equality. One of his greatest supporters was the United States, which exported its culture, technology, and weapons systems to aid Iran's modernization. Part of that modernization was a shift to industrialization and urbanization, the benefits of which were unequally distributed. Furthermore, he maintained close ties to Israel at a time when much of the Islamic world refused to recognize its right to exist. These factors, along with enforced reduction in the power of the Shi'i clerics, increasingly offended a growing Islamic fundamentalist movement.[2]

By the 1970s, a population exasperated with the shah's brutal methods became increasingly drawn to recordings of the exiled Ayatollah Ruhollah Khomeini. His message painted a picture of an alternative society ruled by Islamic law. In January 1978, a newspaper's unflattering portrayal of the Ayatollah sparked protests by religious students that turned into a more generalized protest against the shah. This escalated into a cycle of repression and protests that led to the mass unrest that ultimately drove the shah from the country a year later. In February 1979, Iran's military declared neutrality, thus allowing the ayatollah to return to take control of the government. Under the new regime, the military looked to external threats, while the religious students became a militia force of fanatics later known as the Iranian Revolutionary Guard Corps (IRGC). These militias worked to cleanse Iran of all vestiges of Western influence. When the exiled shah was brought to the United States for cancer treatment that same year, IRGC backlash led to the capture of the U.S. embassy with the taking of 66 hostages.[3] When the United States assembled a rescue mission using Navy RH-53 minesweeping helicopters, mechanical problems in five of eight aircraft caused the mission to be aborted with the concurrence of the president.[4] Though the decision probably saved the involved service members from being added to the list of killed or captured, it was widely scorned domestically. Under the shadow of this incident, reinforced by daily reminders from a hostile media, Carter lost the 1980 presidential election to Ronald Reagan, 489 to 49 electoral votes. As an insult to President Carter, the IRGC released their American hostages only two hours after Reagan's inauguration.[5] Iran would continue its hostility toward the West through the present day (2024 as of this writing).

Iraq—Whereas Iran went the direction of religious fanaticism and anti-Westernism, Iraq took the path of secular military dictatorship. Formed in 1921 from a remnant of the Ottoman Empire post–World War I, Iraq gained its independence from its new British rulers in 1932 under King Fayṣal. In the same year, the British separately granted independence to Kuwait, a move not agreed to by Iraq's government. Iraq developed under the Fayṣal Dynasty until 1958, when a series of military coups installed a line of dictators friendly to the Soviet Union. In 1968, the Ba'ath Party seized control. Over the next 11 years, the country had five presidents. The last one, Saddam Hussein, came into the position in 1979, where he would remain until the country fell during Operation Iraqi Freedom in 2003.[6]

Saddam Hussein came into power in the same year as the Iranian Revolution. Iran's revolutionary zeal and excesses represented both a threat and an opportunity for the new president. The threat was both internal and external. Internally it could inspire radicals in his own country. Externally the risk was that the ayatollah, hostile to the non-religious Saddam, would send his revolutionaries across the border. Also at issue was control of the Shatt al-Arab waterway, the river formed from the confluence of the historically important Tigris and Euphrates Rivers. This waterway was nothing more than a border for Iran, but to Iraq, it was the country's sole access to the world's oceans. Capitalizing on Iran's internal turmoil, Iraq attacked across the river to gain a foothold on its eastern bank in hopes of settling the issue of control over the waterway. By 1982, having seized limited territory, Saddam offered to negotiate peace. The ayatollah refused. An external enemy is a boon to authoritarian regimes, giving them an excuse to continue repressing their citizens when all internal enemies have been purged. Thus, the Iran–Iraq War would continue for most of the decade, featuring fighting on land, attacks on each other's maritime trade, and over a million killed or wounded. In 1987, following the events discussed below, the war would finally end with no territory exchanged.[7]

Libya—While most of this chapter centers on Iran and Iraq, Libya is worth discussing as being one of the original progenitors of mining in the Middle East. Originally, Libya was a long-standing, semiautonomous part of the Ottoman Empire. In 1911, Italy proclaimed its own dominance over the territory as part of the larger European movement to colonize Africa. When Italy lost World War II, it lost its African colonies. In 1951, the United Nations granted Libya its independence under its first and only king. A military coup in 1969 ended the kingdom, with Muammar al-Qadhafi (Gaddafi) taking power as the new dictator. He would reign until 2011, when the country was swept up in the revolutionary wave known as the Arab Spring.[8]

Terrorism became a huge international phenomenon in the 1970s. Between 1970 and 1983, approximately 15,000 major terrorist attacks took place worldwide.[9] Qadhafi envisioned himself as the leader of a movement for Islamic socialism that he hoped would sweep the Third World.[10] Heavily funded by oil revenue, he launched multiple attacks against Chad and supported terrorists (especially Palestinians) throughout the Islamic world. Toward Egypt he cast special vitriol for its acceptance of peace with Israel. He demanded that Libyan expatriates return home and sent assassins to kill those who remained abroad. With Qadhafi's blessing, Libya terrorists committed over 50 acts of terrorism between 1979 and 1985, ranging from assassinations to bombings to the mining discussed below. Included in these were attacks on around five foreign embassies in Libya.[11] This act would be repeated at the U.S. embassy in Benghazi in 2012.[12] Yet, this attack on a U.S. embassy in Libyan territory had historical precedent. In 1801, the Tripolitan Pasha declared war against the United States by chopping down its embassy's flagpole in Tripoli during the First Barbary War.[13]

Middle Eastern Mining

Mines in the Gulf of Suez—In mid-summer 1984, a series of underwater explosions damaged about 16 ships in the Gulf of Suez[14] at three locations ranging from the Suez Canal approaches to the Bab el-Mandeb Strait.[15] Mines were suspected. Soviet-made mines. Mines later deduced to most likely have been laid by Libya to damage and/or bully Egypt, Saudi Arabia, and the greater international community siding against Iran in the ongoing Iran–Iraq War. The Libyan-flagged ship *Ghat* was suspected of the mining during its unusual seven-day delay in the Gulf of Suez starting just three days prior to the first explosion. Having loaded in Tripoli with Libya's mining program manager and other military members onboard, she was later found to have damage to her aft ramp. It was believed that Soviet-made cylindrical mines may have been simply rolled down that ramp into the water under cover of darkness, damaging it in the process. The irony is that the first ship damaged by these Soviet-made mines was Soviet. Given the timing of all the other explosions, it was believed that this had been a premature detonation.[16]

Egypt and Saudia Arabia, having limited MCM capabilities, requested the assistance of the international community to clear the mines threatening international shipping. Though the British did the bulk of the mine hunting, French and Italian ocean minesweepers (MSO) of American build also took part. At Egypt's specific request, the United States sent helicopters from HM-12 and HM-14, which departed within hours on C-5 aircraft. Upon arrival in the area, they embarked on the USS *La Salle* (LPD-3) and the USS *Shreveport* (LPD-12), from which they conducted mine hunting and influence minesweeping. They were helped in their mine hunting by the hydrographic survey ship USNS *Harkness* and an EOD detachment. Though no mines were found in their assigned areas,[17] the decision to embark on amphibs rather than on land proved fortunate. By doing so, they drastically reduced the amount of shore-based support that would need to be flown into the area. It also allowed them to avoid unnecessary complications in negotiating with their local counterparts and allowed them to shift operations with ease.[18]

Though no ships were sunk (and at least one attempt at insurance fraud was attempted),[19] the mining incident and fear of more underwater terrorism caused a bit of a global MCM renaissance. Navies of the world studied the event for lessons learned and reconsidered their approaches to MCM. The main lesson was that mine countermeasures required a balanced force of surface, air, and underwater assets. For the U.S. Navy, the event helped justify its investment in an increase in AMCM aircraft and the building of two new mine countermeasures ship classes then in the early stages of production. For Iran, which vigorously denounced the mining,[20] the attack may have inspired its actions just a few years later. Indeed, it would have helped justify its own ongoing efforts to build mining capabilities.

The Iran–Iraq War

Throughout much of the 1980s, Iran and Iraq continued their war on land and then at sea. Iran sent human-wave attacks across the border several times between 1982 and 1984 but achieved nothing except to fuel the popularity of martyrdom as a source of family pride. In 1984, Iraq started fighting off these attacks using chemical weapons and began an economic campaign to cripple Iran's economy by striking its oil tankers in the gulf. A shift to bombing attacks on civilian populations by both sides began in 1985, to be followed by more human-wave attacks by Iran over the two years that followed. Through all of this, the attacks against each other's sea-based oil industries continued, in often brutal ways. Iran mostly used small boats and gunfire, while Iraq preferred to stay relatively close to home and launch Exocet missiles from a distance. In one instance, one Iraqi strike accidentally hit the neutral American naval ship USS *Stark*, killing 37 sailors. That ships were in the region at all despite such conditions reflects the region's importance in supplying much of the world's energy needs.[21]

The Iranian Navy (IRAN) began building a mine warfare capability as early as 1983, followed by the IRGC Navy (IRGCN), which grew a similar capability the following year. Acquiring simple pre–World War I contact mines from the Soviet Union, their mining programs grew to include storage, assembly, and even a limited mine fabrication capability. By 1984, they were testing innovations like dropping contact mines from their small number of RH-53 helicopters. Lacking purpose-built minelayers, they experimented with methods for deploying mines from other crafts of opportunity such as destroyers, patrol boats, or even troop transport ships. At the time, it was assessed that the country would avoid laying mines outside their own territorial waters for fear of inciting the involvement of the United States and other countries. However, when they extended their ship attacks to Kuwaiti tankers, the United States became involved anyway. At that point, Iran overcame such inhibitions. Its subsequent mining efforts caused it to lose both its navy and the long-standing war in the process.[22]

By 1987, the Iran–Iraq War was heating up at sea. Unable to gain the upper hand against Iraq and its Exocet missiles, Iran turned its attention to the smaller, weaker Kuwait. Situated at the extreme end of the Arabian Gulf, Kuwait was like a referee trapped in a cage with two professional wrestlers who were fighting a no-holds-barred battle royale, folding chairs and all. With battles raging just across the river from its border, Kuwait supported its closer neighbor, Iraq, against the more belligerent Iran.[23] Although this support did not extend to providing military forces, it allowed its infrastructure to be used for the transshipment of foreign military aid. It also loaned Iraq large sums of money that could be used to continue prosecuting the war. In retaliation, Iran first used its network of fanatics to foment acts of terrorism throughout Kuwait. By late 1986, it was directly attacking Kuwaiti ships and neutral ships destined for Kuwait. Desperate for relief, it approached both the United States and the USSR, predicting their help would ultimately bring the war to an end.[24]

The Soviets were the first to send a merchant ship, which Iran promptly attacked. Responding by sending 50 aircraft across their huge, shared border, the Soviets reminded Iran that its campaign in Afghanistan could be easily extended to encompass Iran as well. For the United States, which did not particularly like either belligerent, it was the prospect of the Soviets gaining favorability in the gulf that prompted it to act. Having lost credibility in the region during the Iranian Revolution, the United States wanted to shift favor back in its direction. In January, it accepted a Kuwaiti proposal to reflag all 11 of its tankers under a U.S. company, thereby gaining U.S. protection.[25] During the six months required to complete the reflagging process, Iran began mining the sea-lanes used by these ships. One of the first ships hit was a Soviet-owned tanker leased to Kuwait. On July 22, 1987, the United States formally began convoy escort operations with the newly reflagged Kuwaiti vessels.[26] They got underway that same day, with three warships escorting two reflagged ships including the *Bridgeton*, the sixth-largest tanker in the world.[27]

That Iran would attack the convoy was no surprise. It already knew the convoy's expected timeline because of press reporting. Knowing such information over a week in advance, Iran had plenty of time to prepare an ambush. As the convoy passed into the gulf through the Silkworm-missile-infested Strait of Hormuz, Iranian aircraft made threatening gestures but did not attack. As the convoy passed the Iranian Abu Musa

Island in the southern Arabian Gulf (SAG), the convoy once more braced for an ambush that never came. On the night of July 23-24, AWACS detected significant Iranian activity near Farsi Island. Suspecting that the activity was rehearsal for an Iranian small boat attack, the escort ships once more prepared to fight off a swarm. No gunboats appeared. Nothing appeared. The waters of the Northern Arabian Gulf (NAG) appeared empty. They were not.[28]

At 0655, *Bridgeton* struck a mine. The explosion shot shrapnel from the hull up through intervening decks to burst out of the weather deck. People were thrown off their feet as far away as the bridge, hundreds of feet from the blast. The ship slowed as water quickly flooded four of 31 compartments with over 20,000 tons of seawater. But this double-hulled behemoth was designed to carry far more liquid than that. None of the crew members were injured. They, and the giant vessel that carried them, simply shook off the initial shock and kept on going despite the new 15-foot-by-30-foot hole in the hull below the waterline. Knowing that a similar blast could destroy the smaller escorts, the *Bridgeton*'s captain, a former Navy O-6, advised them to follow in his ship's wake. With no clue as to the extent of the minefield, the escorts heeded his advice. Later that minefield was found to be two lines of M-08 contact mines crossing the deep-water channel, laid the night before by the small boats from Farsi Island.[29]

The event brought instant embarrassment to the U.S. Navy. Despite bringing a cruiser, a destroyer, and a frigate to fight off any attacks on or above the water's surface, nobody had brought anything to look *below* the surface. Considering that Iran had already used mines to attack the Soviet merchant just two months prior, it should have come as no surprise. Helicopter Mine-Countermeasures Squadron (HM) 14 had been on 24-hour alert for months before these mines were dropped in the gulf. A Senate report from less than two weeks prior actually warned that a mine attack was likely and advised advanced preparation for MCM. But those preparing the convoy were no mine warriors. The mine warfare fleet was then in a pitiable state, and service in that fleet was a career death sentence. Any commander raised high enough to prepare high-profile convoy operations would have little if any experience in mine warfare and, thus, little thought for it. Even sending a helicopter airborne to scout for mines might have found the mine lines. But that simple expedient did not happen. When the *Bridgeton* struck the mine, the United States had no MCM platforms anywhere near the gulf. Minesweeping had to take place from Kuwaiti tugboats, equipped with minesweeping boat (MSB) sweep gear and partially manned with U.S. Navy volunteers from Bahrain.[30] In response to the events, the United States rushed HM-14, six ocean minesweepers (MSO), four MSBs, and a special Army helicopter unit, Task Force (TF) 160, to the Arabian Gulf to assist.[31]

In August 1987, Iran once more mined neutral shipping. Knowing that the U.S. convoys formed off the Gulf of Oman side of the United Arab Emirates (UAE), it laid mines in an anchorage near Fujariah, UAE. That an attack was coming was no surprise. Even the approximate area was suspected. On the 10th, a Panamanian-flagged supertanker struck one of these mines, forcing it to offload its oil to avoid an environmental disaster (ironically, it was loaded with Iranian oil). Five days later, another one of these mines destroyed the 156-foot Swedish supply boat *Anita*, killing five sailors. The resulting blowback from the *Anita* mine strike was not what Iran wanted. Whereas the United States had previously been unable to garner MCM support from other nations, this indiscriminate mining drew international condemnation. Soon, other nations began sending warships to the gulf. Iran would eventually admit that it had laid mines in the gulf, though it severely underreported the geographic scope of its activities.[32]

Five weeks after the *Anita* mining, the United States got proof that it was Iran that was mining international waterways. In mid–September, it received an intelligence report that naval mines were being loaded onto the troop transport *Iran Ajr* prior to it getting underway. The suspected drop location was an area where U.S. naval exercises were scheduled for the next day. Late on the night of September 18, TF-160 helicopters flying from the frigate USS *Jarrett* were tasked with observing a ship known to be *Iran Ajr* at the targeted area. Darkened and designed to be unusually quiet, the helicopters allowed the Army pilots to watch through infrared optics as Iranian crew members uncovered mines on the ship's deck. When the pilots reported the first mine splashing into the water, they received the order to fire. Their initial attack drove the crew back from the mines but did not stop the ship. During a lull in the attack while the helicopters returned to *Jarrett* to rearm, the crew pushed more mines overboard. The helicopters attacked again. With excellent precision, they used gunfire and other munitions to drive the crew from the mines, kill the bridge crew, and disable the engines. The surviving crew abandoned ship.[33]

Soon the Navy SEALs arrived to take the ship. What they found on board were M-08 mines and Iranian documents related to the mine-lay. Seven mines had gone into the water. Members of the rescued crew proved to be Iranian military members in uniform.[34] Eventually, the United States repatriated the crew and destroyed the ship. Chastened, Iran waited over half a year before it would once more mine international sea-lanes.

Over that next half year, both the U.S. presence and Iranian aggression increased in the Arabian Gulf. Attacks against tankers continued. Low-level attacks were exchanged between the two countries, resulting in the destruction of Iranian patrol boats and American helicopters. A Silkworm missile attack on the U.S.-flagged ship *Sea Isle City* led to a retaliatory assault on an Iranian oil platform. After that, the convoys continued while Iran continued to make threatening gestures toward the U.S. ships and their charges.[35]

In April 1988, the United States once more encountered Iranian mines. While traveling between duty assignments, the USS *Samuel B. Roberts* (nicknamed the "Sammy B") found itself confronted with what appeared to be mines. As it backed away from the danger and prepared to send a helicopter to investigate, it struck a moored mine astern. With a 20-foot hole and flooding threatening to tear the ship in half, and with fires raging in the engine room, the crew struggled to keep the vessel afloat. After they fought the damage for hours, the ship finally pulled itself out of danger.[36]

Subsequent mine recovery revealed lot numbers from the same series as those found on the *Iran Ajr*.[37] With this evidence in hand, the U.S. Navy set off to affect a proportionate response in OPERATION PRAYING MANTIS. That response was to strike two Iraqi oil platforms being used for military purposes.[38] Iran answered with counterstrikes by its own naval forces. But with the *Samuel B*'s struggles fresh in their minds, the American sailors were in no mood for restrained reactions.[39] In the action that followed, including the only ship-to-ship missile exchange in history, they sank two Iranian frigates, devastated a third ship, and left the two platforms as smoking ruins.[40]

Following PRAYING MANTIS, Iranian mining stopped. The United States upgraded its policy from protecting American-flagged ships to protecting neutral ships as well. Its first attempt to do so on July 2 proved a success, only to be marred the next day by the accidental shooting down of an Iranian airliner. Although periodic skirmishes continued, an exhausted Iran finally agreed to a cease-fire on August 20, 1988. This ended a war lasting nearly a decade, a war that included attacks on 441 ships of many different nations, at a cost of around $2.5 billion. Yet, mines continued to float around the gulf, striking at least three vessels even as the next war loomed on the horizon.[41]

Desert Storm

In 1991, the MCM world encountered its first case of large-scale naval mining since the Vietnam War. Furthermore, it was the most complex minefield encountered since World War II (Wonsan had only three mine types, with only one type of influence mine in small numbers). On the other hand, the coalition MCM force that countered it did so with the newest concepts. Though the U.S. MCM force had hit a nadir since the end of the Vietnam War, a new ship class and new technology helped it join the company of the other professional navies.

On August 2, 1990, Iraq invaded Kuwait. Kuwait City, the capital, fell in hours while the king fled to Saudi Arabia to build a government in exile. Saddam Hussein declared Kuwait to be the 19th province of Iraq. Within days, the United Nations Security Council (UNSC, which includes all of the great powers from whom Iraq might hope to find support) demanded that Iraq vacate Kuwait immediately. Soon afterward, the United States began to amass a coalition force in the region, ready to use military force to drive out Iraq should diplomacy fail. By November, the UNSC delivered the ultimatum for Iraq to leave Kuwait by January 15 or face repercussions. Unmoved and unmoving, Iraq remained. On January 17, OPERATION DESERT STORM began with an aerial bombardment campaign aimed at crushing Iraq's ability to defend against the impending counterinvasion. That invasion, scheduled for February 27, promised to be a grueling desert war as UN coalition forces threw themselves at the dug-in, battle-hardened Iraqi Republican Guard.[42]

Pertinent to a book on naval mine warfare, part of Iraq's defenses included the laying of naval mines off the coast of Kuwait. Knowing that it faced at least one country with expertise in amphibious invasions (the United States), Iraq prepared for a landing. It had been building up a mine inventory during the 1980s, so that by 1984 it already possessed between 500 and 1,000 Soviet mines. By that same year, it also made plausible claims to have laid some mines in Iranian waters.[43] Doubtlessly, the Iraqis viewed Iran's later use of mines as successful and as a valid method for harming stronger navies. Their skills and inventory matured by 1990, likely through further arms imports and through learning something of mining doctrine from their suppliers. Certainly, a perusal of history may have informed their approach as well. The Russian port of Kronstadt during the Crimean War certainly offered a similar geographic problem to that of Iraq.

Desert Shield MCM Buildup—The Desert Shield military buildup began during a transition period in the international balance of power. For decades, much of the world's power resided in the orbits of the two superpowers, the USA and USSR. The Soviet Union's naval doctrine called for heavy use of anti-access, area-denial (A2/AD) in the form of naval mines. In the event of an all-out war with NATO, it expected to extensively mine many of the same harbors and chokepoints mined by the various belligerents during World War II. At the same time, it expected and would aggressively defend against mining attacks on itself. For this reason, it maintained a minesweeping fleet of some 300 ships[44] and invested most of its ships and sailors with mining capabilities.[45] As the countries closest to that threat, the European NATO countries and Japan invested heavily in their mine countermeasures capabilities. With countless unexploded mines sitting in their waters following World War II, they had plenty of real-life targets against which they could practice. After years of working together with the latest equipment, they had developed a highly professional system for cooperating to clear mine threats.[46]

For the United States, things were different. As discussed in the last chapter, its surface mine countermeasures (SMCM) had been drastically reduced under CNO Zumwalt in the 1970s. By 1991, it had just over 10 aged MSOs remaining in varying states of readiness, mostly in the Naval Reserve. Their replacements were being built, however. In 1979–1980, a pair of persuasive senior officers demonstrated to U.S. Navy leadership that the force could not clear a single U.S. port of Soviet mines. This prompted an MCM resurgence, with 31 ships to be built up as part of President Reagan's 600-ship Navy. It included *Avenger*-class mine countermeasures ships (MCMs) and *Cardinal*-class minesweeper hunters (MSH), as well as other ship types. In 1986, the planned MSH program was scrapped due to design flaws, to be replaced by *Osprey*-class coastal minehunters (MHCs) that followed a preexisting Italian design.[47] Both classes were equipped with mine-hunting sonars and underwater robotic vehicles able to remotely identify and neutralize mines. By the time Desert Shield began, several MCMs had been added to the fleet. Two of these would eventually join the three MSOs sent to the Arabian Gulf, along with six MH-53E AMCM helicopters from HM-14. Being ill-prepared for active mine countermeasures, this force spent the next few months training together from the less-used port of Abu Dhabi, UAE. They were joined by five Royal Navy minehunters and a survey ship. The UK ships, along with the Lynx and Sea King helicopters that they provided for aerial spotting, proved the most capable MCM warriors of the war.[48]

Mines in the Northern Arabian Gulf—It began as a proposed amphibious raid on Kuwait's Failaka Island as a diversionary tactic. Though it was unclear whether an actual landing was intended, there was utility in the MCM and gunfire portions of a landing.[49] After all, such actions are clear indicators of an imminent landing and can feasibly divert defenders even longer than a real landing.[50] With those goals in place, mine clearance operations began.[51]

During Desert Shield, intelligence created an estimate of mined areas off the coast of Kuwait. Up to that point, direct surveillance of the area was not possible because naval and air assets stayed at a distance to avoid provoking the war before the coalition was ready. Special forces operating ashore had observed mining at a distance.[52] Other sources reported large numbers of helicopters working at sea for several days.[53] Additional sources may have included satellite imagery, a captured minelayer, underwater scouting by SEALs (in an area without mines), interviewing fishermen, signals intelligence, etc.[54] What was not available was a detailed insight into Iraq's mining doctrine, since the country itself had only limited institutional experience with mining. The intelligence estimate based on available information showed an approximately 20-nautical-mile gap between the suspected minefields. With nothing else to work with, the task force created a clearance geometry in an area where no mining was expected.

From the arbitrarily designated starting point (Point "E") a little northwest of 29N 49E, a path would be cleared toward the coast. Once complete, other areas would be cleared for the bigger ships. Because no mines were expected in *any* of these areas, "clearance" would be merely verifying that expectation.[55]

As it turned out, the minefields did not match the intelligence estimate. There were a total of 10 minefields or mine lines creating a near-continuous barrier across the entire front of the Kuwaiti coast. That fact would not be discovered until two days after mine clearance operations had begun on the other side of those minefields.

When the U.S. and UK naval forces proceeded toward Point "E" on February 16, 1991, they expected to not find mines. Although the UK ships transited with their mine-avoidance sonars operating, they found no mines as they passed through the minefields.[56] Seven U.S./UK minesweepers, along with helicopter carrier USS *Tripoli* (LPH-10) carrying minesweeping helicopters, unknowingly crossed two minefields to arrive at their start point. Upon reaching Point "E," the AMCM helicopters conducted mechanical and influence minesweeping while two UK minesweepers hunted for mines. None were found, despite a mine line nominally crossing the channel near its beginning. After that unknown mine line was passed, no mines were found because there were none to be found. Ironically, while they worked, the rest of the ships in their group, USS *Princeton* (CG-59) and other ships, all unknowingly steamed back and forth in and around three minefields. On the night of the 17th, in response to an intelligence warning of a likely Silkworm anti-ship missile attack, the force moved farther out to sea. They spent the next 11 hours patrolling inside minefields. Worried about drifting mines, they moved slowly and employed mine watches to scan the surface.[57]

Princeton *and* Tripoli—Early in the morning, the force received the "all clear" for the missile attack warning. At 0435, an exploding contact mine rocked the *Tripoli*.[58] In the ensuing response, the executive officer of the accompanying salvage ship, USS *Beaufort* (ATS-2), was flown out to assess the damage. The damage proved to be contained, though the ship transited at minimum speed to avoid overly stressing the surrounding bulkheads. The brand-new USS *Avenger* (MCM-1), its sonar operators vigorously searching for mines that were suddenly everywhere, led the *Tripoli* out of the minefield. Meanwhile, the *Beaufort* arrived, only to be informed by *Avenger* that she was mere feet away from a mine line. After sallying ship and moving away from the mine line, *Beaufort* stood by to assist.[59]

Less than three hours later, at 0716, the problem became even more complicated.[60] Aboard the *Princeton*, the commanding officer was telling his crew to be vigilant in looking for mines because of the *Tripoli* mine strike. Just as the words left his mouth, an explosion tossed the stern in the air. Two seconds later, another explosion 300 yards off the starboard bow whipped the ship side to side.[61] Both explosions were from Manta mines, high-tech influence mines.[62] Damage was extensive. The combined effect of the ship flexing in two directions was like ripping a beer can in half. Huge cracks appeared in the hull and the superstructure. Fuel tanks and the fire main ruptured, flooding spaces with water and fuel oil. Electrical equipment was brought offline when cooling water was lost, effectively taking down the ship's ability to provide anti-air coverage. With their ship crippled with extensive damage, damage-control parties got to work isolating flooding, dewatering, and conducting emergency repairs. Within two hours, though still crippled, the ship could once more protect the fleet against air threats. She remained in this role until relieved by USS *Valley Forge* (CG-50).[63]

The *Beaufort* was sent to the aid of the *Princeton*, supported by the USS *Adroit* (MSO-509) to guide them through the minefield. It took five hours for the two ships to traverse the 12 miles to *Princeton*, as there proved to be dozens of mines to be dodged. Several more hours after reaching her, *Beaufort* commenced towing *Princeton* in the darkness through a minefield so thickly laden with mines that *Adroit* ran out of marking flares. She improvised by using chem lights attached to makeshift floats. After a harrowing ordeal, they eventually reached safety beyond the minefield, where a commercial vessel resumed towing duties.[64]

Following the *Tripoli* and *Princeton* mine strikes, the mission focus shifted. The MCM force concentrated on clearing the approach lane, which was extended 33 miles to well beyond the boundaries of the newly discovered minefields.[65] Once cleared, the lane allowed the USS *Missouri* (BB-63) to provide the sort of overwhelming over-the-horizon gunfire support that only a battleship can give. It also gave the appearance that an amphibious landing was still imminent.[66] Initially, the *Tripoli* (as the only flat deck on station) continued to host the AMCM helicopters conducting precursor sweeps through the minefield. The increased coverage area, as well as limitations caused by *Tripoli*'s tribulations, meant that they could not do all required precursor sweeps. As a result, the SMCM risk increased dramatically, while mitigation measures slowed SMCM efforts to a crawl. On February 24, *Tripoli* and her helicopters departed, being relieved by the USS *LaSalle*, an amphibious ship with a much smaller flightdeck.[67] Afterward, *Tripoli* supported the clearance operations from pier-side in Bahrain. She continued to host the AMCM helicopters, which left each morning to use the *LaSalle* as a lily pad for refueling, meals, and crew rest. Each evening, they returned the hundred-plus miles back to the *Tripoli*.[68]

How Did It Happen?—Whenever damage to naval ships occurs in war, criticism arises. Armchair admirals and historians abound, spouting the kind of wisdom that hindsight makes possible. Most often, though, those critics possess only the tiniest of slivers of understanding of the full situation.

One of the major challenges with any mine clearance operation is figuring out *where* to clear. The safest approach is to start looking for mines the moment a ship enters mineable waters. Unfortunately, a huge part of the world, including much of the Arabian Gulf (as described above), is susceptible to mining. To treat the entire area as if "mined" causes unnecessary restrictions in other areas of warfare. But mines are not usually blanketed uniformly across that entire area. Normally, the miner has some form of restraints that limits mining to some logical balancing between achieving the desired effect while maximizing limited resources. Various tricks can be used to guess the miner's plan, including intelligence analyses. But ultimately, the starting point from which naval forces start to actively watch out for mines is still just an educated guess until the first solid evidence of mining is found.

The best guess of intelligence placed the likely minefields in places that made logical sense based on certain extrapolations. One key assumption was that the miner is competent. In his book *Desert Storm at Sea*, Marvin Pokrant makes a good argument that it was Iraqi incompetence that succeeded in surprising the coalition force. By placing the arc of minefields 50 nautical miles out from Kuwait's rather short (30 miles) coastline, the limited mine assets were far less concentrated than they could have been. They were beyond their ability to observe, and thereby beyond their ability to defend. They were also farther out than necessary to defend against advertised coalition capabilities.

Because a position so far offshore defied logic, it was not guessed at by those applying such reasoning.[69] Another factor may have been that the flood of intelligence reports made it easy to miss something important. Think of it like putting together a puzzle. A 100-piece puzzle is relatively easy to assemble. But if its pieces are mixed with the pieces of 100 other puzzles, the job becomes exponentially more complicated. Some sort of sorting mechanism becomes necessary to ensure that only *relevant* pieces are presented for assembly.

Iraq further revealed its mining incompetence through the sad state of its mine assembly and deployment. Fully 54 percent of its moored mines and 7 percent of its bottom mines were not functional, resulting in only 665 of the 1,157 mines working.[70] With only 57 percent of all mines working and spread too thinly, with perhaps inappropriate ship count settings, it was relatively easy for ships to accidentally miss them. Though this author lacks specific data, it is reasonable to guess that the fields were also not uniform in their effectiveness. Both early growing pains and panic-dumping towards the end could have resulted in lower effectiveness in the beginning and ending of their minelaying efforts. The middle area where the MCM task force operated was most likely one of these. This lack of efficacy, perhaps combined with shapes spread too thinly, made it possible for the coalition ships to operate in the minefield for hours or days without damage. Had the Iraqis been more competent, their mines would have found victims sooner.

Postwar Mine Clearance—Mine clearance began at the tail end of Desert Storm's air campaign, which had begun on January 17. The ground campaign started on the same day that *Tripoli* departed for Bahrain. One hundred hours later, it was all over. Iraq's vaunted Republican Guard, veterans of almost a decade of fighting Iran, had been crushed, with many surrendering without a fight. The United States lost 98 soldiers, while Iraq lost more than 6,600 tanks, armored personnel carriers, and artillery pieces. Decisively ejected from Kuwait, Iraq signed a cease-fire to end the hostilities.[71] As part of that cease-fire, Iraq provided the coalition with detailed information about its mining. Oddly, its reported total of 1,157 mines was a good deal less than the 1,288 mines ultimately uncovered.[72] Perhaps the excess was leftovers from the previous war? On the other hand, sometimes water conditions are such that a non-mine underwater object can be easily mistaken for a mine and so is destroyed. Moored mines that are cut but not tracked properly can be counted again later as unrelated drifting or beached mines. Finally, and least likely, an unsuccessful ship crew may pad its numbers to avoid disgrace.

For the first month after the cease-fire, coalition MCM forces prepared for the final mine clearance. Belgium, France, the Netherlands, Italy, Germany, Japan, Kuwait, Saudi Arabia, and Bahrain all sent MCM vessels to join the U.S. and UK forces already working in the Arabian Gulf. Each country was assigned its own tasking area, and everybody went to work. While the United States and UK continued clearing channels (in mostly mine-free waters) to various Kuwaiti harbors, the rest of the coalition went to work on the minefields. Together, the European and local navies accounted for 874 mines before finishing their tasking on July 20. Following their departure, the United States and Japan continued their tasking through September 10, finding 279 mines between them.[73] The final ship to depart, USS *Guardian* (MCM-5), only recently commissioned, found 29 mines.[74]

Operation Iraqi Freedom

In late 2002, the United States once more built a coalition against the Iraqi regime, under the belief that the ruling Ba'ath Party had "weapons of mass destruction" that it intended to use against civilians and other militaries. Given that it had previously used chemical weapons against both Iranian and Kurdish targets, the claim was not entirely without merit. Though it was never proved that such weapons remained in Iraq, there had been considerable Iraqi resistance to allowing international weapons inspectors to verify their destruction. In the interim, Iraq had been under continuous international sanctions since Desert Storm. Denied the ability to sell its oil internationally, Iraq was hurting far worse economically from having started the war than had it remained at peace. With imports severely restricted, almost 60 percent of the country subsisted on food brought in through the UN Oil for Food program.[75] On March 20, 2003, the invasion of Iraq began, and within five weeks, the Ba'ath Party was defeated.[76]

Mines Discovered at Umm Qasr—On the first night of the invasion, four U.S. patrol craft were conducting maritime security operations in the 41-mile stretch of the Khor Abdullah waterway leading to the main Iraqi port of Umm Qasr. The port was needed for bringing in humanitarian aid supplies as well as for the resupply of military forces. Spotting a suspicious barge preparing to get underway, they stopped it for inspection in the morning. What was found by the Australian special forces who searched it in the morning became instant, worldwide news. Hidden under a layer of oil drum covers were 20 Manta stealth bottom mines and 48 LUGM moored contact mines, laid out on tracks ready to be dropped through a trapdoor. A second barge was found to also contain mines, as did at least one other facility in town.[77]

Mine Clearance—Fearing that the Iraqis had succeeded in mining the waterway, U.S., UK, and Australian forces went to work looking for mines. Using a mix of ships, remote-controlled boats,[78] helicopters, dolphins, unmanned underwater vehicles (UUVs), SEALs, EOD, and other special forces, they spent the next several weeks clearing the waterway. With no underwater visibility, they depended heavily on touch or other nonvisual sensors to identify sonar contacts. When in doubt, underwater objects were blown up.[79] Often, the only way to tell for sure that an object had been a mine was if a placed charge induced a larger, secondary explosion. At least four of HMS *Ledbury*'s sonar contacts yielded these high order secondary explosions when charges were detonated nearby. These sympathetic detonations gave strong evidence that at least some sort of high explosive devices had made it into the water. Soon thereafter, a helicopter scouting at high tide discovered three Manta mines (the same type hit by USS *Princeton*) sitting in the mud. This was clear proof of mining at Umm Qasr.[80]

An important point about the Umm Qasr mine clearance was that it was yet another historical instance of riverine mining. As a river, it had an entirely different visual, acoustic, and physical environment than the harbors and beach areas for which most shallow-water systems were designed. The implication is that mine detection and sweeping systems should be specifically tested in rivers of various types. Tactics should be designed to contend with those environments. If available systems are found incapable of successfully operating in such areas, they should not be used for the purpose. Instead, purpose-designed systems are needed, perhaps looking for inspiration in other industries operating in rivers. Riverine mining will be used again, where a weaker enemy cares to deny their adversary the use of their river.

Lessons Learned from the Middle Eastern Wars

The Middle Eastern mining of the late 20th and early 21st centuries has not entirely stopped. Between Desert Shield and this writing, mines had more appearances in those desert waters. During the Arab Spring revolutions, three mines shut down the port of Misrata in Libya.[81] Houthi rebels laid mines in the Red Sea multiple times over four years, damaging at least two ships. In the Saudi-led MCM effort, at least 171 of these mines were found.[82] During the Syrian civil war, the government dropped old contact mines inside cities as high-yield terror weapons, available at no added cost from 30-year-old stockpiles.[83] The weapon will continue to be used by terrorists for such purposes for as long as asymmetric warfare is practiced.

The lessons to be taken away from the events of this period are many, though not all are subject to an open forum. These pages offer a few that are obviated by the specific types of events discussed:

1. The edges of a minefield are where the mines are located. Intelligence can help narrow down the possibilities, but it is still just a guess. If the task is to attack a live minefield, approach the area with caution. Do not assume that there are *not* mines outside the tasking area.
2. Mines can show up at unexpected times and places. Non-MCM ships should consider adding to their training plans what to do if they unexpectedly find themselves in mined waters. Ships plan for many, many other contingencies (fire, flooding, loss of steering, etc.). Mine strikes can cause all these things and are worth preventing.
3. Expeditionary and organic[84] MCM forces can get basic MCM efforts started until more capable platforms arrive. MCM headquarters staff are sent early and should be staffed, equipped, and trained for expeditionary operations both ashore and on ships.
4. Navies should be prepared for riverine warfare, including mining.
5. The cheapest and most plentiful mines are still moored contact mines. Navies should prepare—and train—for operations against large volumes of contact mines laid at realistic depths. Realism acquaints MCM forces with both the challenges and the opportunities available when fighting this most common type of minefield.
6. Expect to lose ships, including ships from the MCM task force, when your enemy uses mines.

Chapter 12

Post–2022 MIW History and Lessons Learned

At the time of this writing (March 2024), the 2022 Russian invasion of Ukraine (the "Ukraine War") has continued for the past two years. News articles over the past couple of years have addressed naval mines used in that war, but the full cycle is not yet complete. Mines remain in the water, ships are still striking mines, and mine clearance has yet to be a front-of-mind consideration. As in all wars, that will not likely happen until the fighting stops.

This chapter will have three main focuses. First, it will review the mine warfare situation in the Ukraine War. Second, it will look at current trends in naval mine warfare around the war. Finally, it will reiterate lessons learned from throughout this book.

Currently, there is little available information in mine warfare history books past Desert Storm. Anything beyond that point had to be assembled from news articles, essays, reports, asking questions of old colleagues, and even working off personal memory. Two-decade-old events like the 2003 Umm Qasr clearance at least offer the history of completed incidents. Two-year-old events like the Ukraine War have entirely different problems. The full story does not exist at all, and the true details of individual events and programs are not publicly available. What remains are various news articles of varying credibility, all fed by disinformation and modified with misinformation and just plain bad information. Information on mine warfare trends is compiled from similar sources as well as military-sourced news articles and pseudo-databases like Wikipedia (hated by professors everywhere) and enthusiast-assembled collections. The author has done his best to assemble something resembling the truth.

"In war, truth is the first casualty."
—Aeschylus, father of Greek tragedy

Ukraine War

The Ukraine War began with the Russian invasion of Ukraine on February 24, 2022. This was just over eight years after Russia annexed the Crimean Peninsula on February 20, 2014. Possible explanations for these actions vary. One theory is that it was part of a long-term strategy to push the Russian borders to a more defensible position like the Carpathian Mountains or at least the Dnieper River. Certainly, the annexation of Crimea gave it a central position for asserting dominance over the Black Sea. Opponents in NATO proclaimed that it was part of a push to rebuild the old Soviet empire, with the Baltic countries being next. Still others claim that Russia wanted to force Ukraine to stay out of NATO, that it believes that buffer states are important to its security. Russia itself claimed a pretext of saving Russian ethnic people from the deprivations of neo-Nazi forces. The truth is probably somewhere in the middle.

Mining off the Ukraine Coast—In March 2022, approximately one month after the war began, Russia announced that Ukrainian-laid mines were drifting around the Black Sea. Accounts of the number vary as being either 420[1] or approximately 10.[2] Either number is rather suspect, as it is impossible for anybody to determine without knowing the exact position of every mine and then inventorying them in the water. That leaves at least two possibilities: (1) pure speculation, possibly informed by mine mooring characteristics, or (2) an intentional dumping of drifting mines in order to incite international condemnation of Ukraine.

From a few sources, one can get more specific information about the mines used. One states that Russia claimed that a total of 420 mines were laid, including 370 in the Black Sea and 50 in the Sea of Azov.[3] Another says that all 420 were laid at Odessa, Ochakov, and Chernomorsk between the mouths of the Dnieper and Danube Rivers. This source further states that the mines involved were mines made for riverine warfare, the YaM and the YaRM, 70-year-old mines with explosive weights of 20 kilograms and 13 kilograms (respectively).[4] Those weights are correlated in an old, declassified CIA document.[5] Ukraine admits to having laid mines in its own waters, as is its right when defending itself from a possible amphibious landing. However, the Russian theory is that since the old mines were well past their service lives, the mooring cables must have snapped during a storm. It further claims that some mines had been modified with "timers and self-destruct mechanisms" and that one of these struck and sank an Estonian cargo ship.[6]

Again, this appears to be disinformation, misinformation, and an admission of poor Soviet engineering. While it is plausible that Russian intelligence could have detected that 420 YaM and YaRM mines left storage and that some mines were laid, the details are contradictory. An obvious contradiction is that the number laid near Odessa cannot have been both 370 and 420. A second doubtful claim is that Russia could somehow know the specific modifications of those mines, changes that would have turned them into floating time bombs. What value would Ukraine had had in such modifications, which would have resulted in arbitrary explosions at unpredictable places? The third questionable detail is the type of mine used off the coast. It seems implausible that Ukraine would have used old, weak river mines to defend a coastline that was subject to waves and storms. Some credit must be given to their competence and the advice they receive.

It is likely that riverine mines would be effective against the small craft for which they were designed. If used, they were laid in the Dnieper River to prevent a crossing by the Russian forces that had invaded as far as that river. However, they would not have been used as a main battery for deterring a Russian seaborne invasion. The explosive weight of 13 to 20 kilograms is tiny compared with the 115 kilograms used in the M-08[7] that damaged (but did not sink) the USS *Samuel B. Roberts* in the Tanker War as discussed in the last chapter. These weapons would not have deterred Russia from an amphibious landing on the coast, a fact of which both Russia and Ukraine would be aware. By the same logic, if an Estonian cargo ship was really sunk by a mine, it probably was not a YaM or YaRM that did the sinking. A ship robust enough to carry several tons of cargo around Europe and through treacherous and sometimes iced-over northern waters would not be so weak as to be sunk by such a small mine. Finally, if old Soviet mines were still explosive after being cut from their anchors and drifting for over one hour, then Moscow was admitting that the USSR had designed mines without the self-neutralizing features required by the Geneva Convention.

For all these reasons, Russia's claims about the nature of Ukrainian mining are dubious. What seems more plausible is that (1) Ukraine laid sea mines and not riverine mines off its coast; (2) those mines were sufficient to deter a Russian amphibious landing; (3) Ukraine probably laid riverine mines in at least one river; (4) Russia combined this limited data for propaganda; (5) that propaganda *might* have been used to hide Russian use of mines to target commercial vessels, as Ukraine hypothesizes; and/or (6) Russia *might* have set some YaM and YaRM mines adrift to support the propaganda. Of course, it is impossible to tell. Both parties are former Soviet republics, inheritors of the same weapons systems and the same training on mine laying and propaganda. Either side can gain from setting mines adrift and blaming it on its foe.

As 2022 progressed into summer, drifting mines were continuing to be reported in the Black Sea. Ukraine claimed that a 50-year-old man, gone out for his daily swim at a beach in Odessa, was killed by a drifting mine on June 11 as his family looked on in horror. Though there are some details that do not make sense, it is feasible that a mine-related death at a beach did happen. Ukraine subsequently released a statement admitting to having planted mines to prevent an amphibious landing. It went on to prohibit swimming at coastal beaches. It also reported that Russia had laid between 400 and 500 mines in Ukrainian waters, including 372 mines of the "R-421-75 type." This type of mine was reported as being the only type drifting in the Black Sea and one that Ukraine says is not in its inventory.[8] The author could find no open-source technical information on that mine and so cannot analyze claims about it. Information about specific numbers may or may not be accurate, depending on intelligence available to Ukraine.

Elsewhere in the Black Sea, mines continued to drift. By August 2022, 22 mines had been found, including two off the Romanian coast, three off Turkey, 16 in Ukrainian waters, and one in Bulgarian waters. By then, the problem had become sufficiently large enough to require international action. News reports said that U.S., Turkish, Bulgarian, and Georgian explosive ordnance disposal (EOD) divers were training together in dives down to 50 meters.[9] This is probable enough, though the fact that they were diving at all suggests that drifting mines were not their only concern. Drifting mines float on the surface. Furthermore, divers would not dive to look for mines in random locations. There's too much sea for random diving. This likely means that the littoral countries were preparing for general mine clearance as part of the eventual post-conflict cleanup operation. Ukraine was also considering that cleanup as well, suggesting that robots (probably the types of drone technology discussed later in this chapter) would likely be involved.[10]

The war dragged on. Russia destroyed or captured Ukraine's navy early in the conflict. With sea control restricted only by Ukrainian mining, Russia used its freedom of navigation to fire ship-launched land attack missiles into southern Ukrainian cities. Ukraine, however, was not without defenses. Early in the war, it sank the Russian flagship *Moskva* using anti-ship missiles and scored missile hits on other Russian craft. In October 2022, Ukraine began using $250,000 drone surface vessels like fire ships of old to destroy Russian ships and to sow chaos. In essence, they are long-range, remote-controlled torpedoes. One of these blew up a section of the bridge going over the Kerch Strait (same strait mentioned in the Crimean War chapter). Another heavily damaged the replacement Russian flagship, *Admiral Makarov.* This is ironic because that admiral, discussed in the Russo–Japanese War chapter, had been responsible for the first successful torpedo attack. Increasingly, Russian ships began to avoid straying too close to the Ukraine coast.[11]

By September 2023, Ukrainian drones had attacked several more Russian ships and the Sevastopol naval base. The following month, Russia withdrew its fleet east of the Kerch Strait to its naval base at Novorossiysk. Even that was within range of Ukrainian drones, prompting Russia to explore building a new base even farther away in Georgia.[12] A major fleet was forced into hiding by a weapon whose lineage can be traced back to David Bushnell and his 18th-century inventions. What complicated the Russian situation was that the fleet units could not be reinforced from one of its other three ports. The 1936 Montreux Convention, which limited Black Sea access for non–Black Sea navies, also allowed Turkey to close the straits during conflicts to combatants *not* based in the Black Sea. As a result, Russia and the other Black Sea littoral countries were limited to what they had at the war's beginning. Any replacements would have to be built at Black Sea ports or somehow brought downriver from points farther inland.[13]

As all of these actions continued, Ukraine's grain exports suffered. For most countries, this might not matter, but Ukraine produced 10 percent of the world's wheat and 15 percent of its barley. What is more, most of this grain went to the huge populations of food-insecure countries like China, Egypt, Indonesia, and Bangladesh. By blocking this immensely important part of the world's supply chain, Russia's embargo raised food prices around the globe, risking the starvation of millions of people. In July 2022, Turkey sponsored an agreement between the countries that allowed Ukraine to resume exporting this important commodity.[14] Under that deal, a team of UN, Turkish, Russian, and Ukrainian agents would inspect shipments to/from Ukraine for war matériel. When cleared, these ships would be allowed to ship grain and fertilizer (another vital Ukrainian product). It also allowed Russia to bypass UN sanctions on exports for these same products.[15]

In July 2023, Russia withdrew from the grain deal.[16] Its pretext was that it felt that it was not receiving enough benefit from the deal and that Ukraine benefited too much.[17] It then began a bombardment on Odessa, resulting in damage to 130 infrastructure facilities and the destruction of 300,000 tons of Ukrainian grain. It further declared that any ships bound for Ukraine would be treated as potentially bringing war matériel to the country and would be treated accordingly. The UK claimed that intelligence sources postulated that Russia would not want to attack neutral ships directly (as had Iran under similar circumstances in the Tanker War). Instead, it would likely target these ships with mines and blame Ukrainian mines.[18] To be fair, Russia likely considered ships going to Ukraine to be as neutral as ships going to the UK during World War II. With NATO and other countries openly providing weaponry and direct funding to Ukraine while also sanctioning Russia, it had cause to believe that some shipments might not be entirely neutral. On the other hand, the flood of military aid undermined any portrayal that grain sales could be only for the purpose of funding Ukraine's defenses.

With Russia threatening neutral shipping, the Ukrainian and NATO response was to designate a "humanitarian corridor" for safe passage. That corridor avoided the open sea. Instead, it followed the Ukrainian coast where missiles gave some protection from airborne or seaborne attacks. It then continued along the coast through the territorial waters of Romania, Bulgaria, and Turkey (all members of NATO).[19] By using this approach, ships enjoyed some protection from direct Russian attacks where such acts would invite more direct NATO involvement. Despite such protection, reports of potential Russian mining of this channel gained credence when ships reported mine strikes in both November and December.[20] One such case happened in December when a Greek-owned ship struck a mine. Forced to run aground to avoid sinking, the ship had been headed toward the Ukrainian port of Izmail on the Danube River.[21] This was egregious because the Danube River forms the border between Ukraine and Romania. Because several countries border the Danube, any mining of the Danube or its approaches also threatens any shipping destined for those countries.

Mine Clearance for the Ukraine War

With mines drifting throughout the Black Sea, and with rumors of Russian mining in NATO countries' territorial waters, the question turned to mine clearance. While the war rages, it is a balancing act for littoral countries. Freedom of navigation to nonbelligerent countries necessitates an anti-mine effort. Most small countries do not make their own naval vessels. They buy them from more industrialized countries like the United Kingdom. Yet, the Montreux Convention that prevents Russian navy reinforcement also stops the Ukrainian navy from acquiring reinforcements. In December, Turkey used the convention's dictates to prevent the transit of two minehunters promised to Ukraine by the UK in 2022. Those ships would have helped Ukraine to clear mines from its shipping lanes and would have substantially grown the country's almost completely destroyed navy.[22] On the other hand, the convention did not stop Turkey from allowing the transit of the first of two other minehunters that had been recently purchased by Romania.[23]

Those ships would be needed. Mines represent a threat to all Black Sea nations. Some drifted even as far as the mouth of the Bosporus at Rumelifeneri, Turkey, complicating ferry traffic and making fishermen afraid to do their work. Merchant vessels struck mines off the Romanian coast near Costinești and Constantza.[24] This forced the Romanian navy to have its ships and helicopters out visibly patrolling for drifting mines,[25] despite the task being like looking for a needle in a haystack. On January 10, Turkey, Romania, and Bulgaria signed a memorandum of understanding (MOU) that created a trilateral MCM task group (MCMTG). That task group would include three minehunters from each country as well as a command and control (C2) ship (most likely provided by Turkey). This group would be responsible for keeping the humanitarian channel clear of mines.[26]

Multilateral coalitions come with their challenges, not least of which is interoperability. At a most basic level, participants need to be able to communicate and cooperate with one another to deconflict movements and to avoid duplication of efforts. They need to understand one another's abilities so they can understand their limits. They need to prove the credibility of their own reports while also learning to trust their partners' reports. After all, partner navies put their lives in the hands of strangers. On a more advanced level, coalition members need to be able to share information with precise geographic details like contact reports, danger areas, and tasking areas. Military C2 systems are usually designed to be *not* accessible by outsiders. New partners can quickly learn that they lack the technical capability to share information and so must find workarounds. Those workarounds can often be different enough from their normal systems that watch-standers require considerable training and practice to develop the same level of effectiveness that they have with their normal systems. Sometimes, a lack of shared language and even shared alphabets complicates the building of a common operational picture (COP) without the heavy use of multilingual liaison officers (LNO).[27]

With all of these challenges, multilateral exercises become vital for working out the kinks. For that reason, various regional and NATO countries have been holding multilateral MCM exercises to prepare for the eventual mine clearance. For example, in July 2023, ships, EOD, and unmanned systems from nine countries practiced together in an exercise off Constanza, Romania.[28] Events like this likely helped the new trilateral MCMTG to practice working with each other while also allowing them to benefit from the experience of other countries. When the war finishes, other countries may be invited to join this coalition.[29] Whether or not the UN can force a wider coalition is another question that still must be worked out.[30]

The mines in the Black Sea are now a part of mine warfare history. Though much of that history is yet to happen, these mines are shaping it even as this sentence is being written. Two navies, each with a shared heritage in Soviet mine warfare, are exchanging mines in the first major land-and-sea war outside a Middle Eastern desert in half a century. Like the Iran–Iraq War, it is a naval war fought in a confined waterway, where belligerent mines affect other countries directly and threaten the global supply chain of vital commodities. Access by outside ships to the affected geography is restricted. This creates the opportunity to find new ways to fight the threat using expeditionary means to increase capacity, means that include systems that can be brought in by air and by land, such as unmanned systems, small craft, helicopters, EOD, etc. The war has the potential to spark a renaissance in how the world fights mines. It will either prove, or disprove, the concepts behind the trending systems discussed in the next section.

The Many Configurations of the World's MIW Forces

The world's mine warfare forces have shifted considerably since the Cold War, when two factions expected an epic repeat of World War II to break out at any time. Today's mine warfare is more likely to be more small-scale and regional in nature. At the same time, historically rogue nations and nongovernment factions, burdened with paranoia, are more likely to lay mines or launch drones or missiles outside the restraints of international law. As such, MCM forces are more likely to work together as parts of coalitions than to work in their own waters. Russia's invasion of Ukraine has increased the likelihood of conflict with NATO. Yet with few allies and hostile supply lines, any follow-on aggression outside Ukraine would be limited in scope. Furthermore, Ukraine's successful campaign against Russia's Black Sea Fleet has signaled a potential sea change in how navies fight in littoral waters. For Russia, its adversaries, and other rogue nations, these new conditions may result in massively different approaches to how they train and equip themselves for combat in such settings.

Fifty-one of the world's navies have vessels equipped for MIW.[31] The type and quantity of MIW platforms that a country chooses to use appears to depend on several factors:

1. The size of the country's tax base.
2. The proximity of potential adversaries.
3. The likelihood that a country would deploy its MCM force overseas.
4. The impact of mining on the country's military plans and economy.
5. The probability of having one's homeland mined by an enemy.
6. The probability/level of using defensive mining in war.
7. The probability that the country will need to defeat another country's defensive mining.
8. The capabilities of their domestic military-industrial complex.
9. The prestige of the country's navy.

Of the various approaches that many different countries take to mine warfare, there are five basic categories:

1. Countries with large navies and whose neighbors are adversaries that also have large navies. These expect to be on both the giving and receiving ends of heavy mining close to their own shores. As such, they maintain large inventories of mines and MCM assets.

2. Countries with large navies whose adversaries are farther afield. These maintain enough of a high-tech mine inventory to be a credible threat but expect time to ramp up production to achieve large-scale mining. Their MCM forces are designed to be able to take the fight to their adversaries. Because of this, they are most likely to favor riskier technology that offers an outsized technological advantage from limited resources.

3. Countries with medium-sized navies optimized for countering aggressive neighbors. These will often pursue the diplomatic advantage of participating in coalitions. Because of these dual concerns, they build outsized MCM forces and a decent defensive mining capability. They also frequently maintain unique capabilities that can contribute meaningfully to their coalitions beyond mere capacity increases.

4. Countries that do not expect to be involved in the affairs of more belligerent nations but still maintain a "just in case" MCM capacity. Frequently, these are countries that have been mined at some point or whose economies are heavily impacted by maritime trade. They will participate in multilateral naval exercises primarily to gain experience and to be seen. In being *seen* conducting MCM procedures, they project an image of not being helpless against mining attacks, thus deterring such actions. They would expect both to host outside MCM forces and have enough professional knowledge to anticipate and/or translate the needs of those forces.

5. Countries that do not maintain MCM capabilities. Usually these are countries with small economies and little maritime trade. What navies they have are mostly used for policing and revenue collection. While these may temporarily host MCM forces, they neither expect nor prepare for that contingency.

A Review of the World's Current MCM Force Configurations

Mine warfare technology, like any technology, transitions over time. Each new innovation lowers an adversary's effectiveness. Yet, it is easy for a country with established systems to ignore developments elsewhere. Moreover, different countries can tackle the same innovative concept in vastly different ways. If decision-makers pay attention to what other countries are doing, they can often see where their own approach is flawed and inappropriate. Countries can learn from each other. Such synergy was vital for the United States during both world wars when British technology acted as a springboard for its own MIW technology development.

The remainder of this section will act as an overview of the MCM platform inventories of the 51 countries that have MCM vessels as of April 2024. It is intended to identify the various basic approaches, rather than act as a compiled intelligence list. The underlying study was compiled from reviewing around 500 different open-source webpages, with Wikipedia being the starting point.[32] Some of this will be a repeat of the information from the introduction.

To begin with, all MCM platforms do either mine hunting, minesweeping, or both. Where either mine hunting or mechanical minesweeping takes place, there is also some form of mine neutralization.

Mine Hunting: The vast majority of MCM forces do some form of mine hunting. These platforms use some form of active sonar, either forward-looking or side-scan.

Forward-Looking Sonar: Forward-looking sonars are either hull-mounted or are lowered using a winch. The main advantage of forward-looking sonars on ships is they allow the platforms to "see" mines in their path in real time, thus allowing them to avoid hazards. Their second advantage is they allow operators to investigate a contact from multiple angles to determine if they have the consistent properties of mines.[33] Their third advantage is that they allow the ship to maintain continuous sight of the mine while guiding neutralization assets into place.

Side-Scan Sonar: Side-scan sonars are incredibly high-definition. They are always either towed or are part of unmanned underwater vehicles (UUV). As they travel, they send high-frequency sonar signals a relatively short distance down and to the side. The returns are stitched together electronically to form an electronic map of an area. Their main advantage is their highly detailed maps, which operators can review afterward to look for potential mines, and which can be saved for future comparisons. Their main disadvantage is that they show only where the platform has been, not where it is going. Consequently, they offer no safety to the hunting platform. Furthermore, any contacts classified as mine-like must be reacquired later by a different system for identification and neutralization. Finally, the limited number of looks at the contact from a single angle can make it harder to spot distinguishing characteristics. This can lead to either missing mines or unacceptably high numbers of false contacts. Historically, however, operators have had to wait until after the sonar finishes its mission to review the sonar video. As a result, it can be several hours after a drone's run is complete before the team will know whether an area is mined. That delay, without a real-time video uplink, could affect mission timelines and make this sonar type inappropriate for leading ships out of areas not previously hunted.

Mine-Hunting Platforms: Most of the world's MCM platforms are ships, though at least two countries also use helicopters to pull side-scan sonars. Approximately half of these ships are also configured for minesweeping ("minesweepers"), while the rest do only mine hunting ("minehunters"). A growing trend is the use of drones to do the mine-hunting mission.

Mine-Hunting Ships: There are currently five approaches to ship-based mine hunting.

1. Ships using classic forward-looking sonar to hunt for—and neutralize—mines in real time.
2. Smaller vessels pulling only side-scan sonars through the water. With low freeboard, these navies most likely assume that any mine shallow enough to threaten the ship is also shallow enough to be seen by a motivated lookout.
3. Ships using both forms of sonar in concert. A forward-looking, mine-avoidance sonar is used to watch for mines ahead of the ship while it also pulls the side-scan sonar. This is a balanced approach that combines the speed and high definition of side-scan sonar with the safety of forward-looking sonar. Because hull-mounted sonars dedicated solely to avoiding moored mines can be made much more simply than full-service sonars, this combination may offer a cheaper approach.
4. Minehunters that deploy drones as a force multiplier. While drones use side-scan sonar to map the underwater environment of one area, the ship can focus on identifying contacts and neutralizing mines in areas previously hunted by the drone. In some cases, ships can deploy multiple drones at the same time.
5. Ships that are solely motherships for drones. This theoretically keeps these ships out of minefields, which then allows the ships to retain the performance characteristics of normal ships while also bringing MCM capabilities to the fight. Being able to be larger, they also can carry helicopters. The U.S. Navy's two littoral combat ship platforms are larger versions of this concept. The Franco-Belgium "City-class" ships appear to be smaller versions that incorporate many of the lessons learned from the development of the American versions.

Mine-Hunting Drones: Mine-hunting drones have developed along two basic lines. For either one, the data is usually uploaded to a support system once the drone returns to its operator. Potentially these can be upgraded to allow for in-situ decision-making. One approach would be enabling the operator to stream the sonar picture live, along with the ability to adjust the drone's programming to direct it to identify specified contacts. An alternative approach would be to include some form of artificial intelligence that enables it to adjust its own programming to identify contacts that meet certain parameters. Of note, there is also at least one country that does aerial spotting with laser-equipped drone helicopters.

1. *Unmanned Surface Vessels (USV):* These drones act as small, unmanned minehunters. Equipped with a towed sonar, either side-scan or combined forward-looking and side-scan, they are programmed to hunt a predefined area. Once programmed, they transit to the tasking area where they deploy their sonar, hunt per the navigation plan, recover the sonar, and then return to the ship or the pier. They may or may not also have a system for identifying contacts, either in-stride or as a separate mission. Also, some USVs are designed to deploy marker buoys or even neutralization drones. Their advantage is that being larger in size, they can potentially last longer and deploy more/better systems than UUVs. Their disadvantages are that they are visible above the surface and present a hazard to navigation.
2. *Unmanned Underwater Vehicles (UUVs):* MCM UUVs are almost exclusively side-scan sonar platforms designed with a robust GPS navigation system and a comms link. Their main advantages are that their footprint is entirely beneath the surface and they do not require the constant supervision of a human. In combination, this allows clandestine mine hunting. It also allows for routine exploratory hunting of an area without interrupting regular traffic. Potentially, in

wartime these could be based in lower-priority ports with small reserve contingents and assigned to daily exploratory operations. Their disadvantage is that their usefulness is directly correlated to their range and their coverage rate. Another challenge is that they are susceptible to fishing nets and possibly to predators. However, they offer interesting possibilities. For example, they could be designed to be launched from submarines to scout areas being considered for amphibious landings. Another option, for which some countries have opted, is for several to be launched by a single ship. If a ship carries four UUVs, each with the same coverage rate as a minehunter, then that ship could quadruple its own coverage without hazarding its crew in the minefield. Finally, the fact that UUVs do not need to be launched by MCM-specific ships means that they could theoretically launch from shore or from any vessel rigged with a crane or boat davit. This organic approach enables mine hunting in places devoid of purpose-built MCM vessels.

Mine Identification/Neutralization: Mine hunting alone does not remove mines from a minefield, though it at least reveals hazards to be avoided. The next step is to identify sonar contacts and then neutralize any mines found. Most professional MCM forces neutralize mines by using either explosive ordnance disposal (EOD) divers or tethered, remote-operated vehicles (ROV). However, there are also efforts to field mine neutralization drones. An older method, less favored, is to simply shoot floating contact mines in hopes of either sinking them or exploding them. While shooting has historically worked, doing so comes with risks both during and afterward.

EOD Divers: In many ways, EOD divers are the most capable, lowest-maintenance neutralization assets available. Operating independently from small boats, they can usually handle the neutralization task while other units hunt or sweep. They can identify by sight or touch, then decide on the spot what to do next. They recover mines for exploitation, defuse them, or detonate them with explosive charges. In some cases, they even have small ROVs of their own that can be used to identify an object on the bottom without wasting the diver. Their disadvantages are that they are limited in their working depths without additional equipment, and their unique skill sets do not always have shipboard applicability outside a minefield. The world's navies appear to take at least five different approaches to how they organize their MCM EOD forces.

1. *EOD-specific MCM ships:* Small ships crewed primarily by EOD personnel with a small number of other, specialty personnel assigned. For these ships, EOD is the main mission.
2. *EOD as the primary neutralization asset:* These vessels are equipped with hyperbaric chambers and several assigned EOD crew members who form their own division.
3. *EOD as contingency assets:* For these MCM vessels, ROVs form the primary mine neutralization battery, though two or three EOD divers are also assigned. In those cases, EOD work is a collateral duty, and hyperbaric support is provided elsewhere.
4. *EOD-specific, shore-based commands:* EOD divers are stationed and train together. They deploy as detachments to wherever needed.
5. *EOD as part of cross-functional clearance teams:* Units equipped with UUVs, ROVs, EOD divers, and purpose-built small craft able to deploy all three. Other specialties are also assigned. The entire unit, including equipment, is organized to be deployable to anywhere capable of housing and deploying them and their boats.

Mine-Neutralization ROVs: These are remote-controlled underwater robotic vehicles that are tethered to their launch platform. Once one is launched, the operator pilots it remotely, using sonar to get close to the underwater object and then video to identify it. If the object is not a mine, the ROV is returned to the launch platform for later reuse. If the object is a mine, the ROV renders it incapable of firing using either explosives or cutters. There are two main categories for these:

1. *ROV acts as a truck for ordnance:* These large, expensive ROVs deploy ordnance to neutralize mines. Once the ordnance is deployed, the ROV returns to the ship for reuse. The ship maintains only one or two ROVs, which undergo regular shipboard maintenance and repairs.
2. *ROV is also the ordnance:* One of several carried in inventory, these small ROVs are intended to be expendable. If an object is identified as a mine, the ROV drives directly into it. The ROV is tipped with a shaped charge that detonates on impact. Both ships and helicopters can deploy this type of ROV. A vital aspect of them is that the ones doing identification must be recoverable. The bottom of the ocean contains countless objects (rocks, barrels, etc.) that can appear mine-like on sonar. If ROVs are expended every time such objects need to be identified, the MCM platform will quickly run out of ROVs.

Minesweeping: In some environments, mine hunting is impractical because of environmental or time factors. In those cases, minesweeping becomes the more efficient approach. As discussed in the introduction, the two main types of minesweeping are mechanical sweeping for moored mines and influence sweeping for bottom influence mines. Currently, the types of influence sweeps claimed by various navies include acoustic, infrasonic, magnetic, pressure, or some combination of these. Sweeping takes place on four different types of platforms.

Minesweeping Ships: By far the most common type of MCM vessel, the minesweeping ships of all but one navy can both mine-hunt and mine-sweep. Being larger than other platforms, they typically carry the largest minesweeping equipment with the widest sweeping path. Their disadvantage is that their location in the water makes them susceptible to mines, though their design and construction makes them safer than non–MCM ships. Their advantage is that they are typically self-contained, 24-hour, all-in-one MCM platforms with which the world's navies have decades of experience in wartime. When it comes to large-scale mine clearance, these have been and will likely continue to be the primary workhorses. Most countries use a mono-hull construction, though at least four countries (United States, Iran, Indonesia, and Norway) use a small waterplane area twin hull (SWATH) design.

Minesweeping Boats (MSBs): MSBs are designed to sweep mines in shallow, protected waters like rivers and harbors. In the current inventory, two approaches are used. One approach is to permanently install smaller versions of multiple sweep types on the same boat and deploy whichever is needed. The alternative is a modular approach, where specific sweep equipment is installed as needed. Under that approach, at any given time, a vessel might be equipped for mechanical sweeping *or* influence sweeping, but not both. This approach (similar to the argument for the littoral combat ship concept) offers several advantages:

1. Conserves limited deck space.
2. Allows higher-capacity equipment to be installed.
3. Allows equipment to be stored and maintained ashore.

4. Allows operable equipment to be shared among operable boats.
5. Limits equipment loss from mine strikes.
6. Gives the option to use the boats for other missions.
7. Enables the conversion of crafts of opportunity (COOP) for minesweeping duties in remote areas because of the self-contained equipment design, perhaps brought in by truck or aircraft.

Minesweeping Helicopters: Similar in role and approach to MSBs, minesweeping helicopters offer the added bonus of speed, rapid deployment, and relative invulnerability to mines. Their lift capability also makes them useful for non–MCM missions. However, they also require far more support and are more expensive.

Minesweeping Drones (MSD): MSDs balance AMCMs' ability to sweep without risking personnel with the lower support needs of MSBs. Two approaches are used, distinguished by how they are controlled.

1. *Radio-Controlled (RC) MSD*: It was originally used by the United States during the Vietnam War, but China had developed its own version before the war's conclusion. Its current version came out in 2004. Capable of being driven to its operating area by a small crew, supported by a larger minesweeping ship, it can be switched to RC mode once on station. After the crew streams the minesweeping gear, they can operate it remotely from the support ship. The disadvantage is that they rely on humans for precision navigation. That problem is relatively simple for lead-through operations (keep it directly on the bow) but becomes much harder if it is used for wide-area coverage. The advantage of this approach is that being simpler technology, it offers a way to quickly and cheaply scale up MCM capabilities by converting existing small craft. Standalone RC minesweeping kits could conceivably be shipped to a mining hot spot, where they can be used to convert locally acquired small craft. Installed on the boats of surface combatants (such as RHIBs), they could potentially offer a way to sweep moored mines from the ship's path.
2. *Unmanned Surface Vessels (USVs)*: These drones perform a similar role as MSDs, but the sweep deployment and the navigation plan are programmed into the USV. Once programmed, it transits without further operator interface to the tasking area, where it then deploys its sweep, sweeps per the plan, recovers the sweep, and then returns to the ship.

As a final note on ships, most of the world's MCM vessels are armed with large enough guns to be able to defend themselves against similar-sized patrol craft. Some Soviet-made vessels even carry anti-aircraft guns and are almost universally designed to be able to deploy mines. Many other countries located near the former Soviet Union also maintain surface minelayers that usually double as MCM command ships.

Summary of Lessons Learned

1. **Mines are still the cheapest way to sink a ship or to fill drydocks.**
   a. Mines can damage or sink ships that are impervious to other forms of attack.
   b. Mines can be as simple as waterproof IEDs.
   c. Mines can be used by countries without navies.

   d. Defensive mines do not require purpose-built launch platforms. In the past, they have been dropped by fishing boats, helicopters, barges, tugs, ferries, etc.

   Examples: Revolutionary War, Crimean War, U.S. Civil War, Russo–Japanese War, World War I, World War II, Korean War, Vietnam War, Tanker War, Desert Storm, Operation Iraqi Freedom, Houthi rebel mining, Ukraine War.

2. **If a war requires waterborne movement, expect mines.**
   a. Naval mines, like drones, are asymmetrical weapons. They can and will be used by weaker nations as area-denial weapons.
   b. The country most dependent on waterborne movement is most at risk from mines.
   c. Countries with tiny coastlines have more to gain than lose from mining.
   d. Countries with huge coastlines require far more resources to effectively mine.
   e. Countries with lots of alternative ports are less susceptible to small-scale mining.

   Examples: Revolutionary War, Crimean War, U.S. Civil War, Russo–Japanese War, World War I, World War II, Korean War, Vietnam War, Tanker War, Desert Storm, Operation Iraqi Freedom, Houthi rebel mining, Ukraine War.

3. **MCM units cannot be everywhere at once. Usually, non–MCM units are the first to find mines.**
   a. Ship captains tend to treat all areas as not mined until they receive unambiguous information to the contrary.
   b. Commanders must balance the risk of mines against increases in other risk areas that may be caused by imposing mine mitigation measures.
   c. Adding capacity at the local level for rapid, daily exploratory hunting can reduce the chances of non–MCM assets being the first to find new mines.
   d. COOPs or UUVs at the local level can add capacity.

   Examples: Crimean War, U.S. Civil War, World War I, World War II, Vietnam War, Tanker War.

4. **Long, skinny maritime countries are susceptible to strategic mining.**
   a. If that movement is on rivers, expect riverine mines.
   b. If both sides need the river, expect remote-controlled (RECO) mines.
   c. If only one side needs the river, expect every mine in the enemy's inventory.
   d. In a river or harbor, an obstacle needs not be explosive to cause damage.
   e. Be prepared to fight mines in rivers. Sonar, sweeps, chain sweeps.
   f. Riverine mine clearance must consider currents and force protection.

   Examples: U.S. Civil War, Vietnam War, Operation Iraqi Freedom.

5. **The best routes go *around* minefields, not through them.**
   a. When timing matters, it is faster and less risky to find a route clear of mines than it is to clear a route through minefields.
   b. Aerial mine spotting using helicopters, slow-moving aircraft, or even drones can be effective at identifying areas that are *definitely* mined and can add a safety measure for non–MCM ships.
   c. Training should occasionally accommodate decisions to bypass minefields.

   Examples: Most amphibious landings have occurred in places with no mining. At Wonsan, the plan was changed when the primary route was found to be heavily mined. In Desert Storm, the planned route avoided the (incorrect) intel-assessed minefields.

6. **The easiest minefield to clear is the one the miner is prevented from laying.**
   a. Preventing mining requires strong intelligence to provide indications and warnings.
   b. Rules of engagement must address attacking vessels engaged in active mining.
   c. Valuable mining intelligence can be collected from captured minelayers.

   Examples: Crimean War, Korean War, Tanker War, Operation Iraqi Freedom.
7. **MCM task group assets *will* be lost during MCM ops.**
   a. Contingency planning should include in-stride replacement of assets.
   b. For operations where higher risk to assets is necessary, it is better to have available assets that are low-cost, numerous, and attritable.
   c. Do not lay mines that your own forces are not trained and equipped to counter.

   Examples: World War I (UK especially), World War II (Balikpapan is notable), Korean War (Wonsan is notable), Desert Storm, Ukraine War.
8. **Postwar mine clearance takes several months or even years.**
   a. International coalitions can reduce the burden.
   b. Timelines are inversely related to clearance rates.
   c. The MCM mix should *also* include assets with long duration and wide coverage.
   d. Crew rest matters. Consider anchoring at night during postwar cleanup.

   Examples: Russo–Japanese War, World War I, World War II, Vietnam War, Desert Storm.
9. **Mines can be laid most effectively when an enemy moves predictably.**

   Examples: World War I Dardanelles Campaign, Vietnam War; Tanker War.
10. **Prescribed, cleared routes can be both good and bad.**
    a. In home waters, they are vital in areas where mining is heavy and random.
    b. Using them in unmined areas adds predictability without increasing protection.
    c. If used consistently within range of an enemy's sensors, they add predictability.
    d. If used, they must be checked daily. MCM forces should be scaled accordingly.

    Examples: World War I, World War II, Tanker War (USS *Samuel B. Roberts*). Ukraine War?
11. **Small-scale raids show defenders their weaknesses and give them time/motivation to shore up those defenses.**
    a. This is the equivalent of giving an enemy an inoculation *before* giving a disease.
    b. Do not introduce revolutionary new weapons until they can be used at scale.
    c. The best time to introduce them is as a part of a larger, war-winning campaign.

    Examples: Early raid on entry forts at the Dardanelles; early use of magnetic and later acoustic mines in UK waters in World War II. Three successful cases were the first offensive mining during the Russo–Japanese War (turning the tide of the whole war and history) and the two strategic mining campaigns that helped end World War II and the Vietnam War.

12. **Any mining technology should account for electrical storms and solar storms.**

    Examples: Crimean War in Kronstadt; Vietnam War.

13. **Private inventors can often offer a military some revolutionary advances in technology.**
    a. Means to do so should be available.
    b. There will be far more chaff than wheat in the suggestions.
    c. In-the-field innovations by service members should be explored for wider use.

    Examples: David Bushnell, Robert Fulton, Samuel Colt, Immanuel Nobel, the Singer Mine in the Civil War, the antenna mine of World War I.

14. **Moored minefields will be a problem for as long as they are the cheapest, simplest, most prevalent naval mines on the market.**
    a. Training should include realistically laid moored minefields.
    b. Training should include aerial spotting.
    c. Training should include all the actions that the entire MCMTG would take in a live moored minefield.
    d. Any plans for mechanical minesweeping must include plans for how to dispose of large volumes of drifting mines in a short period of time.

    Examples: Crimean War, Russo–Japanese War, World War I, World War II, Korean War, Tanker War, Desert Storm, Operation Iraqi Freedom, Houthi rebels in the Red Sea, Ukraine War.

15. **Surveillance matters.**
    a. It makes the surveyor less at risk of surprise. *Example: Desert Storm mines.*
    b. It gives them a chance to surprise and devastate their enemies at critical times.

    Examples: The 26 mines that sank/damaged four battleships at the Dardanelles; the sinking of Admiral Makarov's ship during the Russo–Japanese War.

16. **Intelligence can get it wrong.**
    a. In the Information Age, the problem is often too much information. Mining information needs to flow to the on-scene MCMTG for correlation and analysis.
    b. Intelligence gives educated guesses, not ground truth.
    c. It is the job of MCM units to find ground truth.
    d. MCMTG units should never assume that MCM geometry is the *only* mined area.
    e. MCMTG units should never assume that known minefields are the *only* minefields.
    f. MCM geometry should be approached with caution.
    g. Aerial spotting in the area surrounding MCM geometry and ahead of the MCMTGs' transit can reduce the risk of accidentally operating in unknown moored minefields.

    Examples: Dardanelles Campaign, World War II, Korean War, Desert Storm.

17. **Non-ship MCM assets add rapid capability but require non–MCM support.**
    a. Nontraditional assets like helicopters, drones, AUVs, EOD, dolphins, etc., can rapidly deploy to a mined area to start the mine-hunting and clearance process.
    b. The farther the clearance geometry is from shore, the farther shore-based units will need to transit.
    c. Shore-based units will require facilities for messing, berthing, maintenance, and C2.
    d. Sea-basing reduces these requirements but may displace other mission sets.
    e. Non-MCM ships operating near minefields incur greater mine risk.
    f. Sea-basing can require the use of several ships, usually many flat-top ships for AMCM.
    g. Non-MCM ship owners usually resist participation in MCM.
    h. Different types of MCM units have different capabilities and coverage rates.

    Examples: Crimean War, U.S. Civil War, Vietnam War, Gulf of Suez mining, Tanker War, Desert Storm.
18. **Established tactics, techniques, and procedures are ever-evolving and never all-inclusive. They are written and reviewed by humans who are doing their best in an evolving field. Anything that reduces the risk of mines, even if not quantifiable in tactical doctrine, still adds value if it (a) removes mines as a threat or (b) stops ships from hitting mines. Commanders should entertain realistic suggestions, and lessons learned should be compiled and frequently reviewed.**

    Examples: Every war in which mines made an appearance.

Abbreviations List

ADM Admiral

AI Artificial Intelligence

AM Auxiliary Minesweeper

AMCM Airborne Mine Countermeasures

AMS Coastal Minesweeper

AN/SLQ Water Countermeasures Special Systems (equipment designator)

AN/SQQ Special Systems (equipment designator)

ANZAC Australian and New Zealand Army Corps

AOU Area of Uncertainty

ASPB Assault Support Patrol Boat

ASW Anti-Submarine Warfare

ATS Salvage Ship

AWACS A Surveillance Aircraft Designator

A2/AD Anti-Access/Area-Denial

BG Brigadier General

CDR Commander

CG Cruiser, Guided Missile

CH Designator for Marine Helicopters

CIA Counterintelligence Agency

CM Mine Countermeasures Command Ship

CMWF Commander Mine Warfare Force

COMINEWARCOM Commander Mine Warfare Command

COMINEWARFOR Commander Mine Warfare Force

CNO Chief of Naval Operations

COA Course of Action

COMNAVFE Commander, Navy Forces Far East

COMNAVFOR Commander, Naval Forces

COOP Craft of Opportunity

COP Common Operational Picture

CSS Confederate States Ship

CTF Commander Task Force

DM Destroyer Minelayer

DMS Destroyer Minesweeper

DMZ Demilitarized Zone

DoD Department of Defense

DPRK Democratic People's Republic of Korea (North Korea)

DST Destructor (a mine designator)

DRV Democratic People's Republic of Vietnam

EOD Explosive Ordnance Disposal

FM Frequency Modulation

GDP Gross Domestic Product

GPS Global Positioning System

HM Helicopter, Mine Countermeasures Squadron

HMS His/Her Majesty's Ship

HRP A helicopter equipment designator

IED Improvised Explosive Device

IRGC Iranian Revolutionary Guard Corps

IRGCN Iranian Revolutionary Guard Corps Navy

IRIN Iranian Navy

JA Japan

JCS Joint Chiefs of Staff

JMS Japanese Minesweeper

LCDR Lieutenant Commander

LCM Landing Craft, Mechanical

LCPR Landing Craft Personnel Ramp

LCS Littoral Combat Ship

LNO Liaison Officer

LOB Line of Bearing

LPD Landing Platform Dock (amphibious ship)

LPH Landing Platform, Helicopter (amphibious ship)

LT Lieutenant

LST Landing Ship Tank

LUGM A mine weapons designator

MCM Mine Countermeasures

MCMTG Mine Countermeasures Task Group

MCS Mine Countermeasures Command Ship

MHC Minehunter, Coastal

MIW Mine Warfare

MK Mark (refers to a weapon designator)

MKB A Soviet weapon designator

MLM Modified Motor Launch

MN1 Mineman, 1st Class

MOP Magnetic Orange Pipe

MOU Memorandum of Understanding

MSB Minesweeping Boat

MSC Minesweeper, Coastal

MSD Minesweeping Drone

MSH Minesweeper Hunter

MSI Minesweeper, Inshore

MSL Minesweeping Launch

MSM River Minesweeper

MSO Minesweeper, Oceangoing

MSR Patrol Minesweeper

MSS Minesweeper, Special

NAG Northern Arabian Gulf

NATO North Atlantic Treaty Organization

NLF National Liberation Front

OPSEC Operational Security

PBR Patrol Boat, River

PGM Patrol Gunboat, Motor

PM Program Manager

POW Prisoner of War

PPR Preplanned Responses

Radar Radio Direction and Ranging

RADM Rear Admiral

RC Remote Controlled (vehicles, boats, etc.)

RECO Remote Controlled (a type of mine)

RH A helicopter equipment designator

RMS Royal Mail Ship

RN Royal Navy

ROE Rules of Engagement

ROK Republic of Korea (South Korea)

ROKN Republic of Korea Navy

ROV Remote Operated Vehicles

SAG Southern Arabian Gulf

SEAL Sea, Air, and Land

SMCM Surface Mine Countermeasures

SMS Seiner Majestät Schiff (His Majesty's Ship)

SONAR Sound Navigation and Ranging

SWATH Small Waterplane Area Twin Hull

TF Task Force

TNT Trinitrotoluene

TTP Tactics, Techniques, and Procedures

UAE United Arab Emirates

UC German Coastal Mining Submarine

UDT Underwater Demolition Team

UMCM Underwater Mine Countermeasures

UN United Nations

UNSC United Nations Security Council

USNS United States Naval Ship

USO Uniformed Services Organization

USQ A sonar equipment designator

USS United States Ship

USSR Union of Soviet Socialist Republics

USV Unmanned Surface Vehicle

UUV Unmanned Underwater Vehicle

VADM Vice Admiral

VC Vietcong

XO Executive Officer

YMS Yard Motor Minesweeper

Chapter Notes

Preface

1. And, yes, he mentions his own experiences, for his career (like that of all mine warriors) helped in its own small way to shape mine warfare.

Introduction

1. Dictionary.com (2002). "Mine." https://www.dictionary.com/browse/mine.
2. Hartmann, G.K. (1979). *Weapons That Wait: Mine Warfare in the U.S. Navy.* Naval Institute Press.
3. Cowie, J.S. (1949). *Mines, Minelayers, and Minelaying.* Oxford University Press.
4. *Ibid.*
5. JCS (2019). *DOD Dictionary of Military and Associated Terms.* Office of the Chairman of the Joint Chiefs of Staff. p. 147.
6. Edwards, J. (2020). Evolving Mine Warfare. *U.S. Naval Institute Blog.* https://blog.usni.org/posts/2020/03/10/evolving-mine-warfare.
7. *Ibid.*
8. Hartmann, G.K. (1979). *Weapons That Wait.*
9. Note: To "channelize" means to force shipping to use a set channel rather than to allow it to go anywhere it likes.

Chapter 1

1. Manstan, R.R., & Frese, F.J. (2010). Turtle: *David Bushnell's Revolutionary Vessel.* Westholme. pp. 243–246.
2. *Ibid.*, pp. 250–251.
3. *Ibid.*, pp. 86–98.
4. Cowie, J.S. (1949). *Mines, Minelayers, and Minelaying.* p. 7.
5. Daughan, G.C. (2008). *If By Sea: The Forging of the American Navy from the Revolution to the War of 1812.* Basic Books. p. 102.
6. Potter, E.B. (Ed.). (1981). *Sea Power: A Naval History (2nd ed.).* Naval Institute Press. pp. 1–10.
7. Fluckey, E.B. (1992). *Thunder below!* University of Illinois Press.
8. Manstan, R.R., & Frese, F.J. (2010). *Turtle.* pp. 129–141.
9. Youngblood, N. (2006). *The Development of Mine Warfare: A Most Murderous and Barbarous Conduct.* Praeger Security International. p. 6.
10. Low, A.M. (1940). *Mine and Countermine.* Sheridan House. p. 18.
11. Manstan, R.R., & Frese, F.J. (2010). *Turtle.* pp. 131–139.
12. Youngblood, N. (2006). *The Development of Mine Warfare.* pp. 13–14.
13. *Ibid.*, pp. 14–15.
14. *Ibid.*, p. 15.
15. *Ibid.*, pp. 15–16.
16. *Ibid.*, p. 16.
17. *Ibid.*
18. Potter, E.B. (Ed.). (1981). *Sea Power.* p. 129.
19. Bradford, R.B. (1882). *History of Torpedo Warfare.* U.S. Torpedo Station. pp. 10–11.
20. Winters, R., Albion, R., & Patterson, A. (1968). *A Brief History of Mine Warfare.* Mine Advisory Committee. p. 17.
21. Youngblood, N. (2006). *The Development of Mine Warfare.* p. 18.
22. Melia, T.M. (1991). *"Damn the Torpedoes": A Short History of U.S. Naval Mine Countermeasures, 1777–1991.* Naval Historical Center. p. 7.
23. Hartmann, G.K. (1979). *Weapons That Wait.* p. 7.
24. Cavendish, R. (2007). "Robert Fulton's Paddlesteamer." *History Today,* 57(8). https://www.historytoday.com/archive/robert-fulton%E2%80%99s-paddlesteamer.
25. Heidler, D.S., & Heidler, J.T. (Eds.). (1997). *Encyclopedia of the War of 1812.* ABC-CLIO. p. 197.
26. Jefferson, T. (1806). Letter: From Thomas Jefferson to Robert Smith, 12 December 1806. *Papers of Thomas Jefferson.* https://rotunda.upress.virginia.edu/founders/default.xqy?keys=FOEA-print-04-01-02-4669.
27. Potter, E.B. (Ed.). (1981). *Sea Power.* p. 93.
28. Note: These same features apply to the drones that featured so heavily in the Russo-Ukrainian war over two centuries later.
29. Youngblood, N. (2006). *The Development of Mine Warfare.* p. 18.
30. Fulton, R. (1810). *Torpedo War and Submarine Explosions.* William Elliott. pp. 15–18.
31. Low, A.M. (1940). *Mine and Countermine.* pp. 46–49.
32. Youngblood, N. (2006). *The Development of Mine Warfare.* p. 20.

33. Bradford, R.B. (1882). *History of Torpedo Warfare.* U.S. Torpedo Station. pp. 24–27.

34. Tucker, S.C. (Ed.). (2012). *The Encyclopedia of the War of 1812: A Political, Social, and Military History,* ABC-CLIO. pp. 473–474.

https://books.google.com/books?id=hVSrJBQYAk8C&dq=War+of+1812+mines+norfolk&source=gbs_navlinks_s.

35. Bradford, R.B. (1882). *History of Torpedo Warfare.* p. 26. Note: As the author, I first read about this event from the above reference and then added this passage while in Sackets Harbor at The Sandwich Bar (whose owner, Jamie, has always been super-supportive) on a minus-10-degree (F) January morning. Living in the neighboring town of Watertown, New York, at the time, I drove the 12 minutes to this town as a change of venue to foster my creativity. While in town I also finished writing chapter 3.

36. Schiffer, M.B. (2008). *Power Struggles: Scientific Authority and the Creation of Practical Electricity Before Edison.* The MIT Press. pp. 122–123.

37. *Ibid.*

38. Colt, S. (1836). "U.S. Patent 9430X Revolver and Revolving Rifle: Improvement in Fire-arms." U.S. Patent Office. https://revolvers.candrsenal.com/timeline/patent-colt-2/.

39. History.com (2023). "Samuel Colt." https://www.history.com/topics/inventions/samuel-colt.

40. Schiffer, M.B. (2008). *Power Struggles.* pp. 122–123.

41. *Ibid.*, pp. 123–127.

42. *Ibid.*, pp. 125–127.

43. *Ibid.*

44. Note: The author personally provided such an objection, which directly led to a proposed weapon system being killed.

45. Youngblood, N. (2006). *The Development of Mine Warfare.* pp. 22–24.

46. Siemens (2023). "Founders: Werner Von Siemens." https://www.siemens.com/global/en/company/about/history/people/founders.html.

47. Tolf, R.W. (1976). *The Russian Rockefellers: The Saga of the Nobel Family and the Russian Oil Industry.* Hoover Institution Press. pp. 1–5.

48. *Ibid.*, pp. 8–9.

49. *Ibid.*, pp. 15–38.

50. *Ibid.*, p. 20.

51. Britannica. (2023). "Alfred Nobel." https://www.britannica.com/biography/Alfred-Nobel.

52. *Ibid.*

53. Tolf, R.W. (1976). *The Russian Rockefellers.* pp. 38–45.

54. *Ibid.*, pp. 45–73.

Chapter 2

1. Figes, O. (2010). *The Crimean War.* Picador. p. 394.

2. *Ibid.*, pp. 160–161.

3. Ffrench, B.R.L.V. (1972). *The Crimean War.* Pen and Sword Books. p. 32.

4. Figes, O. (2010). *The Crimean War.* pp. 175–179.

5. *Ibid.*, pp. 194–199.

6. *Ibid.*, pp. 156–157.

7. *Ibid.*, pp. 157–162.

8. Note: Finland, with its capital at Helsinki, was then a part of the Russian Empire.

9. Bist LLC (2022). *Nautical chart app depicting areas around Sveaborg fortress islands.* Bist LLC. https://fishing-app.gpsnauticalcharts.com/i-boating-fishing-web-app/fishing-marine-charts-navigation.html?title=Port%20of%20Helsinki%20boating%20app&fbclid=IwAR0hwt2aCO-4p1lQPJbFfmNcR5aq8BCtl7ZWZAJ9miQBr3ntCBcQzJNiHuU#11.27/60.1551/24.9819.

10. Menarchik, J.D. (2010). *North Korean Protective Mine Warfare: An Analysis of the Naval Minefields at Wonsan, Chinnampo, and Hungnam During the Korean War.* Air Command and Staff College. https://apps.dtic.mil/sti/pdfs/ADA539219.pdf?fbclid=IwAR0TacYt3TiDqRiOTbU7gntsBBp5KarPdzNHJzCDIJA_i1eJ0GBWsSehtXY.

11. Ffrench, B.R.L.V. (1972). *The Crimean War.* pp. 32–33.

12. Duffy, M. (2011). "The gunnery at Trafalgar: Training, tactics or temperament?" *Journal for Maritime Research* 7(1). pp. 140–169. https://doi.org/10.1080/21533369.2005.9668349.

13. RusNavy.com (2021). *Victories of the Russian Fleet.* MilPress Media Group. http://rusnavy.com/history/victory.htm.

14. Earp, G.G. (1854). *The History of the Baltic Campaign: From Documents and Other Materials Furnished by Vice-Admiral Sir. C. Napier.* Chapel Books. pp. 281–282.

15. Earp, G.G. (1854). *The History of the Baltic Campaign.* pp. 281–282.

16. *Ibid.*, pp. 278–281.

17. *Ibid.*, pp. 281–282.

18. Low, A.M. (1940). *Mine and Countermine.* p. 66.

19. *Ibid.*, p. 208.

20. Laskow, S. (2018). "The sun set off 4,000 sea mines during the Vietnam War." Atlas Obscura. https://www.atlasobscura.com/articles/solar-storm-sets-off-mines.

21. Bist LLC (2022). *Nautical chart app.* Bist LLC.

22. Duckers, P. (2011). *The Crimean War at Sea: Naval Campaigns Against Russia, 1854–6.* Pen and Sword Maritime. pp. 61–66.

23. Samarov A.A., & Petrov, F.A. (1951). *Development of Mine Materiel in the Russian Navy.* Naval Publishing House of the Naval Ministry of the USSR. Translated by Lloyd G. Robbins of the Science Translation Service of the University of Alabama. pp. 152–153.

24. Earp, G.G. (1854). *The History of the Baltic Campaign.* pp. 281–282.

25. Duckers, P. (2011). *The Crimean War at Sea.* p. 149.

26. *Ibid.*, pp. 150–158.

27. Samarov A.A., & Petrov, F.A. (1951).

Development of Mine Materiel in the Russian Navy. pp. 152–153.

28. *Ibid.*, p. 153.

29. Russell, W.H. (1858). *The British Expedition to the Crimea.* Routledge. p. 346.

30. *Ibid.*

31. Google (n.d.). [Google Maps directions to walk overland from Rostov-on-Don, Russia to Sevastopol.] https://maps.app.goo.gl/avpWbKaSJZ9CXRXF6.

32. Russell, W.H. (1858). p. 346.

33. Bradford, R.B. (1882). *History of Torpedo Warfare.* p. 38.

34. Ghosh, D. (2021). Sea of Azov. *WorldAtlas.* https://www.worldatlas.com/seas/sea-of-azov.html.

35. Duckers, P. (2011). *The Crimean War at Sea.* pp. 157–177.

36. Historyplex (2021). *Crimean War Causes, Timeline, and Summary.* Historyplex. https://historyplex.com/crimean-war-causes-timeline-summary.

37. Figes, O. (2010). *The Crimean War.* pp. 394.

38. Pike, J. (2021). *Volga-Don Shipping Canal.* GlobalSecurity.org. https://www.globalsecurity.org/military/world/russia/volga-don-canal.htm.

39. Goncharenko, R. (2019). "NATO steps up naval presence on the Black Sea." DW.com. https://www.dw.com/en/nato-steps-up-naval-presence-on-the-black-sea/a-47732883.

40. Google (n.d.). [Google Map displaying of Sea of Azov region including surrounding countries.] https://maps.app.goo.gl/2azYrRaECAwYimP4A.

41. Russia & Ukraine (2003). *Agreement between the Russian Federation and the Ukraine on cooperation in the use of the sea of Azov and the strait of Kerch.* https://www.ecolex.org/details/treaty/-agreement-between-the-russian-federation-and-the-ukraine-on-cooperation-in-the-use-of-the-sea-of-azov-and-the-strait-of-kerch-tre-149547/.

42. NATO (2021). *Relations with Georgia.* Brussels: NATO. https://www.nato.int/cps/en/natohq/topics_38988.htm.

43. NATO (2021). *Relations with Ukraine.* Brussels: NATO. https://www.nato.int/cps/en/natolive/topics_37750.htm.

44. Zeihan, P. (2016). *The Accidental Superpower.* Twelve Books.

45. Google. (n.d.). [Google Map displaying all countries located within approximately 500 miles of Moscow, Russian Federation.] https://maps.app.goo.gl/qBER7yfHy2fHYxUF9.

46. Note: Oddly enough, this chapter was written only a few months before the Russo-Ukrainian war that began in 2022. During that war, Ukraine famously attacked the Crimean Bridge, which connects Russia to the Crimean Peninsula. That bridge is about10 kilometers from Yeni-Kale. When the author heard of the 2022 bridge, he could not help but immediately think of the parallels to the Crimean War.

47. Google. (n.d.). [Google Map displaying the area around Mykolaiv, Ukraine, out to the mouth of the Dnieper River.] https://maps.app.goo.gl/V96kjSxbz4RMQrc69.

48. Gibbs, P. (1960). *The Crimean Blunder.* Holt, Rinehart and Winston. p. 167.

49. Bradford, R.B. (1882). *History of Torpedo Warfare.* p. 33; Winters, R., Albion, R., and Patterson, A. (1968). *A Brief History of Mine Warfare.* p. 25; Langmaid, K. (1965). *The Approaches Are Mined!* Jarrolds. p. 37.

50. Kinglake, A.W. (1874). *The Invasion of the Crimea: Its Origin, and an Account of Its Progress Down to the Death of Lord Raglan.* William Blackwood and Sons. pp. 150–153.

51. Figes, O. (2010). *The Crimean War.* pp. 203–204.

52. Samarov, A.A., & Petrov, F.A. (1951). *Development of Mine Materiel in the Russian Navy.* pp. 152–163.

53. Figes, O. (2010). *The Crimean War.* p. 417.

Chapter 3

1. Bradford, R.B. (1882). *History of Torpedo Warfare.* pp. 47–48.

2. National Park Service (2023). *Tredegar Iron Works—Ironmaker to the Confederacy.* National Park Service. https://www.nps.gov/articles/tredegar-iron-works-ironmaker-to-the-confederacy.htm.

3. NHHC (2014). "Virginia." Naval History and Heritage Command (NHHC). https://www.history.navy.mil/content/history/nhhc/research/histories/ship-histories/confederate_ships/virginia.html.

4. Coggins, J. (1962). *Arms and Equipment of the Civil War.* Dover Publications. p. 134.

5. Potter, E.B. (Ed.). (1981). *Sea Power.* p. 126.

6. *Ibid.*, pp. 126–129.

7. NHHC (2014). "Manassas." Annapolis, MD: Naval History and Heritage Command (NHHC). https://www.history.navy.mil/research/histories/ship-histories/confederate_ships/manassas.html.

8. Dull, J.R. (2012). *American Naval History, 1607–1865: Overcoming the Colonial Legacy.* University of Nebraska Press. https://www.jstor.org/stable/j.ctt1d9nqgf.

9. Melia, T.M. (1991). *"Damn the Torpedoes."* pp. 8–18.

10. Coggins, J. (1962). *Arms and Equipment of the Civil War.* p. 147.

11. Perry, M.F. (1965). *Infernal Machines—The Story of Confederate Submarine and Mine Warfare.* Louisiana State University Press. pp. 43–44.

12. Bradford, R.B. (1882). *History of Torpedo Warfare.* p. 52.

13. *Ibid.*, p. 55.

14. *Ibid.*, pp. 51–52.

15. Coggins, J. (1962). *Arms and Equipment of the Civil War.* p. 146.

16. Michie, P.S. (1865). *Torpedo Book.* Edited and reprinted in Schiller, H.M. (editor) (2011). *Confederate Torpedoes: Two Illustrated*

19th-Century Works with New Appendices and Photographs. McFarland, 2014. pp. 144–146.
17. Melia, T.M. (1991). *"Damn the Torpedoes."* pp. 8–18.
18. Michie, P.S. (1865). *Torpedo Book.* pp. 106–107.
19. Perry, M.F. (1965). *Infernal Machines.* pp. 110–112.
20. Bradford, R.B. (1882). *History of Torpedo Warfare.* p. 67.
21. Melia, T.M. (1991). *"Damn the Torpedoes."* p. 11.
22. Michie, P.S. (1865). *Torpedo Book.* pp. 26–29.
23. Bradford, R.B. (1882). *History of Torpedo Warfare.* pp. 54–55.
24. Michie, P.S. (1865). *Torpedo Book.* p. 158.
25. *Ibid.*
26. Youngblood, N. (2006). *The Development of Mine Warfare.* p. 37.
27. Perry, M.F. (1965). *Infernal Machines.* pp. 1–12.
28. *Ibid.*, pp. 13–19.
29. *Ibid.*, pp. 28–32.
30. Note: Roman roads allowed large numbers of troops and supplies to be deployed quickly to wherever in the empire they were needed. Featuring tough foundations and sloped to allow swift drainage, they lasted for over a millennium.
31. Perry, M.F. (1965). *Infernal Machines.* pp. 20–27.
32. *Ibid.*, pp. 37–38.
33. *Ibid.*, pp. 43–48.
34. Note: Texas had once been its own independent country and had joined the United States only 16 years prior to the outbreak of the Civil War.
35. Potter, E.B. (Ed.). (1981). *Sea Power.* pp. 135–141.
36. Note: The Mississippi Delta should not be confused with the Mississippi River Delta, the alluvial fan where the river flows into the Gulf of Mexico.
37. Note: A demijohn is a big glass bottle, often measured in gallons, that is used for storing liquids.
38. Perry, M.F. (1965). *Infernal Machines.* pp. 32–33.
39. *Ibid.*, pp. 44–45.
40. Melia, T.M. (1991). *"Damn the Torpedoes."* pp. 12–13.
41. Bradford, R.B. (1882). *History of Torpedo Warfare.* pp. 46–47.
42. *Ibid.*, p. 110.
43. Perry, M.F. (1965). *Infernal Machines.* p. 73.
44. Google. (n.d.) [Google Maps scaled to encompass the James River approaches to Richmond, VA.] https://maps.app.goo.gl/2iwZiC11qMN5YwVP8.
45. Bradford, R.B. (1882). *History of Torpedo Warfare.* p. 46.
46. Stern, P.V.D. (1959). *Secret Missions of the Civil War.* Wings Books. p. 213.
47. Low, A.M. (1940). *Mine and Countermine.* p. 70.
48. Cowie, J.S. (1949). *Mines, Minelayers and Minelaying.* p. 21.
49. Perry, M.F. (1965). *Infernal Machines.* pp. 111–112.
50. *Ibid.*, pp. 112–114.
51. International Conferences (Geneva) (1949). *1949 Geneva convention III Relative to the Treatment of Prisoners of War.* Geneva. Entry into force October 21, 1950. https://ihl-databases.icrc.org/en/ihl-treaties/gciii-1949.
52. International Conferences (The Hague) (1910). *1907 Hague Convention VIII Relative to the Laying of Automatic Submarine Contact Mines.* The Hague. Entry in force January 26, 1910. https://ihl-databases.icrc.org/en/ihl-treaties/hague-conv-viii-1907.
53. Perry, M.F. (1965). *Infernal Machines.* p. 130.
54. Google. (n.d.) [Google Maps scaled to include both Petersburg and Richmond, VA.] https://maps.app.goo.gl/MAsLQCHXw1G8iHFg8.
55. Stern, P.V.D. (1959). *Secret Missions of the Civil War.* p. 234–235.
56. Perry, M.F. (1965). *Infernal Machines.* pp. 130–134.
57. Webster, I. (2022). *CPI inflation calculator: value of $1 from 1864 to 2022.* Calculated at OfficialData.org. https://www.officialdata.org/us/inflation/1864?amount=1.
58. Bradford, R.B. (1882). *History of Torpedo Warfare.* pp. 60–61.
59. Perry, M.F. (1965). *Infernal Machines.* pp. 137–138.
60. *Ibid.*, pp. 139–145.
61. *Ibid.*, p. 177.
62. Boutelle, C.O. (1865). *General map of Charleston Harbor, South Carolina, showing rebel defences and obstructions.* U.S. Coastal Survey. https://www.loc.gov/resource/g3912c.cw0370100/.
63. Potter, E.B. (Ed.). (1981). *Sea Power.* pp. 144–145.
64. Boutelle, C.O. (1865). *General map of Charleston Harbor, South Carolina, showing rebel defences and obstructions.*
65. Perry, M.F. (1965). *Infernal machines.* pp. 50–52.
66. Boutelle, C.O. (1865). *General map of Charleston Harbor, South Carolina, showing rebel defences and obstructions.*
67. Perry, M.F. (1965). *Infernal Machines.* pp. 52–53.
68. *Ibid.*, pp. 54–56.
69. Symonds, C.L. (2009). *The Civil War at Sea.* ABC-CLIO. p. 140.
70. Perry, M.F. (1965). *Infernal Machines.* pp. 81–89.
71. Symonds, C.L. (2009). *The Civil War at Sea.* p. 140.
72. Potter, E.B. (Ed.). (1981). *Sea Power.* p. 129.
73. Melia, T.M. (1991). *"Damn the Torpedoes."* pp. 14–15.
74. NUMA. (2022). *Siege of Charleston.* National Underwater and Marine Agency. https://numa.net/expeditions/siege-of-charleston/.

75. Bradford, R.B. (1882). *History of Torpedo Warfare*. p. 46.
76. Google. (n.d.). [Google Maps scaled to show the St. John's River from Lake George to Jacksonville, FL.] https://maps.app.goo.gl/3vdMqUPz8eKBjkSh7.
77. NPS (2015). *Florida Battles (from Dyer's Compendium)*. National Park Service. https://www.nps.gov/civilwar/florida.htm.
78. Perry, M.F. (1965). *Infernal Machines*. pp. 114–116.
79. Bradford, R.B. (1882). *History of Torpedo Warfare*. p. 46.
80. Perry, M.F. (1965). *Infernal Machines*. pp. 116–117.
81. Melia, T.M. (1991). *"Damn the Torpedoes."* pp. 1–3.
82. American Battlefield Trust. (2022). "Mobile Bay." https://www.battlefields.org/learn/civil-war/battles/mobile-bay.
83. Farragut, D.G. (1864). *Detailed Report of the Mobile Battle*. Reproduced in *Blue & Gray at Sea: Naval Memoirs of the Civil War*, edited by Brian M.T. Tom Doherty Associates Book. p. 385.
84. Melia, T.M. (1991). *"Damn the Torpedoes."* pp. 2–3.
85. Perry, M.F. (1965). *Infernal Machines*. p. 159.
86. Melia, T.M. (1991). *"Damn the Torpedoes."* pp. 2–3.
87. Perry, M.F. (1965). *Infernal Machines*. p. 162.
88. *Ibid*., pp. 177–190.
89. Potter, E.B. (Ed.). (1981). *Sea Power*. p. 153.
90. Perry, M.F. (1965). *Infernal Machines*. pp. 200–201.

Chapter 4

1. Kulik, R.H. (2010). "Scramble for Africa." *Encyclopedia Britannica*. https://www.britannica.com/event/Scramble-for-Africa.
2. Britannica (2023). "Monroe Doctrine." https://www.britannica.com/event/Monroe-Doctrine.
3. National Army Museum (2023). *China: The Boxer Rebellion*.: The National Army Museum. https://www.nam.ac.uk/explore/boxer-rebellion.
4. Potter, E.B. (Ed.). (1981). *Sea Power*. pp. 164–168.
5. Jukes, G. (2002). *The Russo-Japanese War, 1904–1905*. Osprey Publishing. pp. 7–8.
6. *Ibid*., pp. 8–9.
7. Forczyk, R. (2009). *Russian Battleship vs Japanese Battleship: Yellow Sea 1904–1905*. Osprey Publishing. pp. 19–26.
8. *Ibid*., pp. 38–39.
9. Of note, the author did two tours in Sasebo while in the U.S. Navy, for a total of five years.
10. *Ibid*., pp. 14–26.
11. *Ibid*., pp. 34–38.
12. Melia, T.M. (1991). *"Damn the Torpedoes."* pp. 16–25.
13. National Park Service. (2023). "The Endicott Era Defenses." https://www.nps.gov/gate/learn/historyculture/endicotteradefenses.htm.
14. Forczyk, R. (2009). *Russian Battleship vs Japanese Battleship*. pp. 8–9.
15. Jukes, G. (2002). *The Russo-Japanese War, 1904–1905*. pp. 29–30.
16. Forczyk, R. (2009). *Russian Battleship vs Japanese Battleship*. pp. 44–45.
17. Britannica (2023). "Stepan Osipovich Makarov." https://www.britannica.com/biography/Stepan-Osipovich-Makarov.
18. Forczyk, R. (2009). *Russian Battleship vs Japanese Battleship*. pp. 44–45.
19. Jukes, G. (2002). *The Russo-Japanese War, 1904–1905*. p. 27.
20. Albion, R., & Patterson, A. (1968). *A Brief History of Mine Warfare*. Mine Advisory Committee. p. 93.
21. Low, A.M. (1940). *Mine and Countermine*. pp. 78–79.
22. Note: The exact number of crew members lost is unclear. Five different references revealed five different answers. Regardless of the specifics, it was a lot of sailors lost to a previously unknown tactic.
23. Forczyk, R. (2009). *Russian Battleship vs Japanese Battleship*. pp. 45–47.
24. History Channel. (2022). "Russo-Japanese War." https://www.history.com/topics/asian-history/russo-japanese-war.
25. Samarov, A.A., and Petrov, F.A. (1951). *Development of Mine Materiel in the Russian Navy*. pp. 416–419.
26. Jukes, G. (2002). *The Russo-Japanese War, 1904–1905*. p. 40.
27. *Ibid*., pp. 39–46.
28. Forczyk, R. (2009). *Russian Battleship vs Japanese Battleship*. pp. 47–54.
29. Note: When HMS *Dreadnought* was built in 1906, she set the new standard for battleships that made all earlier battleships obsolete. Her main features were a uniform main battery of 10 12-inch big guns and steam turbines that made her the fastest battleship afloat. This combination enabled her to both outrun and outgun any adversary.
30. Jukes, G. (2002). *The Russo-Japanese War, 1904–1905*. pp. 68–69.
31. *Ibid*., pp. 59–61.
32. Forczyk, R. (2009). *Russian Battleship vs Japanese Battleship*. p. 55.
33. Jukes, G. *The Russo-Japanese War, 1904–1905*. p. 68–70.
34. Note: A cruiser typically had the guns of a battleship but had lighter armor. These ships prioritized speed over protection.
35. Potter, E.B. (Ed.). (1981). *Sea Power*. pp. 171–174.
36. International Conferences (The Hague) (1910). *1907 Hague Convention VIII Relative to the Laying of Automatic Submarine Contact Mines*.
37. Note: This may be a problem for Ukraine at the end of the war with Russia that is ongoing at the time of this writing.

Chapter 5

1. Making History Fun (2010). "European Alliances on the Eve of World War I." https://makinghistoryfun.files.wordpress.com/2010/01/-wwi-map.png.
2. The History Place (2009). "1914: War Erupts." http://www.historyplace.com/worldhistory/firstworldwar/index-1914.html.
3. Simkin, J. (2016). "Germany in 1914." Spartacus Educational. http://spartacus-educational.com/FWWinGermany.htm.
4. Van Der Vat, D. (2009). *The Dardanelles Disaster: Winston Churchill's Greatest Failure.* Duckworth Overlook. pp. 9–15.
5. DeVoss, D. (2000). "Searching for Gavrilo Princip." *Smithsonian Magazine.* https://www.smithsonianmag.com/history/searching-for-gavrilo-princip-66287114/.
6. History on the Net (2014). "World War One—Timeline." Regnery Publishing. http://www.historyonthenet.com/world-war-one-timeline/.
7. *Ibid.*
8. Van Der Vat, D. (2009). *The Dardanelles Disaster.* pp. 20–23.
9. *Ibid.*, pp. 23–30.
10. Massie, R.K. (2004). *Castles of Steel: Britain, Germany, and the Winning of the Great War at Sea.* Ballantine Books. pp. 48–49.
11. *Ibid.*, pp. 30–32.
12. Van Der Vat, D. (2009). *The Dardanelles Disaster.* pp. 32–34.
13. *Ibid.*, pp. 34–36.
14. Massie, R.K. (2004). *Castles of Steel.* pp. 44–45.
15. *Ibid.*, pp. 48–49.
16. Van Der Vat, D. (2009). *The Dardanelles Disaster.* pp. 56–60.
17. Youngblood, N. (2006). *The Development of Mine Warfare.* p. 92.
18. Potter, E.B. (Ed.). (1981). *Sea Power.* p. 215.
19. Melia, T.M. (1991). *"Damn the Torpedoes."* p. 29.
20. Youngblood, N. (2006). *The Development of Mine Warfare.* p. 94
21. *Ibid.*
22. Crossley, J. (2011). *The Hidden Threat: The Story of Mines and Minesweeping by the Royal Navy in World War I.* Pen and Sword Maritime. p. 103.
23. Potter, E.B. (Ed.). (1981). *Sea Power.* pp. 216–217.
24. *Ibid.*, pp. 94–95.
25. Note: There is some contention over the originator of the idea to lay the additional mine line.
26. Van Der Vat, D. (2009). *The Dardanelles Disaster.* pp. 1–5. (Note: There is some disagreement on the number of mines laid. British sources say 20, whereas Turkish and German sources say up to 26. Assuming that those who laid the mines know best the number laid, I went with 26.)
27. Massie, R.K. (2004). *Castles of Steel.* pp. 462.
28. Denham, H.M. (1981). *Dardanelles: A Midshipman's Diary.* John Murray, 1981.
29. Milford, D. (2017). "Major Warships Sunk in World War 1." www.worldwar1.co.uk/sunk14.htm (and /sunk__ (15, 16, 17, and 18).
30. Van Der Vat, D. (2009). *The Dardanelles Disaster.* pp. 129–130.
31. *Ibid.*, pp. 130–131.
32. Potter, E.B. (Ed.). (1981). *Sea Power.* pp. 219–221.
33. Erickson, E. (2001). *Ordered to Die: A History of the Ottoman Army in the First World War.* Greenwood. pp. 94–95, 327.
34. Crossley, J. (2011). *The Hidden Threat.* p. 100.
35. Hartmann, G.K. (1979). *Weapons That Wait.* p. 47.
36. Britannica (2023). "The Russian Revolution: Russia History (1917)." https://www.britannica.com/topic/Bolshevik.

Chapter 6

1. Hartmann, G. (1979). *Weapons That Wait.* p. 28.
2. Siemens (2023). "Founders: Werner Von Siemens."
3. Youngblood, N. (2006). *The Development of Mine Warfare.* pp. 64–65.
4. Cowie, J.S. (1949). *Mines, Minelayers, and Minelaying.* p. 27.
5. The Hertz horn is more correctly called the "Herz horn" after its inventor, Dr. Alfred Herz of the German Mine Defense Laboratory. The original spelling changed to "Hertz horn" within the U.S. Navy over a century ago. Because that is the most commonly used spelling within most mine warfare references, the author has chosen to go with that spelling. Thank you to Captain Chris O'Flaherty of the Royal Navy for bringing this discrepancy to light with his 2019 book, *Naval Minewarfare: Politics to Practicalities.*
6. *Ibid.*, p. 37.
7. Crossley, J. (2011). *The Hidden Threat.* p. 21.
8. Cowie, J.S. (1949). *Mines, Minelayers, and Minelaying.* pp. 36–37.
9. Crossley, J. (2011). *The Hidden Threat.* pp. 27–29.
10. Langmaid, K.J.R. (1965). *The Approaches Are Mined!* p. 56.
11. CDSG. (2021). *History of America's Seacoast Defenses.* Coastal Defense Group. https://cdsg.org/forts/ and its subsections.
12. Melia, T.M. (1991). *"Damn the Torpedoes."* pp. 27–28.
13. *Ibid.*, pp. 27–30.
14. Cowie, J.S. (1949). *Mines, Minelayers, and Minelaying.* p. 38.
15. *Ibid.*, pp. 45–64.
16. Massie, R.K. (2004). *Castles of Steel.* p. 97.
17. Crossley, J. (2011). *The Hidden Threat.* pp. 1–3.
18. Langmaid, K.J.R. (1965). *The Approaches Are Mined!* p. 95.

19. *Ibid.*, pp. 77–78.
20. Massie, R.K. (2004). *Castles of Steel*. pp. 140–141.
21. Crossley, J. (2011). *The Hidden Threat*. p. 51.
22. *Ibid.*, p. 56.
23. Massie, R.K. (2004). *Castles of Steel*. pp. 141–143.
24. Youngblood, N. (2006). *The Development of Mine Warfare*. p.84.
25. Crossley, J. (2011). *The Hidden Threat*. pp. 52–63.
26. Cowie, J.S. (1949). *Mines, Minelayers, and Minelaying*. pp. 50–52.
27. Crossley, J. (2011). *The Hidden Threat*. p. 65.
28. Massie, R.K. (2004). *Castles of Steel*. p. 560.
29. Crossley, J. (2011). *The Hidden Threat*. pp. 80–95.
30. Ledebur, G.F.V. (1977). *Die Seemine*. J.F. Lehmanns Verlag. p. 186.
31. Note: It is possible that this openness to the advantages of undersea warfare was inspired in part by the work of Victor von Scheliha. Von Scheliha was a German who worked with Rains in the Confederate Corps of Engineers in building the strong mine defenses at Mobile Bay in the Civil War. After the war, he drafted his *A Treatise on Coast-Defense* (E&F.N. SPON, London, 1868). In that work, he described in detail the effectiveness of mines against Union ironclads, as well as the David torpedo boats that were the precursors to the modern submarine. This work would doubtlessly have been of interest to any nation expecting to one day have to stand up to a stronger, well-armored naval opponent. For Imperial Germany, whose country and navy were both founded just three years later, this would have been an essential primer in naval defenses against the armored surface fleet of the UK.
32. Massie, Robert K. (2004). *Castles of Steel*. pp. 529–530.
33. *Ibid.*, pp. 530–534.
34. *Ibid.*, pp. 536–537.
35. Note: "Unrestricted submarine warfare" meant that any vessel, regardless of nationality, found in the war zone would be sunk on sight without notice.
36. Massie, R.K. (2004). *Castles of Steel*. p. 712.
37. *Ibid.*, pp. 712–714.
38. Zieger, R.H. (2000). *America's Great War: World War 1 and the American Experience*. Rowman & Littlefield. p. 113.
39. Hartmann, G. (1979). *Weapons That Wait*. p. 49.
40. GlobalSecurity.org (2023). "Littoral Combat Ship." https://premium.globalsecurity.org/military/systems//ship/lcs.htm.
41. Hartmann, G. (1979). *Weapons That Wait*. p. 50.
42. Cowie, J.S. (1949). *Mines, Minelayers, and Minelaying*. p. 60.
43. Bruhn, D.D., & Hoole, R. (2018). *Home Waters: Royal Navy, Royal Canadian Navy, and U.S. Navy Mine Forces Battling U-boats in World War I*. Heritage Books. pp. 259–269.
44. Massie, R.K. (2004). *Castles of Steel*. p. 761.
45. Youngblood, N. (2006). *The Development of Mine Warfare*. p. 99.
46. Massie, R.K. (2004). *Castles of Steel*. p. 761.
47. Lozada, C. (2005). "The economics of World War I." *The Digest*, issue 1. National Bureau of Economic Research. https://www.nber.org/digest/jan05/economics-world-war-i.
48. Massie, R.K. (2004). *Castles of Steel*. p. 761.
49. DiGiulian, T. (2023). "United States of America Mines." NavWeaps. http://navweaps.com/Weapons/WAMUS_Mines.php.
50. Cowie, J.S. (1949). *Mines, Minelayers, and Minelaying*. p. 86.
51. Massie, R.K. (2004). *Castles of Steel*. p. 50.
52. Youngblood, N. (2006). *The Development of Mine Warfare*. pp. 99–102.
53. Crossley, J. (2011). *The Hidden Threat*. pp. 63–64.
54. Bruhn, D.D., & Hoole, R. (2018). *Home Waters*. pp. 293–298.
55. *Ibid.*, p. 291.
56. Melia, T.M. (1991). *"Damn the Torpedoes."* pp. 37–38.
57. Bruhn, D.D., & Hoole, R. (2018). *Home Waters*. p. 290.
58. Melia, T.M. (1991). *"Damn the Torpedoes."* pp. 34–36.
59. *Ibid.*, pp. 36–39.
60. *Ibid.*
61. Ledebur, G.F.V. (1977). *Die Seemine*. p. 186.

Chapter 7

1. Elliott, P. (1979). *Allied Minesweeping in World War 2*. Patrick Stephens. pp. 30–31.
2. Melia, T.M. (1991). *"Damn the Torpedoes."* p. 44.
3. Lott, A. (1959). *Most Dangerous Sea: The Story of the Brave Crews Who Swept the Seas of Deadly Enemy Mines*. U.S. Naval Institute. pp. 2–3.
4. Note: Anti-sweeping devices could include bouquet mines, delayed-rising mines, sprockets, etc.
5. Elliott, P. (1979). *Allied Minesweeping in World War 2*. pp. 30–62.
6. Hartmann, G. (1979). *Weapons That Wait*.
7. Lott, A. (1959). *Most Dangerous Sea*. pp. 32–33.
8. *Ibid.*, p. 34.
9. Elliott, P. (1979). *Allied Minesweeping in World War 2*. pp. 48–50.
10. *Ibid.*, pp. 117–120.
11. *Ibid.*
12. *Ibid.*
13. *Ibid.*
14. Grosvenor, J., & Bates, L.M. (1956). *Open the Ports: The Story of Human Minesweepers*. William Kimber & Co.
15. History Channel. (2022). "Lend-Lease Act." https://www.history.com/topics/world-war-ii/lend-lease-act-1.

16. Cowie, J.S. (1949). *Mines, Minelayers, and Minelaying.* pp. 122–123.

17. Helgason, G. (2024). "U-boat Bases: The Bases in France." Uboat.net. https://uboat.net/flotillas/bases/france.htm.

18. The author personally saw a sonar snippet of one of these New Zealand mines in 2014 when it was found during a naval exercise he had attended. Cowie, J.S. (1949). *Mines, Minelayers, and Minelaying.* pp. 133–138.

19. *Ibid.*, pp. 139–144.

20. *Ibid.*, pp. 144–151. The *Manxman* ruse reminds the author of a parade float that he once made in Japan, turning his car into a "Washington Crossing the Delaware" boat using PVC pipe, cardboard, string, a Betsy Ross flag, and borrowed oars.

21. Hartmann, G. (1979). *Weapons That Wait.* p. 152.

22. Note: "Expertise" and "independence" are more important than the "officer" part. A senior enlisted who has a decade of experience in the warfare specifics may at times be more useful for this purpose than a junior officer who has done only a single tour as an occasional watch-stander. Above all, the chosen LNO should be able to establish and maintain credibility with the staff they are meant to assist, if the advantages of this exchange are to be realized.

23. Ledebur, G.F.V. (1977). *Die Seemine.* p. 192.

24. Youngblood, N. (2006). *The Development of Mine Warfare.* p. 130.

25. Mapsofworld.com (2023). "WW2 Map of Europe: Explore the Boundaries of World War 2." MapSherpa Inc. https://www.mapsofworld.com/-world-maps/world-war-ii-map-of-europe.html.

26. Youngblood, N. (2006). *The Development of Mine Warfare.* pp.130–131.

27. Saar, C.W. (1964). *Offensive Mining as a Soviet Strategy.* U.S. Naval Institute. https://www.usni.org/magazines/proceedings/1964/august/offensive-mining-soviet-strategy.

28. DiGiulian, T. (2008). "Russia/USSR Mines." NavWeaps. http://www.navweaps.com/Weapons/WAMRussian_Mines.php.

29. Meister, J. (1957). "The Soviet Navy in World War II." U.S. Naval Institute. https://www.usni.org/magazines/proceedings/1957/august/soviet-navy-world-war-ii.

30. Youngblood, N. (2006). *The Development of Mine Warfare.* pp. 131–132.

31. Lott, A. (1959). *Most Dangerous Sea.* pp. 15–26.

32. Note: The author was once stationed with a German MCM liaison officer who told him about a practice used in his own navy where the commanding officer had one-kilo explosives available to throw over the side, simulating the effect of a nearby mine explosion. He explained that one or two tossed over the side while underway had a way of revealing the ship's state of readiness for conducting MCM operations.

33. Melia, T.M. (1991). *"Damn the Torpedoes."* pp. 47–51.

34. Patterson, A., Jr. (1970). *A Brief History of Mine Warfare.* Mine Advisory Committee, National Academy of Sciences. p. 25.

35. Maher, R.A., & Wise, J.E., Jr. (1998). *Sailors' Journey into War.* Kent University Press. p. 38.

36. Duncan, R.C. (1962). *America's Use of Sea Mines.* U.S. Naval Ordnance Laboratory. pp. 130–131.

37. *Ibid.*

38. Maher, R.A., and Wise, J.E., Jr. (1998). *Sailors' Journey into War.* p. 38.

39. CDSG (2021). *A Short History: Harbor Defenses of the United States of America.* Lenexa, KS: Coast Defense Study Group (CDSG). https://cdsg.org/short-history/.

40. Ledebur, G.F.V. (1977). *Die Seemine.* p. 192.

41. Elliott, P. (1979). *Allied Minesweeping in World War 2.* pp. 83–127.

42. Hartmann, G. (1979). *Weapons That Wait.* p. 70.

43. Of note, none of these landings appear to have taken place at locations previously mined by the Allies. This makes sense. MCM is hardest in places where you know for sure that the waters are mined.

Chapter 8

1. Gavin, P. (1999). "World War Two in the Pacific: Timeline of Events 1941–1945." The History Place. https://www.historyplace.com/unitedstates/pacificwar/timeline.htm.

2. Bada, F. 2019. "How Were the Japanese Islands Formed?" World Atlas. https://www.worldatlas.com/articles/how-were-the-japanese-islands-formed.html.

3. Note: "Combat loaded" describes a packing method where the most-needed supplies like ammunition, water, and radio batteries are loaded on top and in front, while less-immediately needed supplies are farther back. Typically, this is a less-efficient use of available storage space but can drastically reduce the time needed to access critical supplies.

4. *Ibid.*, pp. 95–96.

5. *Ibid.*

6. Note: The author served on two different MIW battle staffs, one for an MIW admiral and one for an MCM commodore.

7. Elliott, P. (1979). *Allied Minesweeping in WWII.* p. 42.

8. *Ibid.*, p. 96.

9. Notehelfer, F.G. (2024). "The Emergence of Imperial Japan." Britannica. https://www.britannica.com/place/Japan/The-emergence-of-imperial-Japan.

10. Johnson, E.A., & Katcher, D.A. (1973). *Mines Against Japan.* U.S. Naval Ordnance Laboratory. p. 23.

11. Lott, A. (1959). *Most Dangerous Sea.* pp. 51–53.

12. Johnson, E.A., & Katcher, D.A. (1973). *Mines Against Japan.* p. 23.

13. Duncan, R.C. (1962). *America's Use of Sea Mines*. pp. 138–141.
14. SUBPAC. (2024). *History of SUBPAC*. Submarine Force Pacific. https://www.csp.navy.mil/About-SUBPAC/subpac-history/.
15. *Ibid*., p. 141.
16. Murrin, M.J. (1945). *Pacific War Diary*. Not published (record kept at Navy Department Library, Washington, DC). pp. 18–19.
17. Note: At the time, minemen were primarily a shore-based rate whose work was normally well behind the front lines.
18. AOM (2003). "In Memoriam—First Mineman Casualties USS Aaron Ward (DM-34)." Association of Minemen (AOM). https://minemen.org/PDFs/Aaron_Ward-Story.pdf.
19. Lott, A.S. (1964). *Brave Ship, Brave Men*. Naval Institute Press.
20. Lund, P., & Ludlam, H. (1978). *Out Sweeps! The Story of the Minesweepers in World War II*. W. Foulsham & Co. pp. 164–167.
21. Melia, T.M. (1991). *"Damn the Torpedoes."* p. 62.
22. Youngblood, N. (2006). *The Development of Mine Warfare*. pp. 135–136.
23. DiGiulian, T. (2008). "Japan Mines: World War II." NavWeaps. http://www.navweaps.com/Weapons/WAMJAP_Mines.php.
24. Youngblood, N. (2006). *The Development of Mine Warfare*. pp. 135–137.
25. Lott, A. (1959). *Most Dangerous Sea*. p. 46.
26. *Ibid*., p. 198; Oregon Coast Beach Connection (2022). "World War II Mines and Explosive Problem on Oregon, Washington Coast in '50s." https://www.beachconnection.net/news/world_war2_mines_exploding_washor.php.
27. *Ibid*.
28. Elliott, P. (1979). *Allied Minesweeping in WWII*. pp. 155–167.
29. Youngblood, N. (2006). *The Development of Mine Warfare*. p. 135.
30. Note: Despite Japan's ability to replace machines, it had lost most of its experienced pilots due to its policy of keeping them deployed. Conversely, the U.S. system was to rotate experienced pilots and seamen to training sites and to new units. This allowed those men to pass the lessons of their experience to the next wave of pilots and seamen.
31. Youngblood, N. (2006). *The Development of Mine Warfare*. pp. 73–79.
32. Note: The author's first ship was homeported in Sasebo. During that tour, he transited the Shimonoseki Strait, participated in exercises in Suo Nada, and made a port visit to Kure with a side trip to Hiroshima. That experience makes it especially visceral for the author to imagine the effect of Phase 1 and the conditions under which the Japanese had to respond.
33. Johnson, E.A., & Katcher, D.A. (1973). *Mines Against Japan*. p. 131.
34. Melia, T.M. (1991). *"Damn the Torpedoes."* p. 61.
35. Johnson, E.A., & Katcher, D.A. (1973). *Mines Against Japan*. pp. 131–133.
36. U.S. Army Air Forces (1945). *Starvation: Phase Analysis of the Twentieth Air Force Strategic Mining Blockade of the Japanese Empire*. 20th Army Air Force. p. 48. https://minemen.org/PDFs/OperationStarvationMiningofJapaneseEmpire.pdf.
37. Johnson, E.A., & Katcher, D.A. (1973). *Mines Against Japan*. pp. 131–133.
38. *Ibid*., pp. 29–34.
39. Youngblood, N. (2006). *The Development of Mine Warfare*. p. 139.
40. Lott, A. (1959). *Most Dangerous Sea*. pp. 166–168.
41. Tillman, B. (2010). *Whirlwind: The Air War Against Japan, 1942–1945*. Simon & Schuster. p. 199.
42. Johnson, E.A., & Katcher, D.A. (1973). *Mines Against Japan*. p. 134.
43. Lott, A. (1959). *Most Dangerous Sea*. pp. 188–189.
44. Lund, P., & Ludlam, H. (1978). *Out Sweeps!* pp. 168–173.
45. Lott, A. (1959). *Most Dangerous Sea*. pp. 188–189.
46. *Ibid*., pp. 198–200.
47. Melia, T.M. (1991). *"Damn the Torpedoes."* p. 64.
48. NAVOCEANO. (1967). *H.O. Pub. No. 110: Danger Areas in the Pacific*. U.S. Naval Oceanographic Office. pp. 1–58.
49. Melia, T.M. (1991). *"Damn the Torpedoes."* p. 68; Elliott, P. (1979). *Allied Minesweeping in WWII*. pp. 181–182.
50. Johnson, E.A., & Katcher, D.A. (1973). *Mines Against Japan*; U.S. Army Air Forces (1945). *Starvation*. pp. 1–35.

Chapter 9

1. Lott, A. (1959). *Most Dangerous Sea*. p. 203.
2. Melia, T.M. (1991). *"Damn the torpedoes."* pp. 69–70.
3. Note: DMSs were small destroyers with minesweeping equipment installed. Fast enough to keep up with the fleet, armed like a destroyer, they were built with a degaussed steel hull. They were valuable against moored contact mines, which remained the most likely mine threat to be encountered. In many ways, their virtue was similar to those of the littoral combat ships (LCSs) of the early 2000s.
4. Note: MSBs were small, lightly manned motor launches rigged with small-scale sweeping equipment. Though their individual sweep paths were small, they could operate in shallow, confined waters unreachable by the sweep gear of AMs or DMSs. They were small and lightweight, and larger ships could carry several of them at once. Their size and crew size also made them far less susceptible to most mines, with less effect on the overall mission if one was lost.

5. Melia, T.M. (1991). *"Damn the Torpedoes."* pp. 69–70.
6. *Ibid.*, pp. 71–72.
7. Menarchik, J.D. (2010). *North Korean Protective Mine Warfare.*
8. Encyclopedia Britannica (1993). "Korea." Robert McHenry (Ed.). *The New Encyclopedia Britannica, vol. 22, macropedia.* pp. 498–499.
9. Note: It is unclear to the author whether or not Kim Il Sung worked with or even met the Chinese communist leader in his guerrilla phase. However, it seems plausible that the two were at least aware of each other's existence.
10. Benson, S.G., & Raffaelle, G.-A. (Eds.). (2001). "Kim Il Sung." *Korean War Reference Library.* Gale. https://www.encyclopedia.com/history/-encyclopedias-almanacs-transcripts-and-maps/-kim-il-sung-1.
11. *Ibid.*
12. *Ibid.*
13. United Nations. (2024). *History of the Korean War.* United Nations Command. https://www.unc.mil/History/1950-1953-Korean-War-Active-Conflict/.
14. DPAA (2024). *Pusan Perimeter.* Defense POW/MIA Accounting Agency. https://dpaa-mil/sites.crmforce.mil/dpaaFamWebInPusan.
15. Lott, A. (1959). *Most Dangerous Sea.* p. 205.
16. *Ibid.*
17. *Ibid.*
18. Lott, A. (1959). *Most Dangerous Sea.* p. 206.
19. *Ibid.*, pp. 206–207.
20. Rollins, E.L. (1950). *UA 452.02.02* USS *Mansfield (DD-728).* Naval History and Heritage Command. https://www.history.navy.mil/content/history/nhhc/our-collections/photography/alphabetical---donations0/r/ua-452-02-lcr-edward-l—rollins-collection0/ua-452-02-02-uss-mansfield—dd-728-.html.
21. Lott, A. (1959). *Most Dangerous Sea.* p. 207.
22. Rear Admiral, Upper Half (RADM) and Commander, Naval Forces Far East (COMNAVFE)
23. Melia, T.M. (1991). *"Damn the torpedoes."* pp. 73–74.
24. Lott, A. (1959). *Most Dangerous Sea.* pp. 205–206.
25. Cagle, M.W., & Manson, F.A. (1957). "Wonsan: The Battle of the Mines." *Proceedings*, vol. 83/6/652. U.S. Naval Institute. https://www.usni.org/magazines/proceedings/1957/june/wonsan-battle-mines.
26. Lott, A. (1959). *Most Dangerous Sea.* pp. 205–206.
27. Cagle, M.W., & Manson, F.A. (1957). "Wonsan: The Battle of the Mines."
28. Lott, A. (1959). *Most Dangerous Sea.* pp. 205–206.
29. Cagle, M.W., & Manson, F.A. (1957). "Wonsan: The Battle of the Mines."
30. Fields, J.A. Jr. (2015). *History of U.S. Naval Operations: Korea.* Naval History and Heritage Command. https://www.history.navy.mil/research/library/online-reading-room/title-list-alphabetically/h/history-us-naval-operations-korea/chapter8-on-border.html.
31. Cagle, M.W. & Manson, F.A. (1957). "Wonsan: The Battle of the Mines."
32. Fields, J.A., Jr. (2015). *History of U.S. Naval Operations: Korea.*
33. Menarchik, J.D. (2010). *North Korean Protective Mine Warfare.*
34. Cagle, M.W. & Manson, F.A. (1957). "Wonsan: The Battle of the Mines."
35. Fields, J.A., Jr. (2015). *History of U.S. Naval Operations: Korea.*
36. *Ibid.*
37. *Ibid.*
38. *Ibid.*
39. Naval Vessel Register. (2024). USS *Gerald R. Ford (CVN 78).* Naval Sea Systems Command. https://www.nvr.navy.mil/SHIPDETAILS/SHIPSDETAIL_CVN_78.HTML.
40. Cagle, M.W., & Manson, F.A. (1957). "Wonsan: The Battle of the Mines."
41. McMullen, C.E. (1950). USS *Pirate War Diary.* Ships History Branch, Naval Historical Center. https://www.history.navy.mil/research/histories/ship-histories/danfs/p/pirate-ii/pirate-am-275-ii-report-of-loss.html.
42. Thomas, C.S. (1950). "USS *Pirate*; Presidential Unit Citation." Secretary of the Navy. https://www.history.navy.mil/research/histories/ship-histories/danfs/p/pirate-ii/pirate-am-275-ii-presidential-unit-citation.html.
43. Cagle, M.W., & Manson, F.A. (1957). "Wonsan: The Battle of the Mines."
44. *Ibid.*
45. Lott, A. (1959). *Most Dangerous Sea.* p. 209; Fields, J.A. Jr. (2015). *History of U.S. Naval Operations: Korea.*
46. Melia, T.M. (1991). *"Damn the Torpedoes."* p. 76.
47. *Ibid.*, pp. 71–76.
48. Cagle, M.W., & Manson, F.A. (1957). "Wonsan: The Battle of the Mines."
49. *Ibid*; Fields, J.A., Jr. (2015). *History of U.S. Naval Operations: Korea.*
50. Edwards, H.W. (1954). "A Naval Lesson of the Korean Conflict." *Proceedings*, vol. 80/12/622. U.S. Naval Institute. https://www.usni.org/magazines/proceedings/1954/december/naval-lesson-korean-conflict.
51. Melia, T.M. (1991). *"Damn the Torpedoes."* pp. 77–78.
52. Menarchik, J.D. (2010). *North Korean Protective Mine Warfare.*
53. Fields, J.A., Jr. (2015). *History of U.S. Naval Operations: Korea.*
54. Google. (n.d.). [Google Maps image of mouth of the Taedong River, North Korea.] https://maps.app.goo.gl/zBpkZRFKpoL9KYw29.
55. Menarchik, J.D. (2010). *North Korean Protective Mine Warfare.*
56. Lott, A. (1959). *Most Dangerous Sea.* p. 211.
57. CDR = Commander, the U.S. Navy's O-5 naval rank.

58. *Ibid.*, p. 212.
59. LCDR = Lieutenant Commander, the U.S. Navy's O-4 naval rank.
60. Note: The reader will recall that in 1950, North Korea had existed as a communist state for only a couple of years. After years of bad treatment by the Japanese, only to be followed by communist leadership, the North Korean leadership did not enjoy unwavering, fanatical support. There would have been plenty willing to work against them. Such a pattern of locals willing to betray local leadership repeats often in wars against harsh dictatorships.
61. Lott, A. (1959). *Most Dangerous Sea.* pp. 212–213.
62. Menarchik, J.D. (2010). *North Korean Protective Mine Warfare.*
63. Lott, A. (1959). *Most Dangerous Sea.* p. 214.
64. Menarchik, J.D. (2010). *North Korean Protective Mine Warfare.*
65. Lott, A. (1959). *Most Dangerous Sea.* p. 215.
66. *Ibid.*, pp. 215–216.
67. Menarchik, J.D. (2010). *North Korean Protective Mine Warfare.*
68. Lott, A. (1959). *Most Dangerous Sea.* pp. 216–217.
69. Blakemore, E. (2020). *The Korean War Never Technically Ended. National Geographic.* https://www.nationalgeographic.com/history/article/why-korean-war-never-technically-ended.
70. Melia, T.M. (1991). *"Damn the Torpedoes."* p. 76.
71. *Ibid.*, pp. 86–90.
72. Note: The author started as a sonar technician on a minesweeper and cross-rated to mineman. Please read his biography for one (admittedly atypical) example of a mineman career path.
73. Melia, T.M. (1991). *"Damn the Torpedoes."* p. 76.
74. *Ibid.*, p. 79.

Chapter 10

1. Duiker, W.J. (2004). "Vietnam." *Encyclopedia Americana* (Michael S.-J., Editor). Scholastic Library Publishing. pp. 110e-110m.
2. *Ibid.*, p. 110o.
3. Green, A.L. (Ed.). (2004). "Vietnam." *Compton's Encyclopedia.* p. 47.
4. Duiker, W.J. (2004). "Vietnam." pp. 110o-110p.
5. Vietnam Embassy (2024). "Geography (of Vietnam)." Embassy of the Socialist Republic of Vietnam in the United States of America. https://-vietnamembassy-usa.org/vietnam/geography.
6. Marolda, E.J. (1987). "The War in Vietnam's Shallows." *Naval History Magazine*, vol. 1, no. 1. U.S. Naval Institute. https://www.usni.org/magazines/naval-history-magazine/1987/april/war-vietnams-shallows.
7. Owen, L. et al. (2024). "Mekong River." *Britannica.* https://www.britannica.com/place/Mekong-River/Climate-and-hydrology.
8. Melia, T.M. (1991). *"Damn the Torpedoes."* p. 93.
9. Bruhn, D.D. (2011). *Wooden Ships and Iron Men, vol. III: The U.S. Navy's Coastal and Inshore Minesweepers and the Minecraft That Served in Vietnam, 1953–1976.* Heritage Books. p. 130.
10. Melia, T.M. (1991). *"Damn the Torpedoes."* p. 92.
11. U.S. Army (1965). *Mines and Booby Traps Used by the VC in South Vietnam (Min Bay Do VC Xu Dung Tai Mien Nam Viet Nam).* Department of the Army. pp. 21–25.
12. Melia, T.M. (1991). *"Damn the Torpedoes."* p. 92.
13. Bruhn, D.D. (2011). *Wooden Ships and Iron Men, vol. III.* pp. 239–249.
14. *Ibid.*, p. 317.
15. Melia, T.M. (1991). *"Damn the Torpedoes."* pp. 95–96.
16. Bruhn, D.D. (2011). *Wooden Ships and Iron Men, vol. III.* p. 125.
17. Melia, T.M. (1991). *"Damn the Torpedoes."* p. 95.
18. Bruhn, D.D. (2011). *Wooden Ships and Iron Men, vol. III.* pp. 132–133.
19. Robinette, H. (1987). "Game Warden: A Report from the Field." Reprinted in *Naval History Magazine* (Barber, J., Editor). U.S. Naval Institute. https://www.usni.org/magazines/naval-history-magazine/1987/april/game-warden-report-field.
20. Melia, T.M. (1991). *"Damn the Torpedoes."* p. 93.
21. MRFA. (1999). "Task Force 116." Mobile Riverine Force Association. https://www.mrfa.org/us-navy/us-navy-mobile-riverine-force/task-force-116/.
22. MRFA (1999). "Task Force 117." Mobile Riverine Force Association. https://www.mrfa.org/us-navy/us-navy-mobile-riverine-force/task-force-117/.
23. Bruhn, D.D. (2011). *Wooden Ships and Iron Men, vol. III.* pp. 125–195.
24. Melia, T.M. (1991). *"Damn the Torpedoes."* p. 93.
25. Note: A mineman is an enlisted specialty in the U.S. Navy. Up until the mid-1990s, they built and maintained mines. Today they concentrate on both mines and mine countermeasures.
Note: To run "hard" aground is a more serious condition than simply running aground. A vessel runs aground when it makes contact with ground under the water on which it is traveling and is stopped from traveling farther. Often the vessel can be refloated with the incoming tide and/or by reducing weight. A vessel is hard aground when it cannot be refloated without help.
26. SECNAV (1969). "Navy Cross Citation for MN1 Cecil Martin." Office of the Secretary of the Navy. https://minemen.org/wp/memories/-address-email-info-tributes/honor-roll-of-minemen/.
27. NHHC (1968). "Damaged Assault Support Patrol Boat (ASPB) Under Fire." Naval History

and Heritage Command. https://www.history.navy.mil/content/history/nhhc/our-collections/photography/numerical-list-of-images/nara-series/usn/USN-1130000/usn-1136903-damaged-assault-support-patrol-boat—aspb—under-fir.html.

28. Bruhn, D.D. (2011). *Wooden Ships and Iron Men, vol. III.* p. 133.

29. Google Earth.

30. Melia, T.M. (1991). *"Damn the Torpedoes."* p. 94.

31. Sherwood, J.D. (2015). *War in the Shallows: U.S. Navy Coastal and Riverine Warfare in Vietnam, 1965–1968.* Naval History and Heritage Command. pp. 119–123.

32. Melia, T.M. (1991). *"Damn the Torpedoes."* pp. 94–95.

33. Bruhn, D.D. (2011). *Wooden Ships and Iron Men, vol. III.* p. 133.

34. Pike, J.E. (2022). "Type 312 Futi Class Drone Minesweeper." GlobalSecurity.org. https://www.globalsecurity.org/military/world/china/futi.htm.

35. Youngblood, N. (2006). *The Development of Mine Warfare.* p. 156.

36. CIA (1967). *Mining of the Inland Waterways of the DRV—Approved for Release on 2006/11/07.* Central Intelligence Agency. pp. 1–6. https://www.cia.gov/readingroom/docs/CIA-RDP78T02095R000600170001-4.pdf.

37. *Ibid.*

38. CIA (1967). *Impact on Logistics of Air Attacks South of the 20th Parallel—Approved for Release on 2006/11/07.* Central Intelligence Agency. pp. 1–6. https://www.cia.gov/readingroom/docs/-CIA-RDP78T02095R000600170001-4.pdf.

39. Bruhn, D.D. (2007). *Wooden Ships and Iron Men, vol. I: The U.S. Navy's Ocean Minesweepers, 1953–1994.* Heritage Books. pp. 162–163.

40. *Ibid.*

41. Melia, T.M. (1991). *"Damn the Torpedoes."* p. 92.

42. Bruhn, D.D. (2007). *Wooden Ships and Iron Men, vol. I.* pp. 163–164.

43. U.S. Navy Minemen (2024). "If there are any Minemen on here who worked with DST mines in the Vietnam War, could you shoot me an IM." Facebook group post. https://www.facebook.com/groups/1623312257946818/posts/3645819592362731.

44. Nalty, B.C. (2005). *The War Against Trucks: Aerial Interdiction in Southern Laos, 1968–1972.* Air Force History and Museums Program. p. 64.

45. *Ibid.*

46. *Ibid.*

47. *Ibid.*, p. 107.

48. NHHC (2024). "Zumwalt, Elmo R., Jr." Naval History and Heritage Command. https://www.history.navy.mil/our-collections/photography/us-people/z/zumwalt—elmo-r---jr-.html.

49. Melia, T.M. (1991). *"Damn the Torpedoes."* pp. 98–99.

50. *Ibid.*

51. *Ibid.*

52. Note: A big challenge in a supply chain for specialty parts like those used on minesweepers is that orders must be placed for them on a regular basis. If all ships in the class are built at the same time, and a huge stockpile of replacement parts are ordered as well, then there is unlikely to be new orders for those parts for several years. Without new orders for a product line, manufacturers often leave the business. This results in searching for new manufacturers, along with a large retooling expense and long delays. The solution is to set a high level for inventory needed to maintain a desired service level (preferably based on data), establish lows based on manufacturing and transportation times, and then reorder whenever those lows are reached. (The author served two tours managing inventories and later earned a supply chain management MBA.)

53. Bruhn, D.D. (2011). *Wooden Ships and Iron Men, vol. III.* pp. 197–254.

54. Bruhn, D.D. (2007). *Wooden Ships and Iron Men, vol. I.* pp. 183–186.

55. Melia, T.M. (1991). *"Damn the torpedoes."* pp. 98–99.

56. *Ibid.*, p. 95.

57. *Ibid.*, pp. 98–99.

58. Marolda, E.J. (Ed.). (1993). *Operation End Sweep: A History of Minesweeping Operations in the DRV.* Naval Historical Center. p. 75.

59. Cox, S.J. (2022). *H-070-1: The Vietnam War Easter Offensive, Part 1.* Naval History and Heritage Command. https://www.history.navy.mil/content/dam/nhhc/about-us/leadership/hgram_pdfs/H-Gram_070.pdf.

60. Youngblood, N. (2006). *The Development of Mine Warfare.* pp. 156–158.

61. PM-19 (1975). *The Mining of the DRV 08 May 1972 to 14 January 1973.* CNO's Mine Warfare Project Office (PM-19). p. 4–1.

62. *Ibid.*, pp. 3–3 through 3–28.

63. *Ibid.*

64. Youngblood, N. (2006). *The Development of Mine Warfare.* p. 158.

65. Melia, T.M. (1991). *"Damn the Torpedoes."* p. 101.

66. Bruhn, D.D. (2007). *Wooden Ships and Iron Men, vol. I.* pp. 166–167.

67. PM-19 (1975). *The Mining of the DRV.* p. 3–41.

68. Youngblood, N. (2006). *The Development of Mine Warfare.* p. 158.

69. Bruhn, D.D. (2007). *Wooden Ships and Iron Men, vol. I.* p. 167.

70. Letzter, R., & LiveScience. (2018). "A Solar Storm Detonated U.S. Navy Mines During the Vietnam War. SCI AM. https://www.scientificamerican.com/article/a-solar-storm-detonated-u-s-navy-mines-during-the-vietnam-war/.

71. *Ibid.*, pp. 3–24 through 3–26.

72. *Ibid.*, p. 7–4.

73. Letzter, R. & LiveScience (2018). "A Solar Storm Detonated U.S. Navy Mines During the Vietnam War."

74. Marolda, E.J. (Ed.). (1993). *Operation End Sweep*. p. 3.
75. Cornish, G.J. (2003). *U.S. Naval Mine Warfare Strategy: Analysis of the Way Ahead*. U.S. Army War College. p. 13. https://apps.dtic.mil/sti/tr/pdf/ADA414973.pdf.
76. *Ibid.*, pp. 119–120.
77. Bruhn, D.D. (2007). *Wooden Ships and Iron Men, vol. I*. pp. 169–170.
78. Marolda, E.J. (Ed.). (1993). *Operation End Sweep*. p. 120.
79. Melia, T.M. (1991). *"Damn the Torpedoes."* pp. 98–99.
80. Bruhn, D.D. (2007). *Wooden Ships and Iron Men, vol. I*. p. 171.
81. Marolda, E.J. (Ed.). (1993). *Operation End Sweep*. pp. 47–50.
82. *Ibid.*
83. *Ibid.*, pp. 57–91.
84. Melia, T.M. (1991). *"Damn the Torpedoes."* pp. 101–111.
85. Marolda, E.J. (Ed.). (1993). *Operation End Sweep*. pp. 57–68.
86. *Ibid.*, p. 99.
87. *Ibid.*, p. 67.
88. *Ibid.*, pp. 66–91.
89. *Ibid.*, pp. 70–77.
90. Bruhn, D.D. (2007). *Wooden Ships and Iron Men, vol. I*. pp. 171–172.
91. Marolda, E.J. (Ed.). (1993). *Operation End Sweep*. pp. 80–81.
92. Melia, T.M. (1991). *"Damn the Torpedoes."* p. 109.
93. Marolda, E.J. (Ed.). (1993). *Operation End Sweep*. pp. 84–91.
94. Melia, T.M. (1991). *"Damn the Torpedoes."* p. 109.
95. PM-19 (1975). *The Mining of the DRV*. p. 4–1.
96. Melia, T.M. (1991). *"Damn the Torpedoes."* p. 110.

Chapter 11

1. Note: The Arabian Gulf is another name for the Persian Gulf. In territorial disputes, legal status is most often conferred by widespread recognition of one's claims to territory. Since a name implies ownership, the gulf is often called the "Arabian Gulf" to deny Iran its claims to territorial waters beyond that allowed by international law.
2. Britannica (2024). "History of Iran." https://www.britannica.com/place/Iran/History.
3. Afary, J. (2024). "Iranian Revolution." *Encyclopedia Britannica*. https://www.britannica.com/event/Iranian-Revolution.
4. Russell, T.R. (2012). *Crisis in Iran: Operation EAGLE CLAW*. U.S. Department of Defense. https://media.defense.gov/2012/Aug/23/2001330106/-1/-1/0/Eagleclaw.pdf.
5. Collins, M.K. (2013). "The Effect of the Iranian Hostage Crisis on the 1980 Presidential Election." *Tenor of Our Times*, 2(6). Harding University. https://scholarworks.harding.edu/cgi/viewcontent.cgi?article=1017&context=tenor.
6. Britannica (2024). "History of Iraq." https://www.britannica.com/topic/history-of-Iraq.
7. Zatarain, L.A. (2008). *Tanker War: America's First Clash with Iran, 1987–88*. Casemate. pp. 1–5.
8. CIA World Factbook (2024). "Explore All Countries—Libya." Criminal Intelligence Agency. https://www.cia.gov/the-world-factbook/countries/libya/.
9. U.S. Naval Institute (1985). *Seminar—The Gulf of Suez Mining Crisis: Terrorism at Sea*. (Seminar at the U.S. Naval Academy, May 30, 1985). Transcript located at Naval Historical Library, Washington, D.C. p. 16.
10. Note: The term "third world" is not meant as a slur. The "first world" was the U.S. and its allies. The "second world" was the Soviet Union and *its* allies, while the "third world" was those countries not in either country's orbit.
11. Mandel, J.E. (1985). "Libya Under Qadhafi: A Pattern of Aggression." Ronald Reagan Presidential Library. https://www.reaganlibrary.gov/public/digitallibrary/smof/nsc-communicationsandinformation/mandel/91721/-libyaunderaqdhafi-apatternofaggression.pdf.
12. Britannica (2023). "2012 Benghazi Attacks." https://www.britannica.com/event/2012-Benghazi-attacks.
13. Huff, E. (2011). "First Barbary War." Monticello.org. https://www.monticello.org/research-education/thomas-jefferson-encyclopedia/first-barbary-war/.
14. Melia, T.M. (1991). *"Damn the Torpedoes."* p. 118.
15. U.S. Naval Institute (1985). *Seminar—The Gulf of Suez Mining Crisis*. p. 3.
16. *Ibid.*, pp. 14–16.
17. Melia, T.M. (1991). *"Damn the Torpedoes."* pp. 118–119.
18. As a sidenote, American helicopters also conducted lead-through operations for the Saudi king's yacht transiting from one port to another. He refused to trust the words of his naval officer second cousin, who insisted that the route had been cleared of mines. Instead, the king preferred to witness the third-party clearance actions himself. U.S. Naval Institute (1985). *Seminar—The Gulf of Suez Mining Crisis*. pp. 7–11.
19. *Ibid.*, pp. 6–7.
20. CIA (1984). *The Iranian Mine Warfare Threat*. Central Intelligence Agency. https://www.cia.gov/readingroom/docs/CIA-RDP85T00314R000300100002-7.pdf.
21. Zatarain, L.A. (2008). *Tanker War*. pp. 1–207.
22. CIA (1984). *The Iranian Mine Warfare Threat*.
23. Youngblood, N. (2006). *The Development of Mine Warfare*. p. 170.
24. Zatarain, L.A. (2008). *Tanker War*. pp. 30–31.

25. *Ibid.*
26. Youngblood, N. (2006). *The Development of Mine Warfare.* p. 170.
27. Zatarain, L.A. (2008). *Tanker War.* pp. 63–65.
28. *Ibid.*, pp. 63–69.
29. *Ibid.*, pp. 63–71.
30. Melia, T.M. (1991). *"Damn the Torpedoes."* pp. 120–121.
31. Zatarain, L.A. (2008). *Tanker War.* pp. 77–86.
32. *Ibid.*, pp. 87–95.
33. *Ibid.*, pp. 106–117.
34. *Ibid.*, pp. 117–124.
35. *Ibid.*, pp. 125–189.
36. Melia, T.M. (1991). *"Damn the Torpedoes."* pp. 126–127.
37. Zatarain, L.A. (2008). *Tanker War.* p. 205.
38. Tangredi, S.J. (2013). *Anti-Access Warfare: Countering A2/AD Strategies.* Naval Institute Press. p. 193.
39. NHHC. (2023). "Operation Praying Mantis." Naval History and Heritage Command. https://www.history.navy.mil/browse-by-topic/-wars-conflicts-and-operations/middle-east/-praying-mantis.html.
40. *Ibid.*
41. Zatarain, L.A. (2008). *Tanker War.* pp. 223–387.
42. Ibrahim, Arwa (2020). "Thirty years on, Iraq's invasion of Kuwait still haunts the region." Al-Jazeera Media Network. https://www.aljazeera.com/news/2020/8/1/thirty-years-on-iraqs-invasion-of-kuwait-still-haunts-region.
43. CIA (1984). *The Iranian Mine Warfare Threat.*
44. Herrick, R.W. (1968). *Soviet Naval Strategy: Fifty Years of Theory and Practice.* U.S. Naval Institute. pp. 134–151.
45. Melia, T.M. (1991). *"Damn the Torpedoes."* p. 72.
46. Marolda, E.J. (1998). *Shield and Sword: The United States Navy and the Persian Gulf War.* U.S. Naval Historical Center. pp. 322–325.
47. Melia, T.M. (1991). *"Damn the Torpedoes."* pp. 115–118.
48. Marolda, E.J. (1998). *Shield and Sword.* pp. 75–77.
49. Pokrant, M. (1999). *Desert Storm at Sea: What the Navy Really Did.* Greenwood Press. pp. 151–153.
50. Marolda, E.J. (1998). *Shield and Sword.* p. 118.
51. Pokrant, M. (1999). *Desert Storm at Sea.* pp. 151–153.
52. Marolda, E.J. (1998). *Shield and Sword.* pp. 148–149.
53. DoD. (1992). *Conduct of the Persian Gulf War: Final Report to Congress.* Department of Defense. p. 200.
54. Pokrant, M. (1999). *Desert Storm at Sea.* pp. 233–239.
55. *Ibid.*, pp. 154–159.
56. Marolda, E.J. (1998). *Shield and Sword.* p. 264. Note: Sonar operators are not robots. They are usually youngsters in their early 20s. Their effectiveness in finding mines is directly associated with their level of motivation and experience.
57. Marolda, E.J. (1998). *Shield and Sword.* p. 264.
58. DoD (1992). *Conduct of the Persian Gulf War.* p. 206.
59. Belt, D.D. (2022). "Damn! The Torpedoes!: Desert Storm mine warfare has lessons for today's great power competition." *Proceedings*, vol. 148/7/1,433 (Editor: Bill Hamblet). U.S. Naval Institute. https://www.usni.org/magazines/proceedings/2022/july/damn-torpedoes.
60. USS *Princeton* (1991). *Desert Storm '91 Cruise Book.* U.S. Navy. p. 28.
61. DoD (1992). *Conduct of the Persian Gulf War.* p. 205.
62. Belt, D.D. (2022). "Damn! The Torpedoes!"
63. DoD (1992). *Conduct of the Persian Gulf War.* p. 205.
64. Belt, D.D. (2022). "Damn! The Torpedoes!"
65. Pokrant, M. (1999). *Desert Storm at Sea.* p. 162.
66. DoD (1992). *Conduct of the Persian Gulf War.* p. 200.
67. Pokrant, M. (1999). *Desert Storm at Sea.* pp.162–167.
68. Marolda, E.J. (1998). *Shield and Sword.* p. 267.
69. Pokrant, M. (1999). *Desert Storm at Sea.* pp. 231–233.
70. Marolda, E.J. (1998). *Shield and Sword.* pp. 322–325.
71. USACMH (2024). *Operation Desert Storm: 17 January to 28 February 1991.* U.S. Army Center for Military Heritage. https://history.army.mil/html/bookshelves/resmat/desert-storm/index.html.
72. Marolda, E.J. (1998). *Shield and Sword.* pp. 322–325.
73. *Ibid.*
74. Note: The author mentions this as the *Guardian* was his first ship starting in 1996. His mentor on that ship was previously stationed on the *Guardian*, where he found his first mine on the day that he checked onto the ship. Under his tutelage, along with the institutional practices of a ship that learned its trade from its NATO counterparts during Desert Storm, the author learned the profession of an MCM sonar operator. He would eventually get "stack time" on hundreds of exercise mines. On the same ship, he began learning naval history from a Naval Academy textbook that his commanding officer allowed him to read on quarterdeck watch. The author passed those lessons on to most minesweepers in the U.S. fleet during his time at NMAWC and COMCMRON 7 between 2009 and 2015, in the U.S. mineman rating manual, and now to you, the reader. Sadly, *Guardian*'s war record was lost in Wikipedia, to be overshadowed by her disastrous ending on the Tubbataha

Reef in the Philippines in 2013. The author watched in agony from the distance of the ship's squadron headquarters in Japan as he received the daily photos of its disassembly. It was an unfit ending to the tale of the storied warrior that plied Far-Eastern waters for two decades.

75. United Nations (2003). *Oil-for-food programme background brief—food.* UN Office of the Iraq Programme Oil-for-Food. https://www.un.org/depts/oip/sector-food.html.

76. NHHC (2023). "Operation Iraqi Freedom." Naval History and Heritage Command. https://www.history.navy.mil/browse-by-topic/wars-conflicts-and-operations/middle-east/operation-iraqi-freedom.html.

77. Think Defence (2020). *Al Faw and the Operation to Open the Iraqi Port of Umm Qasr.* Think Defence. https://www.thinkdefence.co.uk/al-faw-operation-open-iraqi-port-umm-qasr/.

78. *Ibid.*

79. Ryan, P. (2003). "Mine countermeasures a success." *Proceedings,* May 2003. U.S. Naval Institute. https://www.usni.org/magazines/proceedings/2003/may/mine-countermeasures-success.

80. O'Flaherty, C. (2019). *Naval Minewarfare: Politics to Practicalities.* The Choir Press. p. 150.

81. DefenceWeb (2011). "NATO forces clearing sea mines off Libyan coast." DefenceWeb. https://www.defenceweb.co.za/sea/sea-sea/nato-forces-clearing-sea-mines-off-libyan-coast/.

82. Egozi, A. (2021). "Houthis lay sea mines in Red Sea; Coalition boasts few minesweepers." Breaking Defense. https://breakingdefense.com/2021/06/houthis-lay-sea-mines-in-red-sea-coalition-boasts-few-minesweepers/.

83. SHRC (2015). *Naval mines: A new weapon of death.* Syrian Human Rights Committee. https://www.shrc.org/en/?p=25312.

84. Note: "Expeditionary" MCM is specially trained and equipped MCM professionals who can be deployed on cargo planes as a unit to a hot spot. Upon arrival, they can operate from a pier, an airfield, or from any vessels able to provide hotel services. AMCM and EOD fit this definition. "Organic" MCM can be defined as MCM equipment that can be installed on any vessel like a fishing trawler, tugboat, patrol craft, etc. that can provide a basic capability.

Chapter 12

1. Hacaloğlu, H. (2022). "Amid Russia-Ukraine war, Turkey worries about floating mines in Black Sea." Voice of America. https://www.voanews.com/a/amid-russia-ukraine-war-turkey-worries-about-floating-mines-in-black-sea/6521222.html.

2. Teslova, E. (2022). "Russia claims 10 mines planted by Ukrainian navy adrift in Black Sea, Sea of Azov." Anadolu Ajansi. https://www.aa.com.tr/en/russia-ukraine-war/russia-claims-10-mines-planted-by-ukrainian-navy-adrift-in-black-sea-sea-of-azov/2551959#.

3. *Ibid.*

4. Chacko, J.P. (2022). "420 Obsolete Ukrainian sea mines adrift towards Bosphorus Straits and the Mediterranean Sea". *Frontier India.* https://frontierindia.com/420-obsolete-ukrainian-sea-mm mines-adrift-towards-bosphorus-straits-and-the-mediterranean-sea/.

5. CIA (1951). *Soviet Models YaM and R-1 River Mines (Declassified).* Criminal Intelligence Agency. https://www.cia.gov/readingroom/docs/-CIA-RDP82-00457R008300130011-8.pdf.

6. Chacko, J.P. (2022). "Obsolete Ukrainian sea mines adrift towards Bosphorus Straits and the Mediterranean Sea."

7. DiGiulian, T. (2013). "Russia/USSR mines." NavWeaps. https://www.navweaps.com/Weapons/WAMRussian_Mines.php.

8. Tondo, L. (2022). "Sea mines: the deadly danger lurking in Ukraine's waters." *The Guardian.* https://www.theguardian.com/world/2022/jul/11/-sea-mines-ukraine-waters-russia-war-black-sea.

9. EuroNews (2022). "Ukraine war: Drifting mines pose deadly threat in Black Sea waters." https://www.euronews.com/2022/08/11/ukraine-war-drifting-mines-pose-deadly-threat-in-black-sea-water.

10. Tondo, L. (2022). "Sea mines."

11. Mongilio, H. (2023). "A Brief Summary of the Battle of the Black Sea." U.S. Naval Institute. https://news.usni.org/2023/11/15/a-brief-summary-of-the-battle-of-the-black-sea.

12. *Ibid.*

13. Savitz, S., & Courtney, W. (2023). "The Black Sea and the Changing Face of Naval Warfare." RAND Corporation. https://www.rand.org/pubs/commentary/2023/10/the-black-sea-and-the-changing-face-of-naval-warfare.html.

14. Eisele, I. (2022). "Five facts on grain and the war in Ukraine." *Deutsche Welle.* https://www.dw.com/en/five-facts-on-grain-and-the-war-in-ukraine/a-62601467.

15. Wintour, P. (2023). "What was the Black Sea grain deal and why did it collapse?" *The Guardian.* https://www.theguardian.com/world/2023/jul/20/-what-was-the-black-sea-grain-deal-and-why-did-it-collapse.

16. Savitz, S., & Courtney, W. (2023). "The Black Sea and the Changing Face of Naval Warfare."

17. Wintour, P. (2023). "What was the Black Sea grain deal and why did it collapse?"

18. Al Jazeera (2023). "UK warns Russia laying 'sea mines' to deter Black Sea cargo ships." *Al Jazeera.* https://www.aljazeera.com/news/2023/10/5/uk-warns-of-russia-laying-sea-mines-to-deter-black-sea-cargo-ships.

19. *Ibid.*

20. Mongilio, H. (2024). "Turkey to Block U.K. Minesweepers on Loan to Ukraine from Black Sea." U.S. Naval Institute. https://news.usni.org/2024/01/03/turkey-to-block-u-k-minesweepers-on-loan-to-ukraine-from-black-sea.

21. France24 (2023). "Cargo ship hits naval mine in Black Sea en route to Ukraine." *France24.*

https://www.france24.com/en/europe/20231228-cargo-ship-hits-naval-mine-in-black-sea-en-route-to-ukraine.

22. Nye, L. (2023). "Why Ukraine's new mine hunters might not join war." We Are the Mighty. https://www.wearethemighty.com/tactical/-ukraine-new-mine-hunters-might-not-join-war/.

23. Tringham, K. (2023). "First Romanian ex-Royal Navy Sandown-class minehunter arrives in country." Janes. https://www.janes.com/osint-insights/defence-news/sea/-first-romanian-ex-royal-navy-sandown-class-minehunter-arrives-in-country.

24. Tactics Institute (2024). "Addressing the menace of Russian sea mines in the Black Sea." Tactics Institute. https://tacticsinstitute.com/-press-releases/addressing-the-menace-of-russian-sea-mines-in-the-black-sea/.

25. Dumitrescu, R. (2023). "Five sea mines neutralized in the Black Sea so far, Romanian navy says." *Romania Insider.* https://www.romania-insider.com/five-sea-mines-neutralized-black-sea-romanian-navy-august-2023.

26. Kucukgocmen, A., & Hayatsever, H. (2024). "Turkey, Romania, Bulgaria sign deal to clear floating Black Sea mines." Reuters. https://www.reuters.com/world/turkey-romania-bulgaria-ink-deal-clear-floating-black-sea-mines-2024-01-11/.

27. Note: In the past, the author frequently shared watch-floors with other navies, with each side displaying their own information for all to see. However, the lack of a shared language made it difficult to decipher each other's status boards. In one case, the problem was resolved when the author learned enough of the other country's alphabet and language to create a translation guide that allowed basic watch-standers on the U.S. side to read the partner nation's status boards.

28. BNR (2023). "Bulgarian ships take part in international naval exercise in the Black Sea." Bulgarian National Radio. https://bnr.bg/en/post/101851398/bulgarian-ships-take-part-in-internation-naval-exercises-in-the-black-sea.

29. Kucukgocmen, A., & Hayatsever, H. (2024). "Turkey, Romania, Bulgaria sign deal to clear floating Black Sea mines."

30. McLaughlin, R. (2023). "Ukraine symposium—Mine clearance operations in the Black Sea." Lieber Institute. https://lieber.westpoint.edu/mine-clearance-operations-black-sea/.

31. Note: As an example, the author once participated in an MCM exercise in New Zealand that included EOD participants from several countries, including the U.S., Japan, South Korea, and China. The hero was New Zealand. Being well removed from many of the bilateral animosities of participating countries, it could credibly play the role of neutral host. Furthermore, its command ship, optimized for projecting soft power among many island nations, had the strengths of having perhaps the best naval ship's galley (kitchen) afloat, combined with good command-and-control (C2) spaces and being able to launch boats and UUVs. The amazing meals facilitated the building of relationships among participants.

32. Note: Recognizing that Wikipedia is not considered an academic source, the author cross-correlated information elsewhere. That said, Wikipedia had the good property of organizing similar information in a similar manner across multiple countries, making it the most efficient approach. Also, "academic" open-source information on current military inventories across the globe doesn't exist. Most sources are either the passion projects of enthusiasts or the published information of governments, which in turn are often propagandistic in nature.

33. Note: Something that looks like a mine from one angle will often prove to look nothing like a mine from other angles.

Bibliography

Afary, J. (2024). "Iranian revolution." *Encyclopedia Britannica*. https://www.britannica.com/event/Iranian-Revolution.

Al Jazeera. (2023). "UK warns Russia laying 'sea mines' to deter Black Sea cargo ships." https://www.aljazeera.com/news/2023/10/5/uk-warns-of-russia-laying-sea-mines-to-deter-black-sea-cargo-ships.

Albion, R., & Patterson, A. (1968). *A Brief History of Mine Warfare*. Mine Advisory Committee.

American Battlefield Trust. (2022). "Mobile Bay." https://www.battlefields.org/learn/civil-war/battles/mobile-bay.

Association of Minemen. (2003). "In memoriam—First mineman casualties USS *Aaron Ward* (DM-34)." https://minemen.org/PDFs/Aaron_Ward-Story.pdf.

Belt, D.D. (2022). "Damn! ... The Torpedoes!: Desert Storm mine warfare has lessons for today's great power competition." *Proceedings*, Vol. 148/7/1,433. U.S. Naval Institute. https://www.usni.org/magazines/proceedings/2022/july/damn-torpedoes.

Benson, S.G., & Raffaelle, G.-A. (Eds.). (2001). "Kim Il Sung." *Korean War Reference Library*. Gale. https://www.encyclopedia.com/history/-encyclopedias-almanacs-transcripts-and-maps/-kim-il-sung-1.

Boutelle, C.O. (1865). *General map of Charleston Harbor, South Carolina, showing rebel defences and obstructions*. U.S. Coastal Survey. https://www.loc.gov/resource/g3912c.cw0370100/.

Bradford, R.B. (1882). "History of torpedo warfare." U.S. Torpedo Station.

Britannica. (2023). "Alfred Nobel." https://www.britannica.com/biography/Alfred-Nobel.

Britannica. (2023). "The Russian Revolution: Russia History (1917)." https://www.britannica.com/topic/Bolshevik.

Britannica. (2023). "Stepan Osipovich Makarov." https://www.britannica.com/biography/Stepan-Osipovich-Makarov.

Britannica. (2024). "History of Iran." https://www.britannica.com/place/Iran/History.

Britannica. (2024). "History of Iraq." https://www.britannica.com/topic/history-of-Iraq.

Bruhn, D.D. (2007). *Wooden Ships and Iron Men, Vol. I: The U.S. Navy's Ocean Minesweepers, 1953–1994*. Heritage Books.

Bruhn, D.D. (2011). *Wooden Ships and Iron Men, Vol. III: The U.S. Navy's Coastal and Inshore Minesweepers and the Minecraft That Served in Vietnam, 1953–1976*. Heritage Books.

Bruhn, D.D., & Hoole, R. (2018). *Home Waters: Royal Navy, Royal Canadian Navy, and U.S. Navy Mine Forces Battling U-boats in World War I*. Heritage Books.

Bulgarian National Radio. (2023). "Bulgarian ships take part in international naval exercise in the Black Sea." https://bnr.bg/en/post/101851398/-bulgarian-ships-take-part-in-internation-naval-exercises-in-the-black-sea.

Cagle, M.W., & Manson, F.A. (1957). "Wonsan: The battle of the mines." *Proceedings*, Vol. 83/6/652. U.S. Naval Institute. https://www.usni.org/magazines/proceedings/1957/june/wonsan-battle-mines.

Cavendish, R. (2007). "Robert Fulton's Paddlesteamer." *History Today*, 57(8). https://www.historytoday.com/archive/robert-fulton%E2%80%99s-paddlesteamer.

CDSG. (2021). "History of America's Seacoast Defenses." Coastal Defense Group. https://cdsg.org/forts/ and its subsections.

CDSG. (2021). *A Short History: Harbor Defenses of the United States of America*. Coast Defense Study Group. https://cdsg.org/short-history/.

Chacko, J.P. (2022). "420 Obsolete Ukrainian sea mines adrift towards Bosphorus Straits and the Mediterranean Sea." *Frontier India*. https://frontierindia.com/420-obsolete-ukrainian-sea-mm mines-adrift-towards-bosphorus-straits-and-the-mediterranean-sea/.

CIA. (1951). "Soviet Models YaM and R-1 River Mines (Declassified)." Central Intelligence Agency. https://www.cia.gov/readingroom/docs/CIA-RDP82-00457R008300130011-8.pdf.

CIA. (1967). "Impact on Logistics of Air Attacks South of the 20th Parallel—Approved for Release on 2006/11/07." Central Intelligence Agency. https://www.cia.gov/readingroom/docs/CIA-RDP78T02095R000600170001-4.pdf.

CIA. (1967). "Mining of the Inland Waterways of the DRV—Approved for Release on 2006/11/07." Central Intelligence Agency.

https://www.cia.gov/readingroom/docs/CIA-RDP78T02095R000600170001-4.pdf.
CIA. (1984). "The Iranian Mine Warfare Threat." Central Intelligence Agency. https://www.cia.gov/readingroom/docs/CIA-RDP85T00314R000300100002-7.pdf.
CIA World Factbook. (2024). "Explore All Countries—Libya." Central Intelligence Agency. https://www.cia.gov/the-world-factbook/countries/libya/.
Coggins, J. (1962). *Arms and Equipment of the Civil War.* Dover Publications.
Collins, M.K. (2013). "The Effect of the Iranian Hostage Crisis on the 1980 Presidential Election." *Tenor of Our Times*, 2(6). Harding University. https://scholarworks.harding.edu/cgi/viewcontent.cgi?article=1017&context=tenor.
Colt, S. (1836). "US Patent 9430X Revolver and Revolving Rifle: Improvement in Fire-arms." U.S. Patent Office. https://revolvers.candrsenal.com/timeline/patent-colt-2/.
Cornish, G.J. (2003). "U.S. Naval Mine Warfare Strategy: Analysis of the Way Ahead." U.S. Army War College. https://apps.dtic.mil/sti/tr/pdf/ADA414973.pdf.
Cowie, J.S. (1949). *Mines, Minelayers, and Minelaying.* Oxford University Press.
Cox, S.J. (2022). "H-070-1: The Vietnam War Easter Offensive, Part 1." Naval History and Heritage Command. https://www.history.navy.mil/content/dam/nhhc/about-us/leadership/hgram_pdfs/H-Gram_070.pdf.
Crossley, J. (2011). *The Hidden Threat: The Story of Mines and Minesweeping by the Royal Navy in World War I.* Pen and Sword Maritime.
Daughan, G.C. (2008). *If By Sea: The Forging of the American Navy from the Revolution to the War of 1812.* Basic Books.
DefenceWeb. (2011). "NATO forces clearing sea mines off Libyan coast." DefenceWeb. https://www.defenceweb.co.za/sea/sea-sea/nato-forces-clearing-sea-mines-off-libyan-coast/.
Department of Defense. (1992). *Conduct of the Persian Gulf War: Final Report to Congress.*
DiGiulian, T. (2008). "Japan mines: World War II." NavWeaps. http://www.navweaps.com/Weapons/WAMJAP_Mines.php.
DiGiulian, T. (2013). "Russia/USSR mines." NavWeaps. https://www.navweaps.com/Weapons/WAMRussian_Mines.php.
DiGiulian, T. (2023). "United States of America mines." NavWeaps. http://navweaps.com/Weapons/WAMUS_Mines.php.
Duckers, P. (2011). *The Crimean War at Sea: Naval Campaigns Against Russia, 1854–6.* Pen and Sword Maritime.
Duffy, M. (2011). "The gunnery at Trafalgar: Training, tactics or temperament?" *Journal for Maritime Research*, 7(1), 140–169. https://doi.org/10.1080/21533369.2005.9668349.
Duiker, W.J. (2004). "Vietnam." *Encyclopedia Americana.* Scholastic Library Publishing.
Dull, J.R. (2012). *American Naval History, 1607–1865: Overcoming the Colonial Legacy.* University of Nebraska Press. https://www.jstor.org/stable/j.ctt1d9nqgf.
Dumitrescu, R. (2023). "Five sea mines neutralized in the Black Sea so far, Romanian navy says." *Romania Insider.* https://www.romania-insider.com/five-sea-mines-neutralized-black-sea-romanian-navy-august-2023.
Duncan, R.C. (1962). "America's Use of Sea Mines." U.S. Naval Ordnance Laboratory.
Earp, G.G. (1854). *The History of the Baltic campaign: From Documents and Other Materials Furnished by Vice-Admiral Sir. C. Napier.* Chapel Books.
Edwards, H.W. (1954). "A Naval Lesson of the Korean Conflict." *Proceedings*, Vol. 80/12/622. U.S. Naval Institute. https://www.usni.org/magazines/proceedings/1954/december/naval-lesson-korean-conflict.
Edwards, J. (2020). "Evolving Mine Warfare." *U.S. Naval Institute Blog.* https://blog.usni.org/posts/2020/03/10/evolving-mine-warfare.
Egozi, A. (2021). "Houthis lay sea mines in Red Sea; Coalition boasts few minesweepers." Breaking Defense. https://breakingdefense.com/2021/06/-houthis-lay-sea-mines-in-red-sea-coalition-boasts-few-minesweepers/.
Eisele, I. (2022). "Five facts on grain and the war in Ukraine." *Deutsche Welle.* https://www.dw.com/en/five-facts-on-grain-and-the-war-in-ukraine/a-62601467.
Elliott, P. (1979). *Allied Minesweeping in World War 2.* Patrick Stephens.
Encyclopedia Britannica. (1993). "Korea." Robert McHenry (Ed.). *The New Encyclopedia Britannica*, Vol. 22, macropedia.
EuroNews. (2022). "Ukraine war: Drifting mines pose deadly threat in Black Sea waters." https://www.euronews.com/2022/08/11/ukraine-war-drifting-mines-pose-deadly-threat-in-black-sea-water.
Farragut, D.G. (1864). *Detailed report of the Mobile battle.* Reproduced in *Blue & Gray at Sea: Naval Memoirs of the Civil War*, edited by Brian Thomsen. Tom Doherty Associates, 2003.
Fields, J.A., Jr. (2015). *History of U.S. Naval Operations: Korea.* Naval History and Heritage Command. https://www.history.navy.mil/research/library/online-reading-room/title-list-alphabetically/h/history-us-naval-operations-korea/chapter8-on-border.html.
Figes, O. (2010). *The Crimean War.* Picador.
Forczyk, R. (2009). *Russian Battleship vs Japanese Battleship: Yellow Sea 1904–1905.* Osprey Publishing.
France24. (2023). "Cargo ship hits naval mine in Black Sea en route to Ukraine." https://www.france24.com/en/europe/20231228-cargo-ship-hits-naval-mine-in-black-sea-en-route-to-ukraine.
French, B.R.L.V. (1972). *The Crimean War.* Pen and Sword Books.
Fulton, R. (1810). *Torpedo War and Submarine Explosions.* William Elliott.

Gavin, P. (1999). "World War Two in the Pacific: Timeline of Events 1941–1945." The History Place. https://www.historyplace.com/unitedstates/pacificwar/timeline.htm.

Ghosh, D. (2021). "Sea of Azov." In *WorldAtlas*. https://www.worldatlas.com/seas/sea-of-azov.html.

Gibbs, P. (1960). *The Crimean Blunder.* Holt, Rinehart and Winston.

Goncharenko, R. (2019). "NATO steps up naval presence on the Black Sea." DW.com. https://www.dw.com/en/nato-steps-up-naval-presence-on-the-black-sea/a-47732883.

Green, A.L. (Ed.). (2004). "Vietnam." *Compton's Encyclopedia.* Encyclopedia Britannica.

Grosvenor, J., & Bates, L.M. (1956). *Open the Ports: The Story of Human Minesweepers.* William Kimber & Co. Ltd.

Hacaloğlu, H. (2022). "Amid Russia-Ukraine war, Turkey worries about floating mines in Black Sea." Voice of America. https://www.voanews.com/a/-amid-russia-ukraine-war-turkey-worries-about-floating-mines-in-black-sea/6521222.html.

Harris, G. (1973). *USN 711571 Operation End Sweep, 1973.* Navy History and Heritage Command. https://www.history.navy.mil/content/history/nhhc/our-collections/photography/numerical-list-of-images/nhhc-series/nh-series/USN-711000/USN-711571.html.

Hartmann, G.K. (1979). *Weapons That Wait: Mine Warfare in the U.S. Navy.* Naval Institute Press.

Heidler, D.S., & Heidler, J.T. (Eds.). (1997). *Encyclopedia of the War of 1812.* ABC-CLIO.

Herrick, R.W. (1968). *Soviet Naval Strategy: Fifty Years of Theory and Practice.* U.S. Naval Institute.

History.com. (2023). "Samuel Colt." History Channel. https://www.history.com/topics/inventions/samuel-colt.

Historyplex. (2021). *Crimean War Causes, Timeline, and Summary.* Historyplex. https://historyplex.com/crimean-war-causes-timeline-summary.

Ibrahim, Arwa. (2020). "Thirty years on, Iraq's invasion of Kuwait still haunts the region." Al-Jazeera Media Network. https://www.aljazeera.com/news/2020/8/1/thirty-years-on-iraqs-invasion-of-kuwait-still-haunts-region.

International Conferences (Geneva). (1949). *1949 Geneva Convention III Relative to the Treatment of Prisoners of War.* Geneva. Entry into force October 21, 1950. https://ihl-databases.icrc.org/en/ihl-treaties/gciii-1949.

International Conferences (The Hague). (1910). *1907 Hague Convention VIII Relative to the Laying of Automatic Submarine Contact Mines.* The Hague. Entry in force January 26, 1910. https://ihl-databases.icrc.org/en/ihl-treaties/hague-conv-viii-1907.

Jefferson, T. (1806). Letter: From Thomas Jefferson to Robert Smith, 12 December 1806. *Papers of Thomas Jefferson.* https://rotunda.upress.virginia.edu/founders/default.xqy?keys=FOEA-print-04-01-02-4669.

Johnson, E.A., & Katcher, D.A. (1973). *Mines against Japan.* U.S. Naval Ordnance Laboratory.

Jukes, G. (2002). *The Russo-Japanese War, 1904–1905.* Osprey Publishing.

Kinglake, A.W. (1874). *The Invasion of the Crimea: Its Origin, and an Account of Its Progress Down to the Death of Lord Raglan.* William Blackwood and Sons.

Kucukgocmen, A., & Hayatsever, H. (2024). "Turkey, Romania, Bulgaria sign deal to clear floating Black Sea mines." Reuters. https://www.reuters.com/world/turkey-romania-bulgaria-ink-deal-clear-floating-black-sea-mines-2024-01-11/.

Langmaid, K. (1965). *The Approaches Are Mined!* Jarrolds.

Laskow, S. (2018). "The Sun Set Off 4,000 Sea Mines During the Vietnam War." Atlas Obscura. https://www.atlasobscura.com/articles/solar-storm-sets-off-mines.

Ledebur, G.F.V. (1977). *Die seemine.* J.F. Lehmanns Verlag.

Letzter, R., & LiveScience. (2018). "A solar Storm Detonated U.S. Navy Mines During the Vietnam War." https://www.scientificamerican.com/article/a-solar-storm-detonated-u-s-navy-mines-during-the-vietnam-war/.

Lott, A. (1959). *Most Dangerous Sea: The Story of the Brave Crews Who Swept the Seas of Deadly Enemy Mines.* U.S. Naval Institute.

Lott, A.S. (1964). *Brave Ship, Brave Men.* Naval Institute Press.

Low, A.M. (1940). *Mine and Countermine.* Sheridan House.

Lozada, C. (2005). "The Economics of World War I." *The Digest*, issue 1. National Bureau of Economic Research. https://www.nber.org/digest/jan05/economics-world-war-i.

Lund, P., & Ludlam, H. (1978). *Out Sweeps! The Story of the Minesweepers in World War II.* W. Foulsham & Co.

Mandel, J.E. (1985). "Libya Under Qadhafi: A Pattern of Aggression." Ronald Reagan Presidential Library. https://www.reaganlibrary.gov/public/digitallibrary/smof/nsc-communicationsandinformation/mandel/91721/libyaunderaqdhafi-apatternofaggression.pdf.

Manstan, R.R., & Frese, F.J. (2010). Turtle: *David Bushnell's Revolutionary Vessel.* Westholme.

Marolda, E.J. (1987). "The War in Vietnam's Shallows." *Naval History Magazine*, Vol. 1, No. 1. U.S. Naval Institute. https://www.usni.org/magazines/naval-history-magazine/1987/april/war-vietnams-shallows.

Marolda, E.J. (1998). *Shield and Sword: The United States Navy and the Persian Gulf War.* U.S. Naval Historical Center.

Marolda, E.J. (Ed.). (1993). *Operation End Sweep: A History of Minesweeping Operations in the DRV.* Naval Historical Center.

Massie, R.K. (2004). *Castles of Steel: Britain, Germany, and the Winning of the Great War at Sea.* Ballantine Books.

McLaughlin, R. (2023). "Ukraine Symposium—Mine Clearance Operations in the Black Sea." Lieber Institute. https://lieber.westpoint.edu/mine-clearance-operations-black-sea/.

Meister, J. (1957). *The Soviet Navy in World War II*. U.S. Naval Institute. https://www.usni.org/magazines/proceedings/1957/august/soviet-navy-world-war-ii.

Melia, T.M. (1991). *"Damn the Torpedoes": A Short History of U.S. Naval Mine Countermeasures, 1977–1991*. Naval Historical Center.

Menarchik, J.D. (2010). *North Korean Protective Mine Warfare: An Analysis of the Naval Minefields at Wonsan, Chinnampo, and Hungnam During the Korean War*. Air Command and Staff College. https://apps.dtic.mil/sti/pdfs/ADA539219.pdf?fbclid=IwAR0TacYt3TiDqRiOTbU7gntsBBp5KarPdzNHJzCDIJA_i1eJ0GBWsSehtXY.

Michie, P.S. (1865). *Torpedo Book*. Edited and reprinted in Schiller, H.M. (editor) (2011). *Confederate Torpedoes: Two Illustrated 19th-Century Works with New Appendices and Photographs*. McFarland, 2014.

Mongilio, H. (2023). "A Brief Summary of the Battle of the Black Sea." U.S. Naval Institute. https://news.usni.org/2023/11/15/a-brief-summary-of-the-battle-of-the-black-sea.

Mongilio, H. (2024). "Turkey to Block U.K. Minesweepers on Loan to Ukraine from Black Sea." U.S. Naval Institute. https://news.usni.org/2024/01/03/-turkey-to-block-u-k-minesweepers-on-loan-to-ukraine-from-black-sea.

MRFA. (1999). *Task Force 116*. Mobile Riverine Force Association. https://www.mrfa.org/us-navy/us-navy-mobile-riverine-force/task-force-116/.

MRFA. (1999). *Task Force 117*. Mobile Riverine Force Association. https://www.mrfa.org/us-navy/us-navy-mobile-riverine-force/task-force-117/.

Murrin, M.J. (1945). *Pacific War Diary*. Not published (record kept at Navy Department Library, Washington, D.C.).

Nalty, B.C. (2005). *The War Against Trucks: Aerial Interdiction in Southern Laos, 1968–1972*. Air Force History and Museums Program.

National Park Service. (2023). "Tredegar Iron Works—Ironmaker to the Confederacy." National Park Service. https://www.nps.gov/articles/tredegar-iron-works-ironmaker-to-the-confederacy.htm.

NATO. (2021). *Relations with Georgia*. Brussels: NATO. https://www.nato.int/cps/en/natohq/topics_38988.htm.

NATO. (2021). *Relations with Ukraine*. Brussels: NATO. https://www.nato.int/cps/en/natolive/topics_37750.htm.

NHHC. (2014). "Manassas." Annapolis, MD: Naval History and Heritage Command. https://www.history.navy.mil/research/histories/ship-histories/confederate_ships/manassas.html.

NHHC. (2014). "Virginia." Naval History and Heritage Command. https://www.history.navy.mil/content/history/nhhc/research/histories/-ship-histories/confederate_ships/virginia.html.

NHHC. (2023). "Operation Iraqi Freedom." Naval History and Heritage Command. https://www.history.navy.mil/browse-by-topic/-wars-conflicts-and-operations/middle-east/-operation-iraqi-freedom.html.

NHHC. (2023). "Operation Praying Mantis." Naval History and Heritage Command. https://www.history.navy.mil/browse-by-topic/wars-conflicts-and-operations/middle-east/praying-mantis.html.

NHHC. (2024). "Zumwalt, Elmo R., Jr." Naval History and Heritage Command. https://www.history.navy.mil/our-collections/photography/-us-people/z/zumwalt—elmo-r---jr-.html.

Notehelfer, F.G. (2024). "The Emergence of Imperial Japan." *Britannica*. https://www.britannica.com/place/Japan/The-emergence-of-imperial-Japan.

NPS. (2015). "Florida battles (from Dyer's Compendium)." National Park Service. https://www.nps.gov/civilwar/florida.htm.

NUMA. (2022). "Siege of Charleston." National Underwater and Marine Agency. https://numa.net/expeditions/siege-of-charleston/.

Nye, L. (2023). "Why Ukraine's new mine hunters might not join war." We Are the Mighty. https://www.wearethemighty.com/tactical/ukraine-new-mine-hunters-might-not-join-war/.

Owen, L. et al. (2024). "Mekong River." *Britannica*. https://www.britannica.com/place/Mekong-River/Climate-and-hydrology.

Patterson, A., Jr. (1970). *A Brief History of Mine Warfare*. Mine Advisory Committee, National Academy of Sciences.

Perry, M.F. (1965). *Infernal Machines—The Story of Confederate Submarine and Mine Warfare*. Louisiana State University Press.

Pike, J.E. (2022). "Type 312 Futi class drone minesweeper." GlobalSecurity.org. https://www.globalsecurity.org/military/world/china/futi.htm.

Pokrant, M. (1999). *Desert Storm at Sea: What the Navy Really Did*. Greenwood Press.

Potter, E.B. (Ed.). (1981). *Sea Power: A Naval History (2nd ed.)*. Naval Institute Press.

Robinette, H. (1987). "Game Warden: A Report from the Field." Reprinted in *Naval History Magazine* (Barber, J. Editor). U.S. Naval Institute. https://www.usni.org/magazines/naval-history-magazine/1987/april/game-warden-report-field.

RusNavy.com. (2021). *Victories of the Russian Fleet*. MilPress Media Group. http://rusnavy.com/history/victory.htm.

Russell, T.R. (2012). *Crisis in Iran: Operation EAGLE CLAW*. U.S. Department of Defense. https://media.defense.gov/2012/Aug/23/2001330106/-1/-1/0/Eagleclaw.pdf.

Russell, W.H. (1858). *The British Expedition to the Crimea*. Routledge.

Russia & Ukraine. (2003). *Agreement between*

the Russian Federation and the Ukraine on cooperation in the use of the sea of Azov and the strait of Kerch. https://www.ecolex.org/details/treaty/agreement-between-the-russian-federation-and-the-ukraine-on-cooperation-in-the-use-of-the-sea-of-azov-and-the-strait-of-kerch-tre-149547/.

Ryan, P. (2003). "Mine countermeasures a success." *Proceedings*, May 2003. U.S. Naval Institute. https://www.usni.org/magazines/proceedings/2003/may/mine-countermeasures-success.

Saar, C.W. (1964). *Offensive Mining as a Soviet Strategy.* U.S. Naval Institute. https://www.usni.org/magazines/proceedings/1964/august/offensive-mining-soviet-strategy.

Samarov A.A., & Petrov, F.A. (1951). *Development of Mine Materiel in the Russian Navy.* Naval Publishing House of the Naval Ministry of the USSR. Translated by Lloyd G. Robbins of the Science Translation Service of the University of Alabama.

Savitz, S., & Courtney, W. (2023). "The Black Sea and the Changing Face of Naval Warfare." RAND Corporation. https://www.rand.org/pubs/commentary/2023/10/the-black-sea-and-the-changing-face-of-naval-warfare.html.

Schiffer, M.B. (2008). *Power Struggles: Scientific Authority and the Creation of Practical Electricity Before Edison.* The MIT Press.

SECNAV. (1969). *Navy Cross citation for MN1 Cecil Martin.* Office of the Secretary of the Navy. https://minemen.org/wp/memories/address-email-info-tributes/honor-roll-of-minemen/.

Sherwood, J.D. (2015). *War in the Shallows: U.S. Navy Coastal and Riverine Warfare in Vietnam, 1965–1968.* Naval History and Heritage Command.

SHRC. (2015). *Naval Mines: A New Weapon of Death.* Syrian Human Rights Committee. https://www.shrc.org/en/?p=25312.

Siemens. (2023). *Founders: Werner Von Siemens.* Siemens International. https://www.siemens.com/global/en/company/about/history/people/founders.html.

Simkin, J. (2016). *Germany in 1914.* Spartacus Educational. http://spartacus-educational.com/FWWinGermany.htm.

Stern, P.V.D. (1959). *Secret Missions of the Civil War.* Wings Books.

Symonds, C.L. (2009). *The Civil War at Sea.* ABC-CLIO.

Tactics Institute. (2024). "Addressing the Menace of Russian Sea Mines in the Black Sea." Tactics Institute. https://tacticsinstitute.com/press-releases/addressing-the-menace-of-russian-sea-mines-in-the-black-sea/.

Tangredi, S.J. (2013). *Anti-Access Warfare: Countering A2/AD Strategies.* Naval Institute Press.

Teslova, E. (2022). "Russia claims 10 mines planted by Ukrainian Navy adrift in Black Sea, Sea of Azov." Anadolu Ajansi. https://www.aa.com.tr/en/russia-ukraine-war/russia-claims-10-mines-planted-by-ukrainian-navy-adrift-in-black-sea-sea-of-azov/2551959#.

Think Defence. (2020). *Al Faw and the operation to open the Iraqi port of Umm Qasr.* https://www.thinkdefence.co.uk/al-faw-operation-open-iraqi-port-umm-qasr/.

Tillman, B. (2010). *Whirlwind: The Air War Against Japan, 1942–1945.* Simon & Schuster.

Tolf, R.W. (1976). *The Russian Rockefellers: The Saga of the Nobel Family and the Russian Oil Industry.* Hoover Institution Press.

Tondo, L. (2022). "Sea mines: the deadly danger lurking in Ukraine's waters." *The Guardian.* https://www.theguardian.com/world/2022/jul/11/sea-mines-ukraine-waters-russia-war-black-sea.

Tringham, K. (2023). "First Romanian ex-Royal Navy Sandown-class minehunter arrives in country." Janes. https://www.janes.com/osint-insights/defence-news/sea/-first-romanian-ex-royal-navy-sandown-class-minehunter-arrives-in-country.

Tucker, S.C. (Ed.). (2012). *The Encyclopedia of the War of 1812: A Political, Social, and Military History,* ABC-CLIO. pp. 473–474. https://books.google.com/books?id=hVSrJBQYAk8C&dq=War+of+1812+mines+norfolk&source=gbs_navlinks_s.

United Nations. (2003). *Oil-for-food programme background brief—food.* UN Office of the Iraq Programme Oil-for-Food. https://www.un.org/depts/oip/sector-food.html.

United Nations. (2024). *History of the Korean War.* United Nations Command. https://www.unc.mil/History/1950-1953-Korean-War-Active-Conflict/.

U.S. Army Air Forces. (1945). *Starvation: Phase Analysis of the Twentieth Air Force Strategic Mining Blockade of the Japanese Empire.* 20th Army Air Force. https://minemen.org/PDFs/OperationStarvationMiningofJapaneseEmpire.pdf.

U.S. Naval Institute. (1985). *Seminar—The Gulf of Suez Mining Crisis: Terrorism at Sea.* (Seminar at the U.S. Naval Academy, May 30, 1985.) Transcript located at Naval Historical Library, Washington, D.C.

USACMH. (2024). *Operation Desert Storm: 17 January to 28 February 1991.* U.S. Army Center for Military Heritage. https://history.army.mil/html/bookshelves/resmat/desert-storm/index.html.

Van Der Vat, D. (2009). *The Dardanelles Disaster: Winston Churchill's Greatest Failure.* Duckworth Overlook.

Vietnam Embassy. (2024). *Geography (of Vietnam).* Embassy of the Socialist Republic of Vietnam in the United States of America. https://vietnamembassy-usa.org/vietnam/geography.

Winters, R., Albion, R., & Patterson, A. (1968). *A Brief History of Mine Warfare.* Mine Advisory Committee.

Wintour, P. (2023). "What was the Black Sea grain deal and why did it collapse?" *The Guardian.*

https://www.theguardian.com/world/2023/jul/20/what-was-the-black-sea-grain-deal-and-why-did-it-collapse.

Youngblood, N. (2006). *The Development of Mine Warfare: A Most Murderous and Barbarous Conduct*. Praeger Security International.

Zatarain, L.A. (2008). *Tanker War: America's First Clash with Iran, 1987–88*. Casemate.

Zeihan, P. (2016). *The Accidental Superpower*. Twelve Books.

Zieger, R.H. (2000). *America's Great War: World War 1 and the American Experience*. Rowman & Littlefield.

Index

www.ingramcontent.com/pod-product-compliance
Ingram Content Group UK Ltd.
Pitfield, Milton Keynes, MK11 3LW, UK
UKHW051930100926
8556IPUK00022B/1468

9 781476 698625